# Materials Horizons: From Nature to Nanomaterials

**Series Editor**

Vijay Kumar Thakur, School of Aerospace, Transport and Manufacturing, Cranfield University, Cranfield, UK

Materials are an indispensable part of human civilization since the inception of life on earth. With the passage of time, innumerable new materials have been explored as well as developed and the search for new innovative materials continues briskly. Keeping in mind the immense perspectives of various classes of materials, this series aims at providing a comprehensive collection of works across the breadth of materials research at cutting-edge interface of materials science with physics, chemistry, biology and engineering.

This series covers a galaxy of materials ranging from natural materials to nanomaterials. Some of the topics include but not limited to: biological materials, biomimetic materials, ceramics, composites, coatings, functional materials, glasses, inorganic materials, inorganic-organic hybrids, metals, membranes, magnetic materials, manufacturing of materials, nanomaterials, organic materials and pigments to name a few. The series provides most timely and comprehensive information on advanced synthesis, processing, characterization, manufacturing and applications in a broad range of interdisciplinary fields in science, engineering and technology.

This series accepts both authored and edited works, including textbooks, monographs, reference works, and professional books. The books in this series will provide a deep insight into the state-of-art of Materials Horizons and serve students, academic, government and industrial scientists involved in all aspects of materials research.

More information about this series at https://link.springer.com/bookseries/16122

Nabisab Mujawar Mubarak · Sreerag Gopi ·
Preetha Balakrishnan
Editors

# Nanotechnology for Electronic Applications

Springer

*Editors*
Nabisab Mujawar Mubarak
Petroleum and Chemical Engineering
Faculty of Engineering
Universiti Teknologi Brunei
Bandar Seri Begawan, Brunei Darussalam

Sreerag Gopi
Centre for Innovations and Technologies
ADSO Naturals Private Limited
Bengaluru, Karnataka, India

Preetha Balakrishnan
Centre for Innovations and Technologies
ADSO Naturals Private Limited
Bengaluru, Karnataka, India

ISSN 2524-5384 ISSN 2524-5392 (electronic)
Materials Horizons: From Nature to Nanomaterials
ISBN 978-981-16-6021-4 ISBN 978-981-16-6022-1 (eBook)
https://doi.org/10.1007/978-981-16-6022-1



This Springer imprint is published by the registered company Springer Nature Singapore Pte Ltd.
The registered company address is: 152 Beach Road, #21-01/04 Gateway East, Singapore 189721, Singapore

# Contents

# About the Editors

**Dr. Nabisab Mujawar Mubarak** is an Associate Professor in the Petroleum and Chemical Engineering, Faculty of Engineering, Universiti Teknologi Brunei. He serves as a scientific reviewer in numerous journals in the area of Chemical Engineering and Nano Technology. In research, Dr. Mubarak has published more than 207 journal papers, 30 conference proceedings and authored 25 book chapters and the H-index is 42. His area of interest is carbon nanomaterials synthesis, magnetic biochar production using microwave and wastewater treatment using advanced materials. He is a recipient of the Curtin Malaysia Most Productive Research award (2020&2021), outstanding faculty of Chemical Engineering award, Best Scientific Research Award London and outstanding scientist in publication and citation by i- Proclaim, Malaysia. He also has the distinction of being listed in the top two per cent of the world's most influential scientists in the area of chemical and energy. The *List of the Top 2% Scientists in the World* compiled and published by Stanford University is based on their international scientific publications, number of scientific citations for research, and participation in the review and editing of scientific research. Dr. Mubarak is a Fellow Member of the Institution of Engineers Australia, Chartered Professional Engineer (CPEng) of The Institution of Engineers Australia and also a Chartered Chemical Engineer of the Institute of Chemical Engineering (IChemE) UK. He is co-editor for 4 ongoing Elsevier edited books: 1) Sustainable Nanotechnology for Environmental Remediation 2) Nanomaterials for Carbon

Capture and Conversion Technique, 3) Advanced nanomaterials and nanocomposites for Bio electrochemical Systems and 4) Green mediated synthesis-based nanomaterials for photocatalysis.

**Dr. Sreerag Gopi, Ph.D** is a Chief Scientific Officer at Centre for Innovations and Technologies (CIT), ADSO Naturals Private Limited company, India. He also serves as Vice President (Research) at Curesupport Holding B.V, Netherlands. In research, Dr. Sreerag has published more than 50 journal papers, several conference proceedings and authored 15 book chapters. He has more than 7 years of total experience in industrial and academic research in chemistry, material science including nanocomposites for drug delivery, environmental remediation and bioprinting. His area of interests are synthesis and functionalization of nanoparticles from biopolymers, application in drug delivery and liposomal encapsulations. He is a recipient of the prestigious Erasmus Mundus fellowship by European Union during his Ph.D. period. Dr. Sreerag is a member of Royal Society of Chemistry, Member of Royal Australian Chemical Institute and Chartered Chemist at Royal chemical institute, Australia. He is a editor for several books published by Elsevier such as Handbook of Chitin and Chitosan 3 volumes and several on going projects in his hand with the collaboration of RSC, Springer, Apple Academic Press, Wiley and ACS.

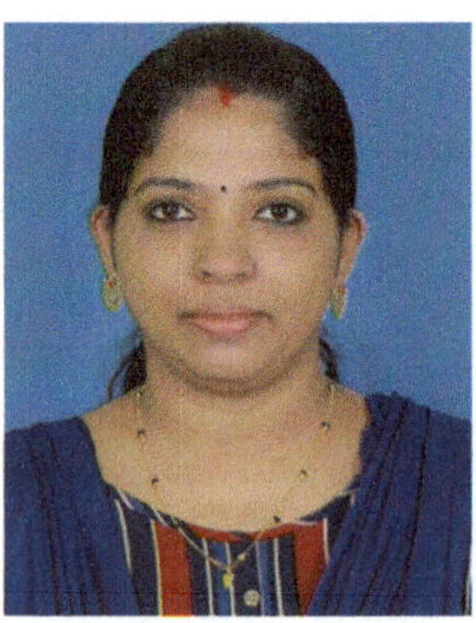

**Dr. Preetha Balakrishnan, Ph.D** is the principal scientist, QA, QC ADSO naturals India and Curesupport Netherlands. She did her graduation in Chemistry from Calicut University Kerala India and post-graduation from Mahatma Gandhi University Kerala with Gold medal and first rank. She is a recipient of prestigious INSPIRE Fellowship from Government of India. She completed her Ph.D. in Chemistry from Mahatma Gandhi University under the guidance of prof. Thomas, vice-chancellor, a renowned scientist in this area. She is an outstanding scientist with sustained international acclaims for his work in Polymer Science and Engineering, Polymer Nanocomposites, Elastomers, Polymer Blends, Interpenetrating Polymer Networks, Polymer Membranes, Green Composites and Nanocomposites, Nanomedicine and Green Nanotechnology. She visited many foreign universities as a part of her research activities and published around 15 research articles, 20+ book chapters in peer reviewed international journals. She edited 10 books with leading publishers like Elsevier, Springer, Wiley, RSC across the globe. Dr. Preetha has received a number of National, international presentation award. She worked as a post-doctoral researcher in the research group of Prof. Thomas and did enormous works in biomaterials for tissue engineering. She also worked as a guest lecturer in Chemistry, at Department of Chemistry, Morning star Home science College Angamaly Kerala, India.

# Chapter 1
# Importance of Nanotechnology, Various Applications in Electronic Field

**Hiba Ghouse, Laith Slewa, Mahd Mahmood, Salman Rehmat, Samia Musharrat, and Yaser Dahman**

## 1 Background

The term nanoelectronic/magnetic is associated with nanotechnology in the electronic/magnetic field. Nanomagnetics have always been in existence but have been recently discovered. Natural existent nanomagnetics originate as magnetite ($Fe_3O_4$), which is found in bacteria, molluscs, and insects. Magnetite is found to be 40–100 nm in length [36]. Currently, there is a great investment done toward nanoelectronics/magnetics throughout the world.

By observing the current technology, it can be seen that the investments are paying off. The size of a transistor has significantly decreased over the past decade. Korea Advanced Institute of Science and Technology (KAIST) has claimed in 2006 that they have developed a 3 nm transistor [10]. "For over 30 years, the industry has been able to double the number of field-effect transistors (FETs) on a chip every 18–24 months" [44]. The purpose behind scaling down the electronics is to significantly reduce the manufacturing cost and to increase the performance of the overall processor.

In the last decade, the size of devices has gone rapidly from 100 to 30 nm or. Also, the thicknesses of critical layers are becoming about 1 nm in the progression of reaching small sizes [45]. For example, a metal-oxide semiconductor field-effect transistor (MOSFET) device's threshold voltage is currently regulated by less than 100 atoms, due to which the requirements for line edge roughness are only a few nanometers now. These progressions in technology have instigated a rise in the demand for atomic level control aimed at patterning, deposition, and characterization.

In the past 10 years, the main cater for economic and technological growth was nanoelectronics. The objective was to increase the capacity of devices by controlling material at the nanoscale. Continuous research will increase the mean for manipulation of magnetic, spintronic, and many other properties of matter which will be at

H. Ghouse · L. Slewa · M. Mahmood · S. Rehmat · S. Musharrat · Y. Dahman (✉)
Department of Chemical Engineering, Ryerson University, Toronto, ON M5B 2K3, Canada
e-mail: ydahman@ryerson.ca

 
M. Mujawar et al. (eds.), *Nanotechnology for Electronic Applications*, Materials Horizons: From Nature to Nanomaterials, https://doi.org/10.1007/978-981-16-6022-1_1

the nanoscale. This will improve the existing products and will also open up areas of new applications such as new sensors, new ultralow power devices, and new flexible electronics. These applications will enhance laptops, cell phones, integral parts of appliances, cars, homes, and healthcare.

# 2 Nanoelectronics

## *2.1 Carbon Nanotubes*

### 2.1.1 Cooling Microprocessors with Carbon Nanotubes

Recently, the Department of Energy in the United States has implemented carbon nanotubes for cooling microprocessor chips. Microprocessors (CPU) are the main hub for all computers, and with increasing demand for faster and smaller microprocessors, the transistors located inside the chips heat up. With technology evolving rapidly, the chips are becoming more densely packed giving rise to the problem of overheating [47]. Carbon nanotubes (CNTs) have demonstrated high thermal conductivity; however, their chemical interaction with other materials has been weak, hampering the idea of combining CNTs and other materials [47]. Physicist, Frank Ogletree, and his team performed a study incorporating organic molecules between carbon nanotubes and metal surfaces.

Researchers developed “covalent bond pathways that work for oxide-forming materials such as aluminum and silicon, and for noble metals, such as gold and copper” [47]. Interaction between the two substances was drastically improved through implementing reactive molecules as a bridge among the CNTs and metal surfaces. “Aminopropyl-trialkoxy-silane (APS) for oxide-forming metals, and cysteamine for noble metals” [47] were used to bridge the two substances together. This was developed by growing vertically aligned CNTs arrays on silicon wafers and thin films of metal (gold or aluminum) were evaporated on cover slips. The metal slips were functionalized with the appropriate reactive molecules allowing bonding with the CNTs. To validate the process, a characterization technique that measures interface-specific heat transport was conducted and provided results with enhanced heat flow.

### 2.1.2 Carbon Nanotubes in Biomaterial Applications

CNTs can be used in various biomaterials for regenerating tissue such as scaffold materials. Scaffold represents an important element in tissue engineering to replace or restore damaged tissues [9]. There is a need for improved composite materials using combinations of synthetic and naturally derived materials. These composites may contain polymers, naturally derived biopolymers or a combination of both. Initial

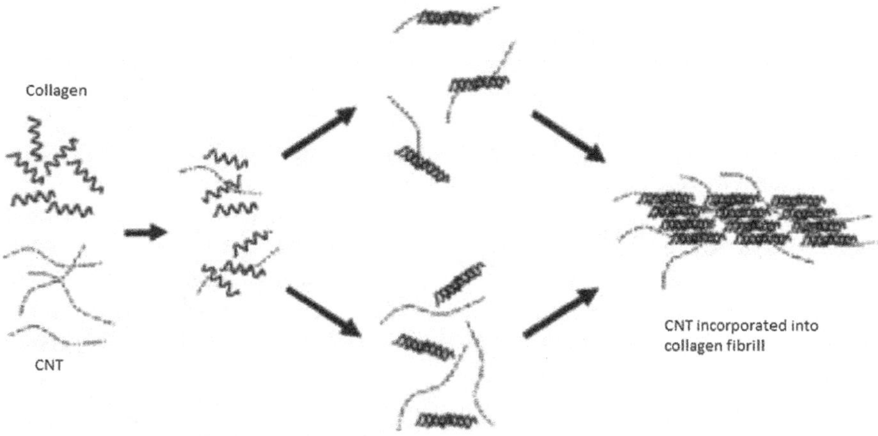


**Fig. 1** Schematic of process by which CNT can be included in collagen fibrils [9]

studies using CNTs with synthetic polymer composites have shown to be promising in improving the mechanical and electrical properties of the material [43].

Figure 1 shows the creation of composite materials for scaffold preparation. This process was done by the combination of CNTs through the incorporation of collagen. CNTs may be attached to the triple-helical collagen molecule (upper middle panel, Fig. 1) or may remain as a separate phase (lower middle panel, Fig. 1). CNT incorporated into collagen fibrils, an output of this process has great mechanical properties and electrically conductive matrix, which is a desirable feature for building scaffolds in which electrical signals can transfer easily between tissue cells [9].

One challenge for replacing or restoring cell tissue is finding the right materials that simulate the environment of the cell. For instance, for tissue to stimulate properly, it requires a scaffold that is electrically conductive to transmit the signal between the cells. This is important to regulate muscle repeated contractions and to improve the mechanical properties of the scaffold. To overcome this limitation, high porous scaffolds are preferred. However, higher porosity may affect the mechanical properties and electrical conductivity of scaffolds, which may aid in directing cell growth [17].

In conclusion, CNTs can be incorporated into polymers to provide novel properties in building scaffolds with interesting physical properties for use in tissue regeneration. However, this area of research needs more time and data to determine the potential of using CNTs in engineering tissue.

### 2.1.3 Carbon Nanotubes in Central Processing Units

Nowadays, all computers are run by Central Processing Units (CPUs). This is a small chip built inside the computer to assign tasks. With faster CPUs, computers can perform complicated tasks in less time. The CPU contains a large number of transistors, and its speed is determined by how fast the transistors can control the current flow between two points. To improve the speed of CPUs, transistors must be made as small as possible. This requires manufacturers to pack more transistors in CPUs, and this, in turn, minimizes the time it takes for a CPU cycle [41]. Small wires made of copper or aluminum connect transistors in CPUs to let electricity flow through them easily. In order to pack as many transistors in CPUs as possible, the length of the wires connecting the transistors needs to be reduced. As the wire length is decreased, the resistance of each wire will increase, therefore shorter wires will not let current flow freely as compared with longer wires.

Components made of CNTs can be beneficial in this case since they are closely connected and easily allow free current flow. In fact, each carbon atom has three strong covalent bonds. Wires that connect transistors in CPUs made of CNTs can endure high temperatures compared with copper or aluminum [41]. Furthermore, CNTs are immune to electromigration effects. This means bonds of CNTs will transfer electrons in one direction without bouncing around. However, the bonds of copper or aluminum will break overtime due to the collisions between the electrons.

In conclusion, CNTs present important properties such as high electrical conductivity and high mechanical strength in which very few other materials can match or even come close to Voldman et al. [41]. Ongoing research with CNTs is a major issue due to high cost, however, it is a promising material, therefore continuous research is being conducted.

#### 2.1.4 Carbon Nanotubes in Medicine

Various developments in the field of nanomedicine are bringing big promises for improved drug delivery and disease diagnosis. CNTs with interesting physical, mechanical, and chemical properties have been extensively explored for a wide range of applications in biology and medicine, such as the delivery of small molecules. Small drug molecules can be loaded on CNTs for drug delivery via covalent conjugation or non-covalent adsorption. Nanotubes are able to be conjugated covalently with small anti-cancer drug molecules. CNTs functionalized via the 1,3-dipolar cycloaddition of azomethine yields to amino groups that can be conjugated with drugs. Animal experiments were first reported by Liu et al. [28] and they indicated that single-walled carbon nanotubes (SWNTs) were attached to branched polyethylene glycol (PEG) via paclitaxel (PTX) conjugation (as shown in Fig. 2).

The results showed that residence time for the blood circulation of the SWNT-PTX complex was found to be longer compared with the free PTX. This indicated that drug accumulation increased in the tumor due to the enhanced and improved therapeutic efficiency for the delayed tumor growth.

Recent studies by Wu et al. [46] discovered that Multi-Walled Nanotubes (MWNTs) could conjugate covalently to the antitumor agent, 10-hydroxycamptothecin (HCPT), using hydrophilic diaminotriethylene glycol as the spacer. The resulting MWNT-HCPT conjugates were found to help prevent the formation or growth of tumors both in vitro and in vivo compared with the clinical HCPT formulation [46].

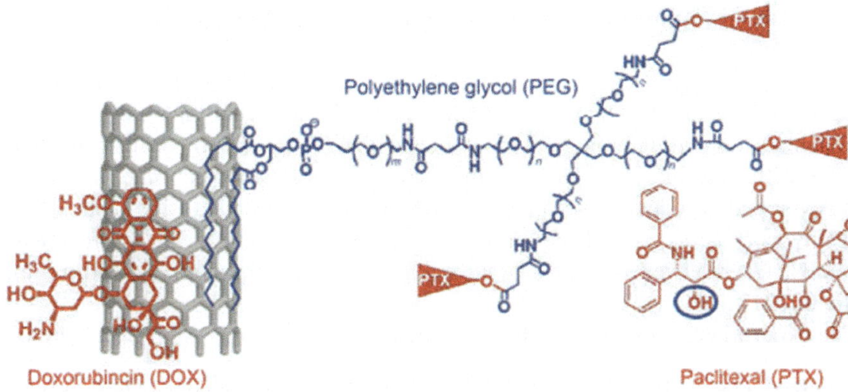


**Fig. 2** Scheme showing drug delivery with SWNTs, covalent conjugation of paclitaxel [28]

In vivo treatment, SWNT drug delivery was also investigated, tumors in mice intravenously injected with SWNTs were almost eliminated after irradiation with NIR laser (808 nm, 1 W/cm$^2$, 5 min) [5].

SWNTs were co-conjugated with cisplatin (a chemotherapy drug) and Epidermal Growth Factor (EGF), to specifically target cancer cells as shown in cancer treatment in the animal experiment in Fig. 3. Tumors were completely eliminated when they were injected with the SWNTs, this occurred without any serious side effects or the return of the tumor for over 6 months.

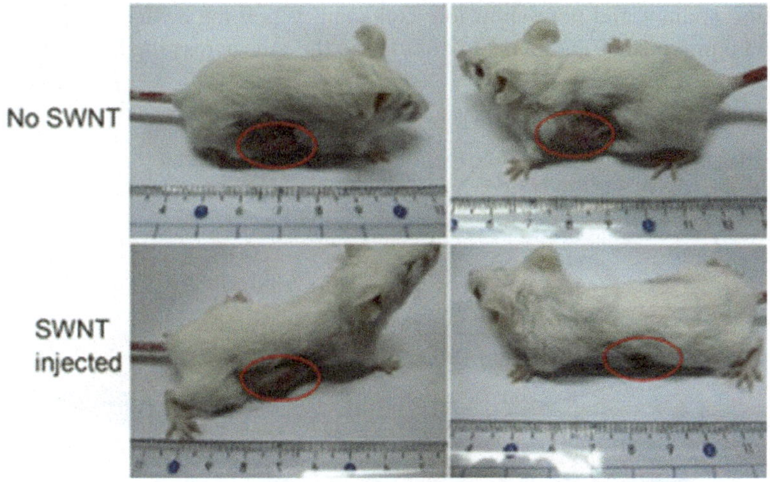


**Fig. 3** Reprehensive photos of tumors after various treatments indicated [5]

### 2.1.5 Carbon Nanotube Transistors Controlled by a Biological Ion Pump Gate

Aleksander Noy and his team at the University of California have developed an approach for integrating biological molecules with CNT devices by coating the layer of CNT with lipid bilayers. Through this property, a bionanoelectronic device has been developed, this device uses protein machine powered by adenosine triphosphate (ATP) hydrolysis, to control the output of a nanotube transistor. An ion transistor protein Na+/K+-ATPase is used to control and maintain potassium and sodium ion gradients [21].

The experimental procedure follows below:

- Using a CNT platform (as shown in Fig. 4) where single-walled CNTs combines two metal electrodes named as S and D for source and drain, respectively.
- S and D are insulated LOR3A photoresist material, leaving only the middle section exposed to the ATP solution.
- The device surface that is exposed is coated with a lipid bilayer, which consists of Na+/K+ molecules, using vesicle fusion.
- The ATP solution was delivered to the pump through a PDMS microchannel.

After the pump has been started, the lipid bilayer covering the CNT would lead to an increase in ion transport and change the electric field around the CNT. This change in the field would then regulate the transistor's current. With a flow 10 mM ATP molecules provided to the lipid bilayer, the conductance in the CNT was observed to increase by 35%, as shown in Fig. 5.

Embedded in this is an ATP-powered pump, this pump helps in spreading and mediating sodium and potassium ion exchange. When the device is on, the ATP-powered pump pushes and helps in circulating the ions across the lipid bilayer, this

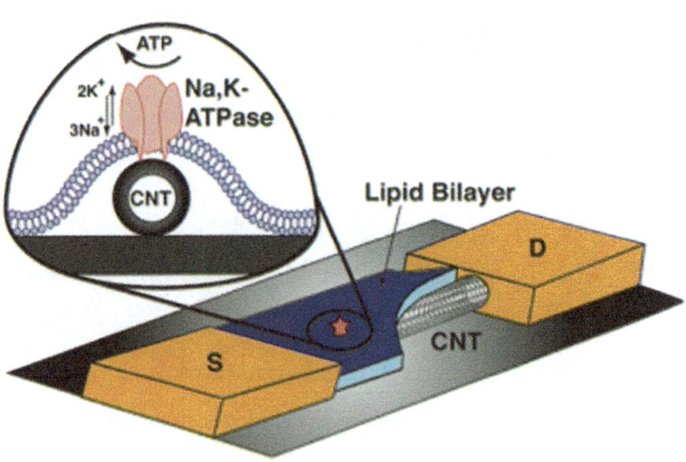


Fig. 4 Carbon nanotube platform [21]

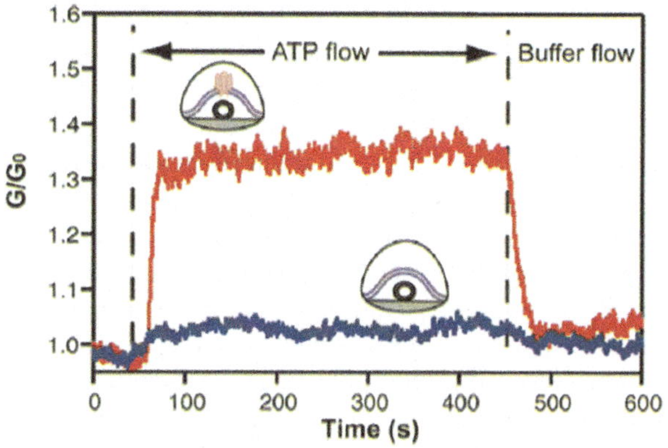


**Fig. 5** Conductance change with regards to ATP flow [21]

would, in turn, change the conductance of CNT and also boosts the transistor's current output by approximately 40% [21].

## 2.2 *Sensors*

### 2.2.1 Carbon Dioxide Sensors

Carbon dioxide is present in almost every aspect of life; therefore, sensing carbon dioxide is vital. Currently, $CO_2$ sensors are used for monitoring air quality in the environment and capnography. Recently, various simplified $CO_2$ sensors are being developed, however, these fail to address problems such as direct detection and high power consumption. Field-effect transistors fabricated from CNT's (NTFET's) have shown promising results as chemical sensors. The response of the device to the chemical analytes "occurs via charge transfer between the NTFET and the analytes" [38]. To implement this technology as $CO_2$ friendly, a supramolecular approach has been employed, the non-covalent functionalization of CNTs by using polymer coating has been used. The functionalizing that has previously been used consisted of covalent bonds and modifying this resulted in interference with CNT physical properties, leading to loss of conductance. Polymer coatings provide versatility and can be modified for $CO_2$ applications and are easily processed for different coating procedures such as dip coating and spin coating [38].

Next, a $CO_2$ recognition layer is attached, consisting of a mixture of polyethyleneimine (PEI) and starch polymers. The $CO_2$ sensor would use PEI/starch-coated NTFET's and through adsorption $CO_2$ gas would come in contact with the polymer coating, establishing equilibrium between water and PEI amino groups. PEI consisting of amino groups is a highly branched molecule capable of adsorbing $CO_2$ from gas mixtures. The results described were "The chemical reaction lowers the total pH of the polymer layer and alters the charge transfer to the semiconducting nanotube channel resulting in the change of NTFET electronic characteristics" [38].

The $CO_2$ reaction with PEI and starch polymers results in an n-type NTFET device characteristic, also, reaction with PEI/starch polymer coatings results in the formation of carbamate, therefore, reducing the electron-donating effect of PEI. As the device characteristic is modified, deformations in the polymer layer may occur resulting in scattering sites in CNTs and reduced conductance.

These overall NTFET characteristic changes due to the presence of $CO_2$ gas can be used for sensor design. The response and recovery occur at a minimal time (few seconds), with high sensitivity and precision even at low $CO_2$ concentrations. The small size and low power consumption will be beneficial for medical $CO_2$ sensors. As shown in Fig. 6, a calibrated curve displays the $CO_2$ with regards to the change in conductance [38]. The conductance of the sensor increases as the $CO_2$ concentration also increases as shown in Fig. 6. In Fig. 7, the graph shows the response time of the sensor. At low concentrations, the response time is rapid, whereas as the $CO_2$ increases to 1%, the response time increases as well.

The experiment that the researchers carried out is as follows [38]:

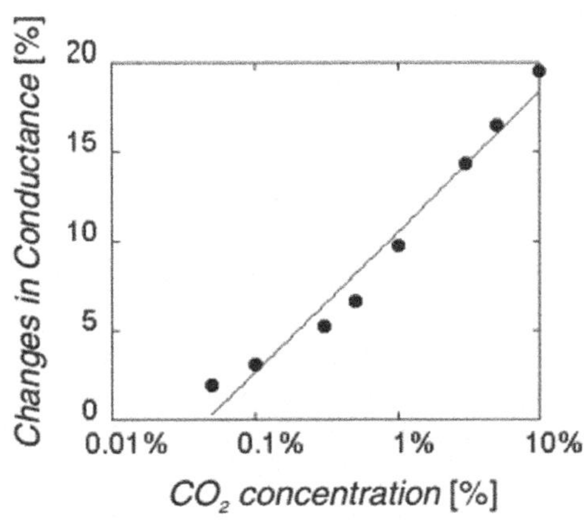


**Fig. 6** $CO_2$ concentration versus conductance [38]

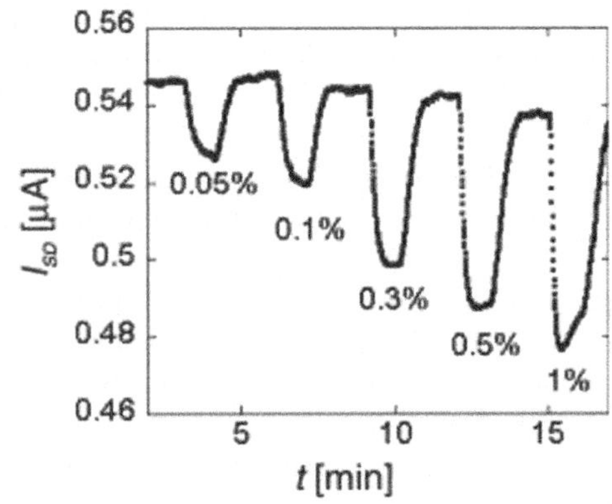


**Fig. 7** $CO_2$ detection time [38]

- NTFET devices were fabricated using SWNT through a chemical vapor deposition technique that was conducted at 900 °C. To increase growth, iron nanoparticles with a mixture of methane/hydrogen gas were used.
- Using titanium films 30 nm thick, electrical lead patterns were formed on the nanotubes and also a 120 nm thick gold layer.
- The substrate was submerged in a 10wt% solution of poly(ethyleneimine) overnight then later washed with water. Following, a thin layer <10 nm was observed.
- Measurement of current flow was obtained by using a semiconductor parameter analyzer.
- For this study, NTFET devices were wire bonded and packaged in a 40 pin CERDIP package, then functionalized with PEI and starch polymers. These were assembled in a flow cell where air and $CO_2$ gases can penetrate.

### 2.2.2 Silicon-Based Nanoelectronic Field-Effect pH Sensor with Local Gate Control

Nanostructures such as nanowires, nanotubes, and quantum dots are being implemented in applications such as biosensors due to their size; this is mainly due to the fact that the size of the sensor is comparably the same as the biological or chemical species being detected. This increases the detection sensitivity due to a large surface to volume ratio [8]. The authors of this article have researched using a conventional silicon nanowire as a sensor for hydrogen ion concentration or a pH sensor. This has been achieved by modifying the surface of the wire with 3-aminopropyltriethoxysilane (APTES) resulting in the surface acting as a receptor for hydrogen ions. The surface charge of the silicon wire would change, and this would consequently change the conductance of the nanowire [8].

An n-type silicon nanowire would experience an increase in negative charge on the surface when the pH increases. This would cause "the channel of the charge carrier to deplete" decreasing the net conductance [8]. The side gates on the nanowire sensor would provide the control of the accumulation of charge carriers, which would allow field-effect control. The change in the concentration of pH can be detected by the change in differential conductance. The engineered silicon nanowires are fabricated through the standard semiconductor processing technique whereas, the nanowire can be controlled with side gates "to induce inversion or depletion layers" [8].

This research has demonstrated that the silicon nanowire can be controlled through nanoscale side gates; one of the benefits from this research is that during functionalizing, the selective gate technique can individually turn on or off the nanowire's array. This method can lead to a detector that is able to simultaneously detect multiple chemical or biological substances.

## 2.3 *Water-Soluble Quantum Dots for Multiphoton Fluorescence Imaging In Vivo*

For biological purposes, the use of quantum dots is being thoroughly examined. Water-soluble quantum dots can be used for multiphoton fluorescence imaging in a living organism. This particular article investigated photo physical properties of water-soluble CdSe-ZnS nanocrystals organized by a synthesis technique [25]. These nanocrystals are compressed within an amphiphilic polymer.

These water-soluble quantum dots were measured to have a 14 and 9 nm radii. These are 2–3 times the dry radius, which indicates a solid interaction between the polymeric surface coatings of the nanocrystals and the solvent [25]. The design of the actual two-photon focal volumes indicates that at the order of cross-sections of 200 Goeppert-Mayer units (GM), the volume surges because of saturation effects. The saturation effects are at less than 10% for powers <5–10 mW, but at cross-sections of 30,000 GM, the focal volume increases at only ~1 mW [25]. This, in turn, reduces the optical resolution to fuzzy images.

The water-soluble quantum dots were tested in vivo by injecting mice with the dots. Due to the size of the dots, it makes them convenient for angiography. The dots were inserted in the mice within the skin and adipose tissue. The imaging was found to be clearly visible through the dermal layer in the vivo mouse. Using a high concentration of the quantum dots (1 μM), the blood velocity was found using line scans, and the heart rate was easily detected using temporal undulations of the capillary wall directly through the skin [25].

The quantum dots are removed from the body before the amphiphilic coating breaks down. The cadmium in these quantum dots is known to be toxic but no ill effects were shown during the imaging tests in vivo [25]. They are, however, continuously being investigated for any long-term toxicity effects.

## 2.4 *Zinc Oxide-Based Transparent Thin-Film Transistors*

The possibility of the most transparent applications is dependent upon the quality of the materials and performance of the devices. Therefore, circuit, system, and manufacturing issues need to be explored for a realistic assessment of nanotechnology of transparent applications [19].

The article "ZnO-based transparent thin-film transistors" mainly discusses about the succession of its unique nanostructure, quality of materials and performance of nanoelectronics [19].

Thin-film transistors (TFT) are fabricated with optical transmission, highly transparent ZnO-based, enhancement mode 75% in the visible portion of the electromagnetic spectrum. It is *n*-channel based, has low light sensitivity, and displays electrical characteristics, which are suitable for performance as a transparent select-transistor, which is in each pixel of an active-matrix liquid crystal display called

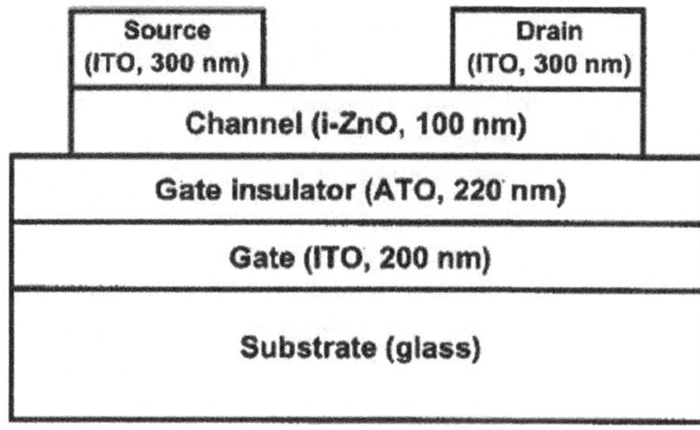


**Fig. 8** Typical TTFT structure [19]

AMLCD. Furthermore, fabrication of this device is simple and compatible along with inexpensive glass substrate technology [19].

The control electrode facilitates the achievement of logic, amplification, memory, other types of signal conditioning and processing functions, whereas a transistor is an important development in transparent electronics and plays a great role in the modern information society. Transparent Thin-Film Transistor (TTFT) devices are an example of that and Fig. 8 displays the structure [19].

Its structure consists of a glass substrate that is blanket coated with a 200-nm-thick layer of sputtered indium tin oxide (ITO) and a 220 nm thick layer of aluminum–titanium oxide (ATO) deposited by atomic layer deposition. ITO is a highly transparent and n-type conductor that serves as the TTFT gate [19].

ATO consists of a superlattice of alternating layers of $Al_2O_3$ and $TiO_2$, which is capped on either end by an $Al_2O_3$ layer which is an engineered insulator. The ATO layer operates as the gate insulator. The ZnO channel and ITO source/drain electrode films are deposited through ion beam sprayed in 1024 Torr of $Ar/O_2$ (80%/20%) while the substrate is unheated throughout the deposition. ZnO channel and ITO source/drain electrodes are pattered by shadow masks. The channel width and length are 15,000 and 1500 um with the width-to-length ratio of 10:1. The source/drain contact dimensions are 15,000 $\mu$m × 31,500 $\mu$m. We know 1 $\mu$m is equal to 1000 nm [19].

A rapid thermal anneal (RTA) usually at 600–800 °C in $O_2$ is utilized to amplify resistivity of ZnO channel resistivity after deposition of the ZnO layer. This improves the electrical quality of the ATO/ZnO interface and enhances the crystallinity of the ZnO layer. Subsequently, for the deposition of the ITO source/drain electrodes, a 300 °C RTA in $O_2$ is carried out to develop the transparency of the ITO layer [19].

## 2.5 *Thin Film Solar Cells*

Thin film solar cells are a renewable energy conversion device where nanotechnology has played an important role. Traditionally, thin film solar cells were made of large grain n- and p-type semiconductors. This was done to avoid any harmful effects of grain boundaries (it is a defect in the crystal structure and inclines toward decreasing electrical and thermal conductivity of the material) and other imperfections [35].

There is a limit of increasing efficiency of solar conversion depending on factors such as characteristics of semiconductors to fabricate the cell. Thin film solar cells are based on polycrystalline and amorphous Si, CdTe, and $CuInSe_2$ (CIS), which are the most examined. Regarding CIS, the efficiency has been enhanced by the integration of Ga and small amount substitution of In. The band gap can be tuned for effect efficiency for ($Cu(In,Ga)Se_2$) (CIGS), while CdTe has an appropriate band gap for future optimum solar conversion [35].

The utilization of nanostructured components in thin film solar cells increases the possibility of even further efficiency of solar conversion. The objective of efficient solar adsorption in p-type semiconductors and maximum light transmittance through n-type material has been examined by opening up the band gap in both kinds of semiconductors. CdTe and CIS have been examined as absorber thin films for further efficiency. Higher band gap CdS forming in the nanostructured thin film has been reported numerous times [35].

The study of such design of devices based on nanostructured semiconductors includes complex characteristics and phenomena. It is usually not compatible with common analytical methods like x-ray diffraction (XRD). The structural and morphological properties of the nanostructured semiconductors are comprised of optoelectronic devices. Primarily, these devices require to be examined by highly technological equipment such as transmission electron microscopy (TEM), scanning transmission electron microscopy (STEM), and atomic force microscopy (AFM) [35].

Nanoscale studies need to be completed for comprehension of photo and electrical responses of nanostructured semiconductors and devices. It has led to a new era in the study of inorganic semiconductor materials to appear. Primarily in photovoltaic applications, there is a possibility for improvement for all the physicochemical characteristics of the semiconductors. The most important challenges for nanotechnology in nanostructured semiconductors design are the improvement of charge carrier generation/recombination ratio and charge transfer mechanisms [35].

These developments in nanotechnology applications such as thin film solar cells indicate the immense opportunity of a lower amount of material utilization, low processing costs, large area thin film solar cells, and consequently the possibility of cost reduction for photovoltaic power [35].

## 2.6 Cadmium Sulfide Nanowires: Ultra-long Growth and Enhanced Field Emission Properties

Semiconductor properties in the chalcogenide are extremely important in the fabrication of optoelectronic devices such as optical switches, field emitters, sensors, and lasers. Cadmium sulfide (CdS), which is a chalcogenide, has a band gap of 2.4 eV at room temperature [7]. CdS is expected to be a good field emitter since it has a 4.2 eV, which is lower than carbon nanotubes: 5.0 eV, zinc oxide: 5.3 eV, and zinc sulfide: 7.0 eV to name a few. The semiconductors can have improved field emission properties, which are attainable by modifying the electronic properties by tailoring the geometry of the emitter [7].

CdS nanowires are already being produced, however, synthesis of ultra-long CdS nanowires by thermal evaporation technique resulted in an aspect ratio of approximately $8 \times 10^4$. Compared with the previous procedures, this advancement is cost effective and easier due to the high degree of condensation [7].

Synthesis consists of high purity CdS powder undergoing thermal evaporation to synthesize single crystalline ultra-long CdS nanowires. The CdS powder is placed in the quartz tube and into the furnace. The silicon substrate used was cleaned in acetone (20 min) and then sputter coated with a thin layer of gold film (approximately 30 Å) [7]. The reaction chamber was flushed with argon for an hour at 100 sccm. These tubes were then heated to 1000 °C at a rate of 50 °C per minute and held at 1000 °C for 30 min. After each run, they were cooled to room temperature and cleaned with dilute (40%) hydrofluoric acid and returned to the furnace for the next run [7]. A yellowish wool-like substance was found on the entire surface of the gold-coated silicon substrates.

Characterization involved a Scanning Electron Microscope (SEM) for the surface morphology. The crystalline nature was confirmed by using the Transmission Electron Microscopy (TEM) and High Resolution TEM (HRTEM) [7]. XRD was used for phase identification and the photoluminescence (PL) spectra investigated the optical properties. The Keithley 485 picoammeter and a Spellman high voltage DC power supply (0–40 kV) was used to measure the field emission current at room temperature, which was 0.17 $\mu A/cm^2$ as shown in Table 1 [7].

## 2.7 Electroless Deposition of Silver Thin Films on Gold Nanoparticle Catalyst for Micro and Nanoelectronic Applications

To produce micro and nanoelectronic mechanical systems, electroless deposition is a method used to enhance the properties of the required catalysts. Thin silver films are deposited onto the silicon oxide/silicon ($SiO_2/Si$) surface. This surface was modified by the self-assembled monolayer (SAM) and activated by gold nanoparticles [22]. Three separate stages were analyzed such as compact and isolated islands, coalesce

**Table 1** Turn on field values for CdS and ZnO nanostructures reported in literature [7]

| Material | Morphology | Turn-on field (V/μm) (for 0.1 μA $cm^{-2}$) | References |
|---|---|---|---|
| CdS | Ultra-long nanowires | 0.17 | Chavan et al. [7] |
| | Quasi aligned Nanowires arrays | 12.2 (~10μA $cm^{-2}$) | Zhai et al. [49] |
| | Aligned Nanowires | 0.68 | Datta et al. [13] |
| | Multipods | 7.2 | Zhang et al. [50] |
| | Nanotubes | 1.4 | Pan et al. [32] |
| | Nanotips | 5.28 (~10μA $cm^{-2}$) | Zhai et al. [48] |
| | Nanocombs | 0.26 | Chavan et al. [6] |
| | Nanobelts | 3.7 (~10μA $cm^{-2}$) | Li et al. [27] |
| ZnO | Nanowires | 6.0 | Lee et al. [26] |

equilibrium shapes, and continuous hole-free films. These stages were analyzed throughout the growth of the sliver films, and the observations mainly occurred by high resolution scanning electron microscopy (HRSEM).

The experimental method is described briefly below [22]:

- Samples cleaned in hot (80 °C) $NH_4OH$:$H_2O_2$:deionize water (DIW) (1:1:6) to remove any possible metallic contamination
- Rinsed in hot (80 °C) HCl:$H_2O_2$:DIW (1:1:5) for residual organics removal
- For improved metal adhesion (Silanization): silane-based amino-terminated SAM with 1–2 nm thickness was coated on the hydroxylate $SiO_2$/Si samples and was done in 1% animopropyl-trimethoxysilan (APTMS) in ethanol solution for 3.5 h at 65 °C.
- To attain a cross-linkage of the chains and densification of the SAM, samples were placed in an air oven at 100 °C for 30 min.
- Surface was activated by immersion of the samples in an aqueous colloid solution of gold nanoparticles (size: 15–25 nm)
- Deposition occurred at room temperature (from amino-acetate complex-based solution and hydrazine hydrate as the reducing agent.
- Minute quantities of additives are added as surfactant and to level and brighten the deposit
- Alpha-step 500: measured the thickness of the silver films
- In-line Four Point Probe Resistivity Tester: measured the resistance of the deposits
- HRSEM (0.9 nm resolution): characterize the gold nanoparticle precipitation on the $SiO_2$/Si substrate and to observe the morphology and the microstructure of the thin film
- AFM: measures the topography and roughness
- UV/Vis/NIR Spectrophotometer: was used to measure the optical properties

Figure 9 displays the microstructure via the HRSEM imaging of the thin silver films that have different deposition times. The first two (**a** and **b**) have a deposition time of 10 and 15 s, respectively, and are both considered to the early stage where

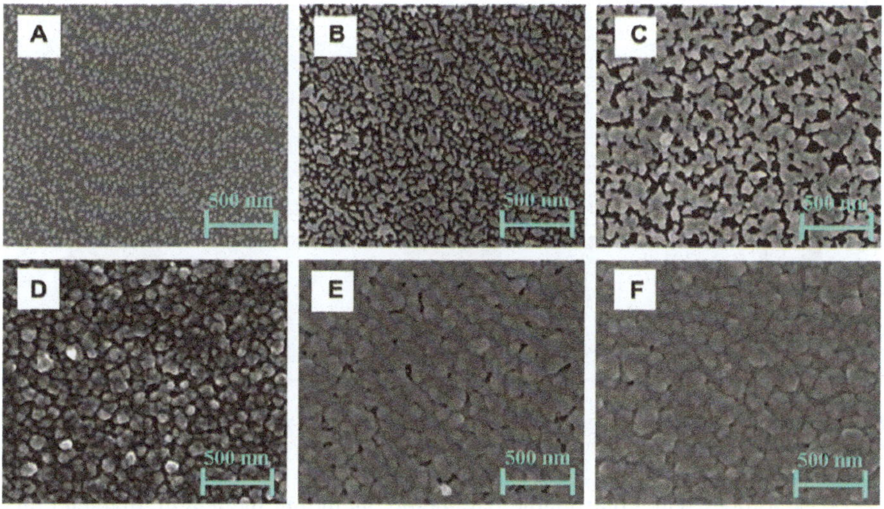


**Fig. 9** HRSEM images of silver films. Deposition time(s): **a** = 10, **b** = 15, **c** = 20, **d** = 45, **e** = 60 and **f** = 80 [22]

the film is extremely thin, it is referred to as isolated compact islands. As the process continues and the time is increased, the growth continues to a point where the islands coalesce into near equilibrium compact shapes that form the percolating structures (**c** and **d**). As time increases further the channels between the small structures are filled up and a free of holes, continuous deposit is formed (**e** and **f**) [22].

The results clearly portray the successful deposition of these films. Currently, even though the same process kinetics were followed, the properties that result such as the optical and electrical are slightly lower than the properties of the Ag films electroless deposition, which was on top of Pd activated substrates. As a next step, the application of minute gold nanoparticles in terms of a catalyst could be used for successful metallization [22].

## 3 Nanomagnetics

### *3.1 Single Crystalline Magnetite Nanotubes*

Magnetite nanotubes are on a potential breakthrough to serve as adjustable networks for magnetic particles, data storage devices, and scanning tips for magnetic force microscopes [29]. Researchers from the Department of Electrical Engineering-Electrophysics, University of Southern California have figured out a way to synthesize a single crystalline $Fe_3O_4$ nanotube. This is done by wet-etching inner cores of the MgO of the MgO/$Fe_3O_4$ core–shell nanowires. This practice allows homogenous

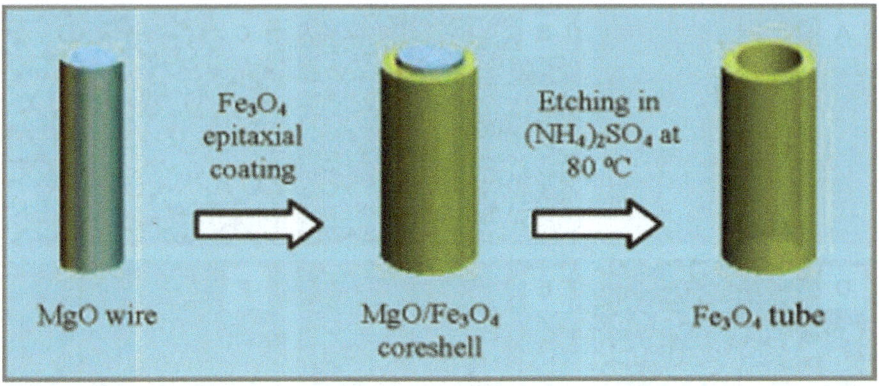


**Fig. 10** Process of growing magnetite nanotubes [29]

$Fe_3O_4$ nanotubes with controllable parameters including the length, diameter, and wall thickness [29].

The MgO crystalline nanowires were initially grown on $Si/SiO_2$ substrates. Then a conformal layer of $Fe_3O_4$ was deposited onto the nanowires with pulsed laser deposition techniques. Once the $MgO/Fe_3O_4$ core–shell nanowires were achieved, the MgO cores were etched in $(NH_4)_2SO_4$ solution at 80 °C for ~1.5 h [29] as shown in Fig. 10. Finally, deionized water was used to rinse the nanotubes to remove the residual contaminants. This process produced $Fe_3O_4$ nanotubes with an outer diameter of ~30 nm, thickness of ~7 nm, and a uniform sidewall. The nanotubes had a cubic reverse spinel structure with a latticed constant of $a = 0.840$ nm and a resistivity of $4 \times 10^{-2}$ Ω cm at room temperature [29].

## *3.2 Nanomagnetics for High-Performance and Miniaturized Power*

An article published in IEEE Nanotechnology Magazine focuses on designing magnetic structures at the nanoscale to achieve high-frequency permeability; low loses and improved power handling. Nanomagnetics are present in natural metals such as iron and nickel, however, aren't suitable for high-frequency applications due to their high electrical conductivity [34].

A copper or iron-based nanocomposite shows "higher permeability and frequency stability at microwave frequencies than those obtained from bulk copper or iron metal or their microscale composites" as shown in Fig. 11 [34]. This phenomenon is due to the exchange interaction between the molecules, the interaction of magnetic orderings within, which extends to the neighboring molecules through spin polarization to a distance of 25 nm. This interaction results in cancellation of magnetic anisotropy

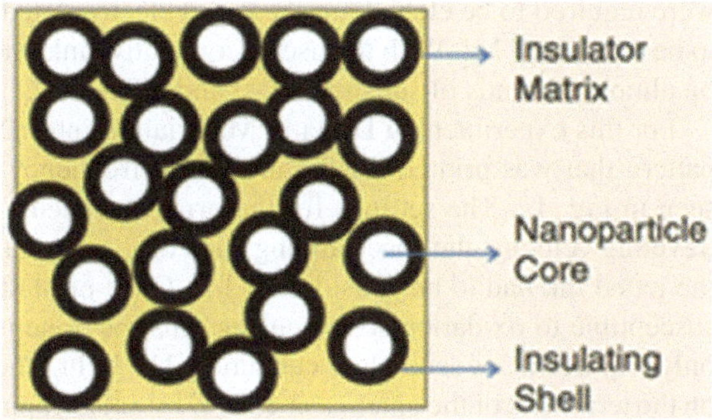


**Fig. 11** Nanomagnetic composite structure [34]

**Table 2** Comparison of different nanomagnetic materials [34]

| Material | Permeability | Frequency | Power handling |
|---|---|---|---|
| Nanomagnetic | 100–500 | 1–10 GHz | Good (50–100 Oe) |
| Ferrites | 120 | 100–500 MHz | Moderate (10–30 Oe) |
| Metal compacts | 125 | 10 MHz | Moderate (10–20 Oe) |

and due to the size, the production of eddy's current is small. This leads to superior magnetic properties such as high permeability (100–500), increased frequency (1–10 GHz), and good power handling. This is implemented for radio frequency inductors and power convertors, where volumetric density can be enhanced up to 10 times without compromising power efficiency. Antennas have always been large in size and to "decrease the antenna size, it has to be surrounded by a material with either high permittivity or permeability" [34]. Nanomagnetics metal polymers such as Cobalt polymer nanocomposites can address this situation, leading to enhanced bandwidth and reduced size.

This shows that magnetic materials at the nanoscale can be used as a means to miniaturize antennas and other radio frequency devices; while achieving high frequency with increased power handling. Table 2 compares various nanomagnetic materials for their permeability, frequency, and power handling [34].

### *3.3 Nanocopper Conductive Ink for RFID Application*

To enhance the Radio Frequency Identification (RFID) applications, a nanocopper conductive ink that is made of anti-oxidation copper particles at the nanoscale can be used. This new technique is proposed to save money and time during the production of RFID devices as well as broaden the scope of applications. Initially, RFID devices were required to be etched, masked, and electroplated, which deemed these devices to be corrosive [14]. With the use of conductive ink, the RFID devices can be printed on almost all kinds of substrates fast and efficiently.

For this experiment, a Dimatix Materials Printer DMP-2800 was used [14]. The pattern that was printed for the ultra-high frequency (UHF) RFID antenna can be seen in Fig. 12. The settings for the printer needed to be precise. The ink volume, traveling velocity, the space among each droplet, occurrence, and the temperature of the jetted ink had to be considered [14]. Once printed, the nanocopper particles are susceptible to oxidation so the curing must be done in a vacuum oven. The pattern only required a few seconds to cure under UV light. The RFID chip was then mounted on the terminals of the antenna at 150 °C by anisotropic conductive adhesive bonding [14].

The copper particles that were used were less than 50 nm in diameter. Once printed, a single layer was over 2 μm [14]. The resistivity was initially high but once multiple layers are added, the resistivity drops. With a cross-cut test, the ink film remained intact a 100% showing good adhesion. The measured resistivity was about $4 \times 10^{-5}$ Ω cm, which was comparable to that of bulk copper $1.75 \times 10^{-6}$ Ω cm [14]. The printing process was also compared to mask lithography and screen printing shown in Table 3.

This new process of printing the RFID inlay was a success. It showed a good reading performance and was easily/quickly/economically prepared using ink-jet printing and UV light curing. The cross-cut test showed strong adhesion and optic microscopes showed a good conductivity [14].

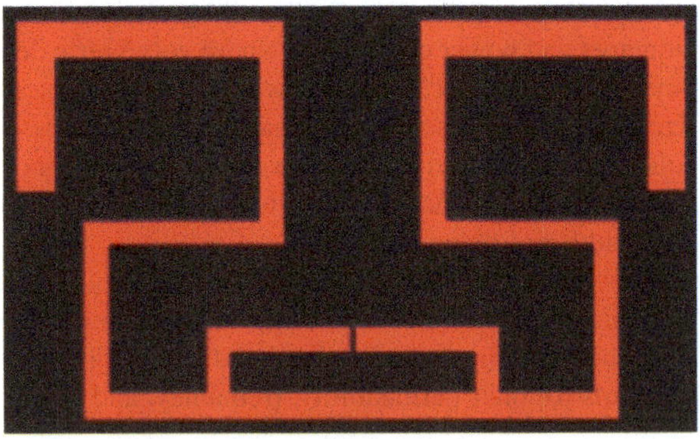

**Fig. 12** Pattern for UHF RFID printing [14]

**Table 3** Comparison of different printing techniques [14]

| Process | Thickness (μm) | Resistivity (Ω cm) |
|---|---|---|
| Screen printing | 5.0 | $2.3 \times 10^{-5}$ |
| Mask lithography | 1.5 | $5.0 \times 10^{-5}$ |
| Ink-jet printing | 2.6 | $4.1 \times 10^{-5}$ |

## *3.4 Drug Delivery System Using Nanomagnetic Fluid*

Using magnetic particles as a system for drug delivery is constantly being researched. The article focuses on a drug delivery system that incorporates a ferrofluid cluster composed of magnetic nanoparticles. The advantages of using this method are that the magnetic fields don't invade the body and is a painless process; also, it consists of no side effects and can provide fast and effective treatment due to the nanoscale. The fluid used was made by the researchers, sizing at 8 nm, with a strong sensitivity to external magnetic fields [30]. The particles were gathered together to make a cluster since it increases magnetism and were placed in an insoluble liquid, Canola oil [30]. The cluster is 0.5 mm, and since it is placed in Canola oil (High viscosity), the motion of the cluster is slow and easily observed. The motion of the particle in two monopoles q and −q was calculated whereas the velocity of the cluster was calculated to be approximately $1.0 \times 10^{-8} \frac{1}{x^3}$ [30]. The velocity versus the distance between the magnet and the cluster is shown in Fig. 13.

In addition, two theoretical motions are studied, dragging in the artery system and pulling toward the artery wall. The dragging in the artery system motion consisted of evaluating the viscosity of the blood and the velocity; this resulted in a velocity of 0.1–0.65 m/s [30]. The drag condition obtained is L <0.1 m, whereas the results that were obtained are displayed in Table 4, where A is the drag force radius.

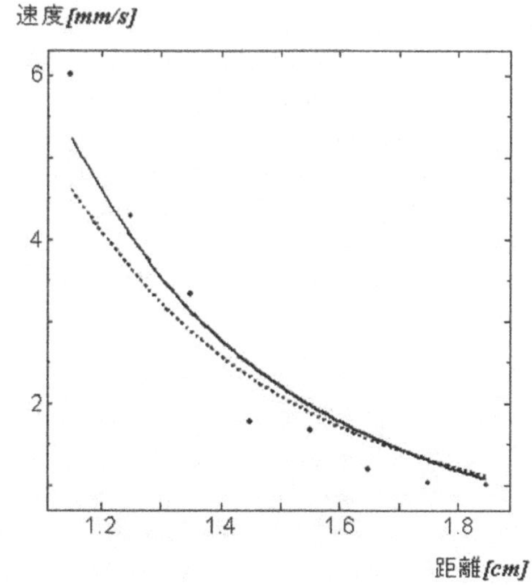


**Fig. 13** The velocity versus the distance between the magnet and the cluster [30]

**Table 4** Velocity change with respect to cluster radius [30]

| V(m/s) | A (μm) | L (mm) |
|---|---|---|
| 10 | 250 | 250 |
| 50 | 50 | 50 |

The pulling toward the artery wall motion study showed that the magnetic force is sufficient to pull a cluster of a radius of 0.1 mm; hence it's important to design a strong magnetic field [30]. The results show that magnetic fluids in blood vessel can be controlled through external magnetic fields. If the magnet is strong enough, it can move a cluster from a distance of 10 cm [30].

### 3.5 Semi-biosynthesis of Magnetite Gold Composite Nanoparticles Using an Ethanol Extract of Eucalyptus Camaldulensis and Study of the Surface Chemistry

Magnetic particles are being thoroughly researched in the nanotechnology field because of their uses in the biomedical sector. By creating nanocapsules with magnetic particles, it becomes easier to target certain cells or bacteria in the human body. The problem that has remained, however, is protecting the magnetic core of the nanocapsule against oxidation. Also, while protecting the magnetic core, the surface must be able to chemically bond with DNA, proteins, amino acids, and so forth [18]. The current products in the field use protective layers of polyethylene glycol, dextran, or silica [18]. The problem with these materials is that they can create hydrophobic surfaces that do not allow proper protein bonding. The other protective layers have non-metallic coatings that decrease the magnetic properties of the magnetite core [18].

A solution that has been found to some of these issues is using gold-coated magnetic nanoparticles. TEM imaging of the gold particles can be viewed in Fig. 14. The gold acts as an inert element in terms of its chemical reactivity and is great for protecting the magnetic core. Kinoshita et al. have synthesized a gold–iron nanocomposite powder by reverse micelle method [18]. Chen et al. [8] have synthesized gold magnetite nanoparticles by a laser process in water, and others have created these structures by reducing agents. The problem that arose was that they use harmful

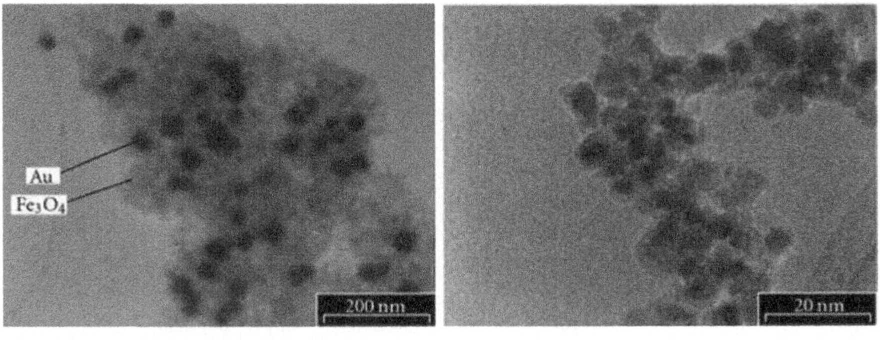


**Fig. 14** TEM images of magnetic-gold composite nanoparticles. Dark particles represent gold and grey particles represent magnetite [18]

surfactants or organic solvents, and it was difficult to maintain magnetic nanoparticles in liquid media [18].

To overcome this, researchers have prepared Au-$Fe_3O_4$ with a combined chemical and biological reducing process. $Au^{3+}$ was attached to the surface of the magnetite by using an ethanol extract of Eucalyptus camaldulensis for the reduction of gold [18]. This created a functionalized gold magnetite nanocomposite particle. The magnetite particle averaged 10 nm and ranged between 6 and 20 nm and using UV visible spectra a peak of 608 nm was seen for the composite material [18]. It was determined by XRD method that these particles were face-centered cubic, and the dominant crystal plane was (111) [18]. These nanocomposites showed a carboxyl group on the surface of the Au-$Fe_3O_4$ [18]. This makes the nanocomposite suitable for rapid separation and purification of biomolecules by magnetic field and drug delivery. Also, the gold outer layer prevents the magnetic core from oxidation during the usage while keeping the magnetic properties intact.

## *3.6 Micro- and Nanomagnetic Particles for Applications in Biosensing*

This review article examines the enhancements of electrochemical biosensors particularly by the addition of magnetic particles. These electrochemical biosensors are primarily used for detecting biomolecules such as nucleic acids and proteins as well as most cells if not any [20]. The high surface area and the paramagnetic or superparamagnetic characteristics of these tiny magnetic particles provide an alluring technological platform for the design and production of these electrochemical biosensors. More research is being conducted to incorporate the electrochemical labels to the magnetic particles and the specific processes that can attain the dual functions of electrochemical detection and magnetic separation [20].

This article considered the synthesis and surface modification of magnetic particles. Many different sized particles were analyzed, and Table One in the article depicts the description of each of the approaches. Biosensors are a major technological advancement in the medical industry [20]. Sequence-specific electrochemical sensors utilizing magnetic separation of DNA allow the collection of targeted nucleic acid from any bone, blood, marrow, cultured tissue, and any other regions.

Another aspect of this article was electrochemical protein sensors utilizing magnetic separation. This is extremely important since biological threat defense and drug development are directly linked to targeted disease diagnosis [20]. This can be, in turn, be used to analyze DNA sequencing.

This was a review article that had an immense amount of information and results which concluded that surface modification will enhance the properties and allow the material to attain the specific properties of labeling, sensing, and detection.

## 4 Research Trends

Research in electronics has often been related with Moore's law, which states that the performance of devices doubles every 18–24 months. This increase in performance is related to the miniaturization of devices in electronics and magnetics. Research trends primarily focus on three parameters; emerging technologies, architecture for integration of nanoelectronics, and the fabrication of these devices [12]. The invention of novel devices is mainly based on elementary processors, logic, and memory cells whereas the issues that are being addressed are the operation of nanodevices at high temperatures, low power consumption, and high performance. Furthermore, trends in research on the architectural component of nanoscale devices have been observed. It is vital for the proper integration of these devices to provide efficient information storage and processing. Lastly, the manufacturing techniques of nanoelectronics are also being researched, the main challenge is to create a fabrication technique that is fast, efficient, and cost effective. Also, the integration of both top-down and bottom-up techniques is being researched whereas the concept of self-assembly has an immense focus [12].

**Emerging Devices**

Complementary metal-oxide semiconductors (CMOS) have been the major technology used for nanoelectronics. However, numerous investigations into other devices are constantly being investigated. Single electron tunneling (SET) is one of the prevailing technologies aimed at high density and low power applications. Research into SET has shown promising results in memory storage in the form of single electron transistors. These would replace the conventional flash type memory devices with a nanoflash memory [12].

Another replacement for CMOS that is currently being studied is the resonant tunneling diodes (RTDs); this technology has potential in a niche market such as the digital to analog convertors (DAC), clock quantizers, and low power SRAM's. These devices, however, require a uniform layer of wafer and tolerance in the monolayer regime. Currently, research is being done on how to implement RTD circuits on non-uniform surfaces [12].

Rapid single flux quantum (RSFQ) logic is another technology being researched for potential electric circuits and conductors. RSFQ is a superconductor circuit technology that can be implemented in the market where Si CMOS are unable to achieve good frequencies. RSFQ is high speed and is micrometer in size, however, is costly due to cryogenics [12]. RSFQ also needs constant cooling and because of which complex structures haven't been produced yet. The bulk of the research is being performed on the above-mentioned drawbacks. Molecular electronics is another technology being explored. However, it's a futuristic approach therefore, many obstacles still need to be researched and addressed. Molecular electronics tries to implement chemical synthesis and self-assembly concepts.

**Architecture**

Nanoelectronic architecture consists of features such as wiring, the degree of parallelism, and data transfer range. Currently, most nanoelectronic devices are based on CMOS; therefore, the architecture related to these devices is relatively similar. As the research trend is moving toward finding alternatives to CMOS, alternative architecture approaches are being studied for an efficient device. In terms of RTD, the architecture for the device would consist of a lower amount of wires and less complex then CMOS architecture. Whereas, using a SET would require more complex architecture but will be more compact [12]. The current research in this parameter is mainly focused on researching and developing a local architecture for different devices, making it a more robust device, by creating a defect and fault tolerant device architecture.

**Fabrication**

The current fabrication that is being used is the top-down approach; this relies mainly on two techniques which are [12]:

1. Optical lithography (high throughput), provides a dimension of 180 nm
2. Electron beam lithography (low throughput), provides a dimension of 30 nm.

Currently, the major portion of the research is based on finding a fabrication technique that provides high throughput with dimensions lower than 100 nm. The viable techniques that can create a device in the range of 100–50 nm are extreme ultraviolet lithography (EUV) and X-ray lithography. X-ray lithography has been more economically viable since it provides a single resist process. The throughput and size of different techniques are compared and shown in Fig. 15.

From Fig. 15, the throughput versus the resolution of various techniques shows that a technique with high throughput is unable to provide a good resolution. E-beam lithography has shown good potential, however, it is not suitable for direct fabrication with a high throughput. Currently, the research is focused on establishing a technique to increase the throughput of electron beam lithography and provide direct writing or imprinting for nanoelectronics.

## 5 Recent Accomplishments

Today, nanotechnology is gaining public attention. Many researchers have developed various technologies for commercial purposes. The Canadian government has begun funding educational institutions for research in this area. This section will summarize briefly the recent accomplishments of nanoelctronics and magnetics:

**Nanomanipulators**

Nanomanipulator system allows researchers to look into materials at the atomic scale. As a result, it has led to new discoveries in biology, material science, and engineering

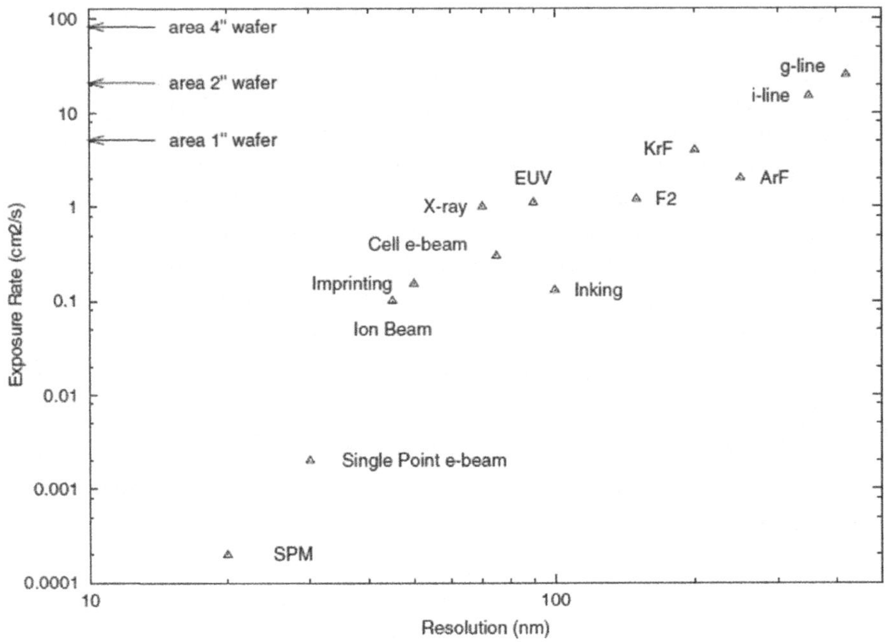


**Fig. 15** Comparison of various fabrication techniques [12]

[15]. For instance, researchers have used nanomanipulator systems to determine the mechanical and physical properties of carbon nanotubes. The recent development of nanomanipulators is the ability to build efficient nanomachines unless the physical principles at nanoscales are not understood.

## Nanofabrication

Nanofabrication is the most recent development in the research community. Researchers have gained the ability to control individual atoms directly. The next step is to manufacture 100 nm structures. Nanofabrication methods can be divided into two categories: top-down methods and bottom-up methods.

A top-down method has been used in developing electronics, such as a photolithography process. This technology can be used in the electronic industry to fabricate microprocessor chips. This process is expensive. Therefore, researchers try to develop soft lithography, which creates an elastic stamp to transfer a structure to a surface. Nevertheless, this technique is still in research and development because it is not ideal for manufacturing complex multi-layered structure electronic devices [1].

On the other hand, bottom-up technology represents a new production method. This technique may replace many devices in the future. One material created with the bottom-up method is a CNT. Several structures of CNTs varying in length and thickness have been developed for various purposes in building nano-devices. Another important material that is made by the bottom-up method is quantum dots. The electrical and mechanical properties are determined by the size of the dot. Therefore, the size and composition are key in production. The reflection of light by quantum dots can be used in the medical industries. It is hoped by the research community that the next stage is to fully expand the effectiveness and speed of the future of medicine treatments with fewer risks and costs [1].

## 6 Future Expectations

There are vast opportunities in the nanotechnology field, especially in nanoelectronics and nanomagnetics. Expansion in the nanoelectronics field has a lot of potential since it is still a fairly new and developing area, improvement and innovation are required in the actual device, the circuitry, architecture, and the final application (result). There are a few prime goals in terms of advancement in the nanoelectronics and magnetics field some that are going to revolutionize the industry. Fabrication, devices, and architecture are the major categories for these revolutionary advancements [45].

Considering fabrication one of the perceivable advancements in the near future includes the ability to achieve near-atomic-level control of reduced-dimensional materials at a 3D level. This is basically more control over the fabrication process in order to have increased control directly on the thickness and the lateral dimensions specifically during the growth and etching process. Having this control would improve the new materials' functionality, thus creating an advanced material [45]. Another possible expectation in the fabrication process is increasing the precision to 1 nm by combining lithography and self-assembly in order to pattern the semi-arbitrary structures. This combination will be extremely important for arbitrary pattern formation since currently, self-assembly alone has many defects as well as a repetitive structure and usually these qualities do not allow versatility for many applications. Therefore, self-assembly defect rates need to be improved as well as lithographic patterning which would allow for effective arbitrary pattern formation and in turn improved fabrication of the nanomaterials [45].

Expectations in the near future concerning devices also have two major perceivable advancements; one of these is reducing energy dissipation caused by devices for logic and memory [45]. The major aspect recognized for advancements is by identifying all the parameters that can be manipulated such that a reasonably attainable energy dissipation can be achieved. Exploiting and manipulating spinning parameters for memory, logic and new and improved functionality is also an advancement regarding devices that may soon be attainable in the future. The manipulation of these parameters shall, in turn, allow for the creation of devices that contain novel properties and may act as new materials. The behavior of single and collective spin will greatly affect the dynamics of the new and advanced devices [45].

The final major category for future advancement in the nanoelectronics and nanomagnetics field is the architecture. The actual integration of all the electrical and magnetic components even at the nanoscale needs to be organized in such a manner as to optimize productivity, efficiency, and cost and this is exactly what architecture is based on. This is known as integrating architecture and nanoscale device research for unique computational functionality [45]. Another important aspect of architecture is increasing the focus on emerging non-information technology applications. Current research is continuously advancing in the information technology (IT) sector; however, there is room for vast improvement in different industries and non-IT applications, which require unique necessities such as high temperature, ultra-low power, biologically friendly packaging, etc. Therefore, it is imperative that research is conducted on these processes from the very beginning to the end, rather than focusing on it as a general problem or even considering it as just improvement of computational analysis methods or improving the memory in traditional IT environments [45].

Many other possible future advancements can be explored; however, these were the major goals set for the next few years and they do seem attainable.

## 7 Conclusion

This literature review paper closely looked into different aspects of nanoelectronics and nanomagnetics. A brief overview was given followed by the review in nanoelectronics and nanomagnetics. In nanoelectronics, carbon nanotubes, nanowires and sensors as well some miscellaneous categories were focused upon. Nanomagnetics had a variety of aspects that were equally focused upon except for RFID; since RFID is such an upcoming prospect in the nanotechnology field, and it was focused on greater detail. Currently, in the fields of nanoelectronics and nanomagnetics architecture, devices and fabrication methods are of great importance. This is the most focused area and even in the near future, these particular areas are the main focus for further advancement. All in all, this literature review paper considered a variety of nanoelectronics and nanomagnetics aspects since it is a promising industry.

## References

1. Adleman L (1994) Molecular computation of solutions to combinatorial problems. Science 226(5187):1021–1024
2. Ajayan PM, Endo M, Song L et al (2012) Fabrication and characterization of single-walled carbon nanotube fiber for electronics applications. Carbon 50(15):5521–5524
3. Barett JT (n.d.) Examples of nanotechnology applications in electronics. http://science.opposingviews.com/examples-nanotechnology-applications-electronics-18240.html
4. Baughman RH, Zakhidov AA, Heer WA (2007) Carbon nanotubes—the route toward applications. Science's Compass. http://www.gel.usherbrooke.ca/beauvais/documents/Science_297_787_2002.pdf
5. Bhirde AA, Patel V, Gavard J et al (2009) Targeted killing of cancer cells in vivo and in vitro with EGF-directed carbon nanotube-based drug delivery. ACS Nano 3:307–316

6. Chavan PG, Badadhe SS, Mulla IS et al (2011) Synthesis of single crystalline CdS nanocombs and their application in photo-sensitive field emission switches. Nanoscale 3:1078–1083
7. Chavan PG, Kashid RV, Badhade SS et al (2014) CdS nanowires: ultra-long growth and enhanced field emission properties. Vacuum 101:38–45
8. Chen Y, Xihua W, Shyamsunder E et al (2006) Silicon-based nanoelectronic field-effect PH sensor with local gate control. Appl Phys Lett 89(22):223512
9. Cheng Q, Katy R, Ehsan J (2013) Carbon nanotube–poly(lactide-co-glycolide) composite scaffolds for bone tissue engineering applications. Bio-Med Eng Soc 41:904–916
10. Clarke P (2006) KAIST claims size record with 3-nm FinFET. EE Times. http://www.eetimes.com/document.asp?doc_id=1160025
11. Cleland AN, Roukes ML (2008) Silicon nanowires: advanced fabrication methods. J Appl Phys 92:2758–2769
12. Compañó R (2001) Trends in nanoelectronics. Nanotechnology 12(2):85–88
13. Datta A, Chavan PG, Sheini FJ et al (2009) Growth, optical, and field emission properties of aligned CdS nanowires. Cryst Growth Des 9(9):4157–4162
14. Deng D, Jin Y, Chen Y et al (2012) Preparation of copper nanoparticles with low sintering temperature. Paper presented at 14th international conference on electronic materials and packaging (EMAP), pp 1–4
15. Drexler KE (2001) Machine-phase nanotechnology. Sci Am 285(3):74–75
16. European Commissions (2009) FP7 workshop on advanced nanoelectronics technologies. Nano-Sil, 11 Sept 2009. Web, 6 Apr 2014. http://www.nanosil-noe.eu/data/document/report-advanced-nanoelectronics.pdf
17. Gentleman E, Lay AN, Dickerson DA et al (2003) Mechanical characterization of collagen fibers and scaffolds for tissue engineering. Biomaterials 24:3805–3813
18. Haratifar E, Shahverdi HR, Shakibaie M et al (2009) Semi-biosynthesis of magnetite-gold composite nanoparticles using an ethanol extract of Eucalyptus camaldulensis and study of the surface chemistry. J Nanomater 5:1687–4110
19. Hoffman RL, Norris BJ, Wager JF (2003) ZnO-based transparent thin-film transistors. Appl Phys Lett 82(5):733–735
20. Hsing I, Xu Y, Zhao W (2007) Micro-and nano-magnetic particles for applications in biosensing. Electroanalysis 19(7–8):755–768
21. Huang S-CJ, Artyukhin AB, Misra N et al (2010) Carbon nanotube transistor controlled by a biological ion pump gate. Nano Lett 10(5):1812–1816
22. Inberg A, Livshits P, Zalevsky Z et al (2012) Electroless deposition of silver thin films on gold nanoparticles catalyst for micro and nanoelectronics applications. Microelectron Eng 98:570–573
23. Johnson RC (2002) Nanoscale metal deposition eyed for MRAMs. EE Times. https://www.eetimes.com/nanoscale-metal-deposition-eyed-for-mrams/
24. Krane J (2001) Bell labs builds nano-transistor. BC Cylce. http://www.lexisnexis.com/hottopics/lnacademic/?verb=sr&csi=7911&sr=HEADLINE(Bell_Labs_builds_nano-transistor)+AND+DATE+IS+2001-10-17
25. Larson D, Zipfel W, Williams R et al (2003) Water-soluble quantum dots for multiphoton fluorescence imaging in vivo. Science 300(5624):1434–1436. http://www.sciencemag.org/content/300/5624/1434.full
26. Lee CJ, Lee TJ, Lyu SC et al (2002) Field emission from well-aligned zinc oxide nanowires grown at low temperature. Appl Phys Lett 81(19):3648
27. Li L, Wu P, Fang X et al (2010) Single-crystalline CdS nanobelts for excellent field-emitters and ultrahigh quantum-efficiency photodetectors. Adv Mater 22(29):3161–3165
28. Liu Z, Chen K, Davis C et al (2008) Drug delivery with carbon nanotubes for in vivo cancer treatment. Cancer Res 68:6652
29. Lui Z, Zhang D, Han S et al (2004) Single crystalline magnetite nanotubes. Unpublished raw data, Department of Electrical Engineering-Electrophysics, University of Southern California, Los Angeles, California. http://pubs.acs.org.ezproxy.lib.ryerson.ca/doi/pdf/10.1021/ja0445239

30. Nakano M, Matsuura H, Dong-ying J et al (2008) Drug delivery system using nano-magnetic fluid. Presented at 3rd international conference on innovative computing information and control, p 338, 18–20 June 2008
31. Nanocyl SA (2009) Carbon nanotubes. http://www.nanocyl.com/en/CNT-Expertise-Centre/Carbon-Nanotubes
32. Pan H, Poh C, Zhu Y et al (2008) Novel CdS nanostructures: synthesis and field emission. J Phys Chem 112(30):11227–11230
33. Qingsu C, Rutledge K, Jabbarzadeh E (2013) Carbon nanotube–poly(lactide-co-glycolide) composite scaffolds for bone tissue engineering applications. Bio-Med Eng Soc 41:904–916
34. Raj PM, Sharma H, Mishra D et al (2012) Nanomagnetics for high-performance, miniaturized power, and RF components [nanopackaging]. IEEE Nanotechnol Mag 6(3):18–23
35. Sebastian PJ, Gamboa SA (2005) Nanotechnology applied to thin film solar cells. Sol Energy Mater Sol Cells 88(2):129–130
36. Skomski R (2003) Nanomagnetics. J Phys: Condens Matter 15(20)
37. Smith D (2012) Nano-transistor breakthrough to offer billion times faster computer. The Sydney Morning Herald. http://www.smh.com.au/technology/sci-tech/nanotransistor-breakthrough-to-offer-billion-times-faster-computer-20120220-1thqk.html
38. Star A, Han T-R, Joshi V et al (2004) Nanoelectronic carbon dioxide sensors. Adv Mater 16:2049–2052
39. Thiess A, Mokrousov Y, Heinze S et al (2009) Magnetically hindered chain formation in transition-metal break junctions. Phys Rev Lett 103(21):217201
40. Torrisi F, Daping C, Tawfique H et al (2012) Inkjet-printed graphene electronics. ACS Nano 6(4):2992–3006
41. Voldman S, Hui D, Warriner L et al (2000) Electrostatic discharge (ESD) protection in silicon-on-insulator (SOI) CMOS technology with aluminum and copper interconnects in advanced microprocessor semiconductor chips. J Electrostat 49(3):151–168
42. Wang C (2011) Carbon nanotube nanoelectronics and macroelectronics. Dissertation, University of Southern California
43. Webster TJ, Waid MC, McKenzie JL et al (2004) Nano-biotechnology: carbon nanofibres as improved neural and orthopaedic implants. Nanotechnology 15:48–54
44. Welser J, Wolf S, Avouris P et al (2010) Applications: nanoelectronics and nanomagnetics. http://www.cein.ucla.edu/new/file_uploads/chapter08.pdf
45. Welser J, Wolf S, Avouris P et al (2011) Applications: nanoelectronics and nanomagnetics. Nanotechnology research directions for societal needs in 2020. Springer, Netherlands, pp 375–415
46. Wu W, Wieckowski S, Pastorin G et al (2005) Targeted delivery of amphotericin B to cells by using functionalized carbon nanotubes. Angew Chem Int Ed 44:6358–6362
47. Yarris L (2014) Cooling microprocessors with carbon nanotubes. Berkeley Lab News Center. http://newscenter.lbl.gov/news-releases/2014/01/22/cooling-microprocessors-with-carbon-nanotubes/
48. Zhai T, Fang X, Bando Y et al (2009) Characterization, cathodoluminescence, and field-emission properties of morphology-tunable CdS micro/nanostructures. Adv Funct Mater 19(15):2423–2430
49. Zhai T, Fang X, Bando Y et al (2009) Morphology-dependent stimulated emission and field emission of ordered CdS nanostructure arrays. ACS Nano 3(4):949–959
50. Zhang M, Zhai T, Wang X et al (2010) Arm-length-controlled synthesis, field emission, and luminescence of CdS multipods. Cryst Growth Des 10(3):1201–1206
51. Zhang D, Ryu K, Liu X et al (2006) Transparent, conductive, and flexible carbon nanotube films and their application in solar energy. Nano Lett 6:1880–1886
52. Zhirnov V, Herr JC, Muirhead G (2008) New frontiers: self-assembly and nanoelectronics. J Nanoelectron 2008:11
53. Zhirnov V, Herr JC (2001) New frontiers: self-assembly and nanolectronics. Computer 34(1):34–43

# Chapter 2
# Nanotechnology for Batteries

**Ashish Bhatnagar, Manoj Tripathi, Shalu, and Abhimanyu Prajapati**

## 1 Introduction

A battery is a device that converts chemical energy into electrical energy. The chemical reaction takes place between the anode and cathode via electrolyte, and electrical energy is stored during charging which is subsequently released during discharging. Availability of battery with higher charge storage capacity, high life cycle, low cost is talk of the day now. After the ground-breaking work of Sony Inc. in 1991 and Kasei and Toshiba in 1992, Li ion battery is continued to be the center of research due to their relatively high charge storage capacity, long cycle life [1]. Since 1992, the Li ion battery research has seen tremendous achievement, thanks to manipulation of matter at a scale of 1–100 nm, without which many advances in battery technology may not be possible [2]. The advancement in the field of battery materials (anode, cathode and electrolyte) relies heavily on dimensionally altered nanomaterials and nanotechnology, to improve conductivity and to suppress irreversible side reactions [2].

Following are the advantages of nanomaterial-based lithium-ion batteries:

1. Short diffusion pathways for electronic and ionic motion lead to quicker reactions and faster charging and discharging. In addition, the electrode's high surface area allows for better interaction and facilitates quicker ion (Li) transfer from and to the electrolyte by providing better interaction between electrode and electrolyte. It does, however, have the downside of making more electrolyte-side decomposition reactions at the electrode surface. This type of modification is generally associated with modification of kinetics [2].

A. Bhatnagar (✉) · M. Tripathi (✉) · A. Prajapati
Department of Physics and Materials Science and Engineering, Jaypee Institute of Information Technology, Noida, Sector-62, Noida 201309, India

Shalu
Department of Physics and Astrophysics, University of Delhi, New Delhi, India

 
M. Mujawar et al. (eds.), *Nanotechnology for Electronic Applications*, Materials Horizons: From Nature to Nanomaterials, https://doi.org/10.1007/978-981-16-6022-1_2

2. When synthesized at the nanoscale, the potential of some of the materials reaches the value, which makes their application in battery feasible. One of the prominent example for this are $Li_2O$ and LiF (potential reaches to the value of Li) which cannot be used as electrode in battery but can be used as electrode when synthesized in nano form. The following properties can be placed under the category of thermodynamics destabilization.
3. During the chemical reaction between electrodes, there is major variation in the volume of electrode, which has an adverse effect on battery capacity. Due to faster stress relief times, nanomaterials compensate for the volume changes and, therefore, improve the battery performance and increase its cycle life. For example, consider the case of bulk silicon in which the storage capacity of 4200 mAh/g can't be maintained after 30 cycles as its volume expands by nearly 300%. Interestingly, this high storage capacity can be maintained up to 100 cycles just by using Si nanowires in place of bulk Si. These changes can be associated with the mechanical stability of materials.

In the coming section, we will highlight the part of nanotechnology for improving the properties of critical components of battery. The focus will be on the nanomaterial-based approach for improving the battery characteristics.

## 2 Nanostructured Anode for Battery Application

Anode (besides cathode and electrolyte) is one of the very important constituents for the battery and contributes largely to its efficiency. The negative electrode of a rechargeable battery cell is commonly known as an anode. Recent advances in the use of batteries for electric cars require much higher energy capacity, greater density of power and long life cycles (compared with 0.240 Wh/kg usable commercial cells) [2].

Practically, viable anode should, therefore, have a high potential and a low voltage. However, it is necessary to remember that an anode's electrical potential should be less than the electrolyte. The absence of this chemistry in potential will corrode the anode by the formation of passivating solid electrolyte interphase (SEI) film on anode and, thus, will affect batteries' performance [3, 4].

A large number of studies over the last decades have concentrated on finding efficient anode material, which have most/all of the properties discussed above [4, 5]. The most investigated and sorted anode in commercial battery cell is carbon [6]. In addition to carbon-based material, IA (Li), IIA (Mg), IVA (Si, Ge, Sn, Pb), V A (P, As, Sb, Bi), IB (Ag, Au) and IIB (Zn, Cd) components were used in different rechargeable battery based on the required anode properties. These elements have been investigated based on one or more relevant properties. Unhappily, they do not meet all specifications in bulk form. Graphite (carbon), for instance, has good electronic conductivities and can be easily synthesized but it offers a low capacity 372 mAh/g, which can be attributed to storage of only 1 Li atom with 6 C atom

(LiC6). In comparison, silicone-based material may have more than 4200 mAh/g of specific capacity and sufficient Li/Li+ potential (alternatively 0.5 V) [7]. However, the volume expansion/contraction problem as discussed limits their application as battery anode.

The design of material properties on nanoscale has received considerable attention in order to achieve a high anode efficiency. Li+ is, for instance, anisotropically diffused and has sluggish kinetics in rutile $TiO_2$ at low temperature, which leads to lower capacity when $TiO_2$ is used as anode in battery [6]. Nanocrystalline $TiO_2$ rutile of 15 nm particles, on the other hand, increases cell reversibility to 263 mAh/g [6]. In the section below, we addressed the improvement in battery capacity by altering dimension of anodes, i.e. shifting it toward the nano regime.

## *2.1 Low-Dimensional Nanostructured Anode*

Low-dimensional structures have several unique favorable feature, primarily related to the kinetics and mechanical stability of the transport, [e.g. 0D nanoparticles] (Fig. 3.1A), 1D nanoparticles, 2D nanosheets thus have higher length to diameter ratio, which is beneficial for ion diffusion. Nanorods and nanowires are solid within whereas nanotubes are hollow inside. In contrast with the nanowire, the nanotubes are less in length and bigger in diameter and offer later more flexibility. Nanobelts are just rectangular elongated structures [2]. We will now proceed to discuss the geometrical aspects of dimensionally altered anode.

Tang et al. [8] synthesized long (30 m μm) $TiO_2$(B) by mechanically driven hydrothermal techniques and reported a capacity of 114 mAh/g at 8.4 A/g, for an impressive 10,000 cycles. The authors concluded that the pseudocapacitative load storage, short diffusion and wide area resulted from condenser-like performance and battery-like high cycling capacity. In addition to kinetics of preferable transmission and diffusion, 1D products provide eases stress relaxation and maintain structural integrity during the repeated loading/unloading process. In addition, Chan et al. [9, 10] investigated as synthesized Si nanowire remains intact even after cycling (10 cycles) and high capacity of 3500 mAh/g has been observed. The 1D structure of nanowire accommodates the strain and provides good contact with the current collector. With little change, the achieved capacity was maintained until 10 cycles. The stable capacity was as high as 2100 mAh/g.

Usually, 2D nanostructurings are very fine structures similar to sheets. These sheets vary from a few to ten nanometers in thickness. Various studies have shown that nanosheets can be considered as practical anode with improved 2D directional kinetics. Figure 3.1C displays the geometric and kinetic features of the nanosheet. 2D nanosheets are capable of achieving exceptional capacity, which improved the rate of charging and discharging because of facile electron transport through its 2D surface. A single-step solvothermal method was used by Kim et al. [11] to prepare crystalline nanosheets based on $SnS_2$. The retention of these nanosheets was 95% at current rate of 323 mA/g after 50 cycles. In $TiO_2$ (B), the structure based on nanosheets showed better rates than nanoparticles and nanotubes [12–14]. In addition, 2D structures can also provide volume-changing strain relaxation and versatility for anodes [13, 14]. The long-cycle life [15] of the anode, as discussed for 1D structures, is due to these characteristics. In addition, 2D structures offer the possibility of exploring certain crystallographic facets to increase an anode's storage capacity owing to the abundance of the exposed area [13, 14].

## 2.2 *High Dimensional Nanostructured Anode*

High-dimensional nanostructures are typically three-dimensional structures consisting of the 1D/2D connections together with the interrelationship and vacuum spaces. This theory is the basis for most of the recent nanostructures. These structures offer all the benefits of small-scale 1D/2D structures. Furthermore, the resultant anodes show improved performance as they were designed to correct the limits of their small-size structure and material properties. Examples are porous, hollow, heart, hybrid and heterogeneous nanostructures [2].

After the ground breaking work of Oumellal et al., hydride ($MgH_2$) emerged as an anode materials for battery application but there practical use beset largely due to its certain volume expansion/shrinkage [16] during repeated charging/discharging processes, which badly effects there performance as battery material. After only 10 cycles, the discharge capacity of $MgH_2$ electrode decreases from 1480 mAh $g^{-1}$ to less than 200 $mAhg^{-1}$. In addition, the absorption of the lithium and electrolyte for reformation of solid electrolyte interface layer on active materials leads to shortage of lithium, which eventually results in battery failure.

To solve these problems, $MgH_2$ nanoparticles (NPs) attached to graphene (GR) were investigated as Li ion battery anode with PMMA binder was also added to $MgH_2$-Gr to improve its conductivity. The $MgH_2$ NPs are synthesized on ultrathin and adjustable GR, forming a functional composite structure. Such formulations and strong graphene interactions obtained at nanoscale lead to improved lithium storage capacity of $MgH_2$/GR/PMMA [16]. The prepared electrodes deliver a maximum lithium storage capacity of 946 $mAhg^{-1}$ per 100 $mAg^{-1}$ (100 full cycles), and capacity of 589 $mAhg^{-1}$ or with a maximum current capacity of 2000 $mAg^{-1}$. Therefore, the $MgH_2$/GR combination promoted by the PMMA binder ensures that it is efficient nanostructures high-performance anode material.

## 2.3 *Hybrid Nanostructured Anode*

The mixture of 1D/2D/3D structures is hybrid structure. Hybrid structures can offer excellent electrochemical and mechanical properties by bringing in all the desirable features of the 1D/2D/3D structures.

Active nanostructures assisted by 2D-reduced graph oxide (RGO) (e.g., nanoparticles, nanospheres, core–shell-tures) are excellent examples of hybrid nanostructured anode. Such 2D sheets may contain multiple structures and by stacking face to face, form layers and may also be referred to as alternating layered or sandwich structures [17]. In one of his pioneering works, Zhong et al. [13, 14] reported a layered composite structure, i.e. long continuous graphene sheets consisting of uniformly distributed $MnCO_3$ nanoparticles.

This layered composite structure of metal oxide/carbon is superior as compared with either metal oxide or graphene in terms of electrochemical performance [18]. Keeping the above in view self-sacrificial property of MOFs has been reported to synthesize transition metal oxide (TMOs) as anode for LIBs [19–21], and so on. Microcubes, microboxes and hierarchically arranged $Fe_2O_3$ microboxes were synthesized by Zhang et al. [20] at 350, 550 and 650 °C, respectively. These molecular structures demonstrated exceptional lithium storage abilities and improved cycling ability when tested as anode material for Li-ion batteries [22].

On the other hand, TMOs synthesized using traditional inorganic methods have poor cycling stability and rate capability, which limits their application in rechargeable battery even after having utmost potential. MOFs owing to its self-sacrificial models can be used for producing nanostructures that satisfy the criteria listed above. Pyrolysis of MOFs in air can be utilized to have different types of TMOs with improved lithium-ion storage efficiency. The MOF network formed thermochemically decomposed and subsequently nucleated to obtain porous $Co_3V_2O_8$ metal oxide in the presence of atmospheric oxygen [23]. This electrode had outstanding cyclability and rate efficiency (940 mAh/g after 100 cycles and 650 mAh/g at 1 and 5 A/g after 400 cycles, respectively). In addition, the same electrode content, i.e. $Co_3V_2O_8$, was recorded by this research group through a simple solvothermal synthesis method, exhibiting performance nearly identical to that thermochemically produced $Co_3V_2O_8$ [24]. Novel $CoFe_2O_4$/GNS nanocomposite from the bi-metal organic system has also been designed and published, showing improved discharge ability and reversible capacity. This improvement can be attributed to the synergistic effect between CO and Fe [25].

The above facts and discussions point toward the general understanding of MOF-derived nanocomposites should offer better charge (Li/Na) storage characteristics.

Miao et al. [26] proposed the novel concept of regulating the size of active materials by using graphene oxide and synthesize 3D reduced graphene oxide aerogel (rGOA). The particle size of oxide has been regulated by the active functional group of GO in-situ due to Ostwald ripening. The modulation of active functional group has been done by regulating the gelation degree of GA (Fig. 1A, B). Interestingly Miao et al. hypothesized that a large number of open pores in 3D rGOA facilitate electrolyte transport transfer channel. On the other hand size regulated nanooxide shorten the ionic diffusion pathways and also buffer the volume expansion and enhance the kinetics od transfer process of Li+/e and enhance the kinetics of the transfer process of Li+/e. In order to synthesize other tunable "guest/substrate" functional nanohybrids, the logical strategy of modulating the gelation degree of GO substrates can be expanded. The reversible capacity of 1038.5 $mAhg^{-1}$ at a current density of 1000

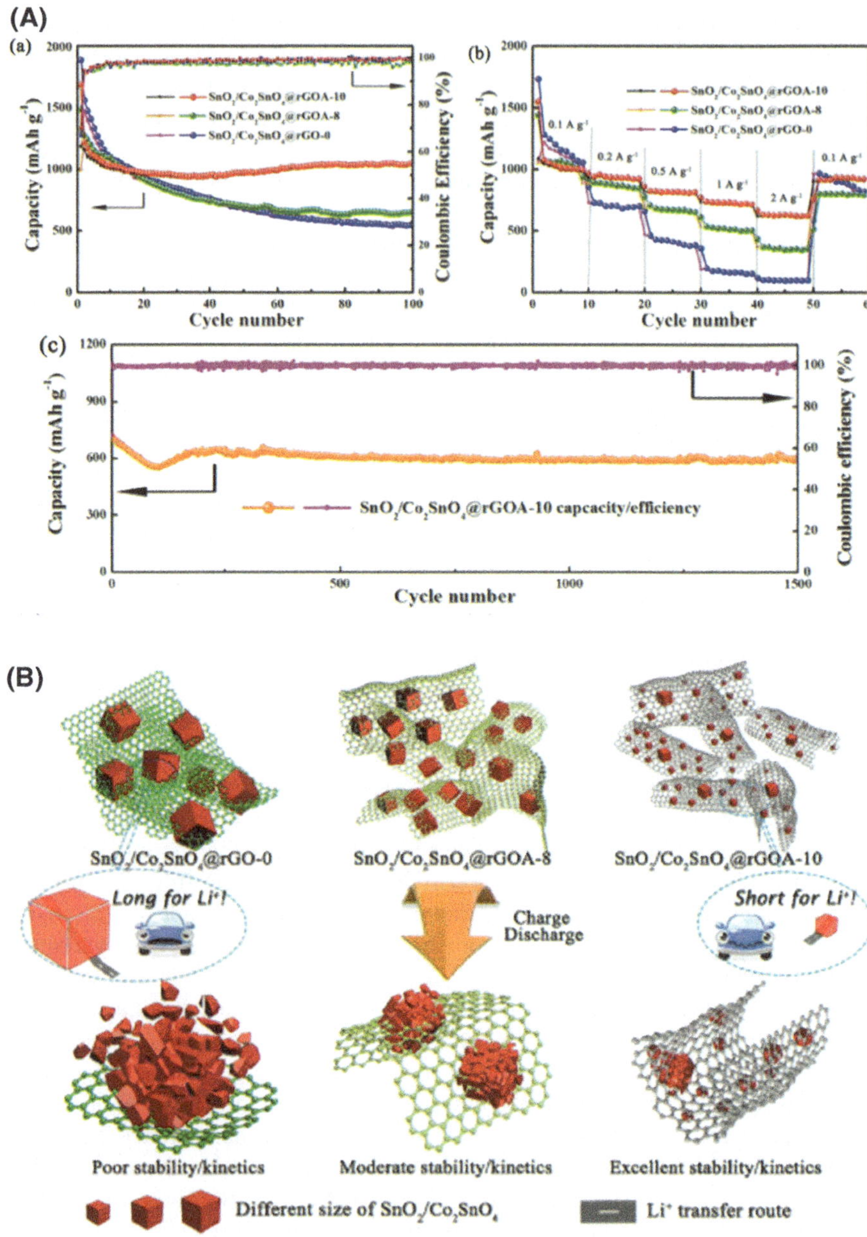


**Fig. 1** [**A**] (**a**) Cyclic performance of as-synthesized different electrodes at a current density of 100 mA/g between 0.01 and 3.0 V. (**b**) Rate performances of as-synthesized different electrodes. (**c**) Long cycling performance of $SnO_2/o_2$ $SnO_4$ @rGOA-10 electrode @ a current density of 1.0 A/g. [**B**] Schematic diagram of the cycling process of $SnO_2/Co_2$ $SnO_4$ @rGOA (Electrochimica Acta, 356 (2020), 136769.)

$mAhg^{-1}$ has been observed even after 1500 cycles in case of $SnO_2/Co_2SnO_4$@rGO) (Fig. 1A).

It is important to have structural aspects of nanostructured anode. As time progresses, new systems with specific advantages are reported. However, the efficiency of an anode does not rely on the structural aspects alone. One structure may be suitable for a specific material or group of materials but may not be suitable for other materials. Many material and structural combinations are possible, with a wide range of new possibilities available. Hybrid structures can simultaneously fulfill several requirements.

More and more works based on the hybrid structure are, therefore, being done. However, it is generally difficult to synthesize hybrid structures with accurate control, and hence, it is difficult to achieve a balance of the high performance anode and synthesis process complexity and eventually to update the entire device. Nonetheless, it is safe to say that nanostructured anodes will be directed in the future to hybrid structures. We will now focus and discuss the effect of nanostructured materials and their dimension in improving the performance of cathode.

## 3 Nanostructured Cathode for Battery Application

Nanotechnology can basically deliver new avenues of designing and manufacturing cathode materials for battery application [27]. We will now proceed forward to discuss the nanotechnological dimensional effect on cathode material performance.

### *3.1 Low-Dimensional Nanostructured Cathode*

For synthesizing LiFePO4 (LFP) nanoparticles, solvothermal method has been used by Zhu et al. [49]. According to Zhu et al., LFP nanoparticles offer improved electrochemical properties (100 mAh/g) at C/10. It has been further demonstrated that diamond-like polycrystalline LFP nanoparticles show enhanced capacity as compared with polygonal-shaped single crystal. Sarvanan et al. [50] used solvothermal technique in order to synthesize nanoplate of LFP-carbon (LFP/C). The morphology of LFP/C has been altered by varying the precursor. For example, LFP/C nanoplates, spindle, diamond and thick plate morphology have been achieved by using iron oxalate, iron acetate and iron gluconate. Out of all the above morphologies, LFP/C nanoplates show improved electrochemical performances owing to their lowest plate thickness. Xiao et al. [73] synthesized nanostructured $LiCOO_2$ by combining co-precipitation and solid-state method. $LiCOO_2$ synthesized with exposed (0001) plane show enhanced Li+ diffusion owing to its perpetual dicular plane. $LiCOO_2$(001) shows 140 mAh/g charging capacity at 100 mA/g and discharging capacity of 113 mAh/g.

## 3.2 High and Hybrid Dimensional Nanostructures

Vu and Stein [28] synthesized 3D hierarchical porous composite, LiFePO4-carbon (LFP/C) with interconnected macrospores and mesoporous. This hybrid geometry of pores facilitates electrolyte movement, offers wide surface area for Li+ exchange between the anode and cathode via electrolyte. The other advantage of this hybrid nanostructure geometry is that Li+ has to diffuse through shorter (180 nm) path.

The combination of the above properties resulted in 3D composite cathode with high capacity (150 mA h/g) at C/5 and 99% columbic efficiency with no capacity fading over 100 cycles. The composite supported current density was recorded to be 2720 mA/g (at 16 C rate) [28].

Wang et al. [29] obtain highly efficient LFP/C composite micro–nano hierarchical structure using hydrothermal process. The synthesized composite LFP/C showed a high ability of 120, 110 and 90 mAh/g at 5, 10 and 20 C, respectively.

Zhang et al. [30] synthesized a peanut-shaped LMR-NMC, using solvo-thermal method. Precursor oxide materials were first developed by a solvothermal process that showed a peanut-like morphology. The larger primary particles (100–300 nm) and pores (>50 nm), after the calcination with $LiOH{\cdot}H_2O$, are still the same. The LMR-NMC cathode prepared by a solvothermal method showed a high initial capacity of 229.9 mA/g.

Abou-Rjeily et al. [31] reported the formation of spinel $LiMn_2O_4$ by solid-state reaction using $\beta$-$MnO_2$ precursors followed by thermal treatment under atmospheric air impregnation. A variety of morphological and structural changes in flake size, bond vibration and mass loss were studied (Fig. 2A). The electrochemical studies of these materials show the initial specific discharge capacities to be in the 95 mAhg $^{-1}$ range. The material was able to cycle at high current rates, while maintaining 50% of its initial discharge capacity, with just a 7% decrease in capacity (Fig. 2B). For 500

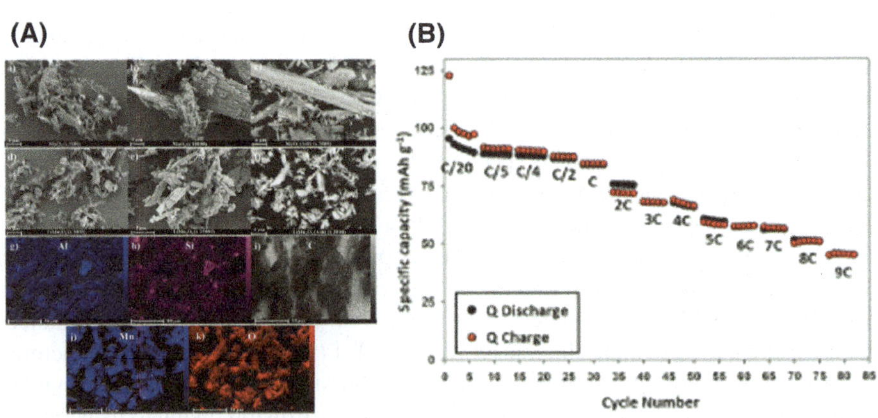


**Fig. 2** [A] (**a–c**) SEM micrographs of $MnO_2$ and $LiMn_2O_4$ (**d–f**) EDS mapping for (**g–i**). [**B**] Plot of specific capacity versus cycle number of $LiMn_2O_4$-based cathodes at different current rates (Energy Storage Materials, 26, (2020) 423–432)

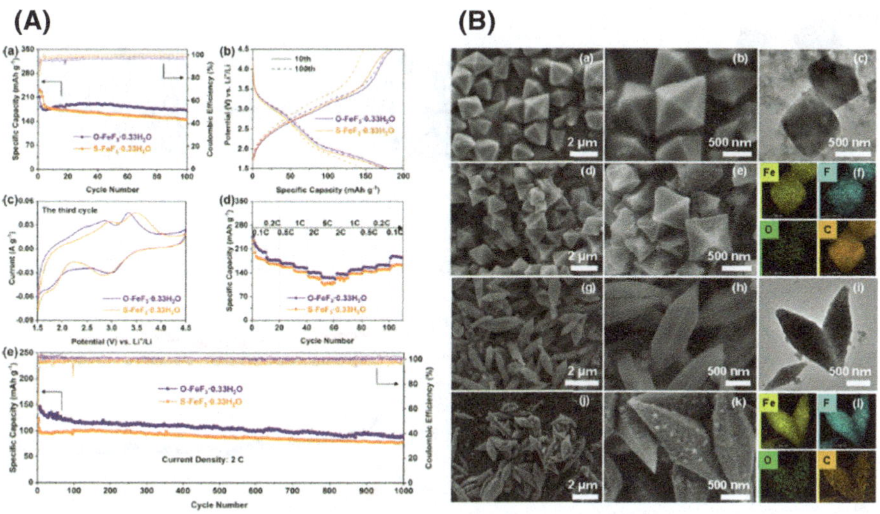


**Fig. 3** [**A**] (**a**) Cyclic performances. (**b**) Charge/discharge profiles. (**c**) CV curves. (**d**) Plot of specific capacity versus cycle number. (**e**) Cycling stability curve of O-$FeF_3{\cdot}0.33H_2O$ and S-$FeF_3{\cdot}0.33H_2O$ and [**B**] (**a, b**) SEM micrograph. (**c–e**) TEM and SEM micrographs. (**f**) EDS of $FeF_3{\cdot}0.33H_2O$. (**g, h**) SEM micrograph. (**i–k**) TEM and SEM micrographs. (**l**) EDS of S-$FeF_3{\cdot}0.33H_2O$ (Inorg. Chem. 59 (2020) 12700–12710)

cycles at a combined rate of charge/discharge, a stable battery was observed with an irreversible capacity loss and a drop in capacity of 3%.

Cheng et al. [32] reported two nanostructured cathode $FeF_3O{\cdot}3H_2O$ (named O-$FeF_3{\cdot}0.33H_2O$ and S-$FeF_3{\cdot}0.33H_2O$ here O implies octahedral and S-spindle) with hierarchical porosity and Skelton of graphite. These materials were prepared by in-situ co-pyrolysis of morphologically regulated different types of Fe-based metal-organic framework. Among the above two materials, owing to its octahedral architecture, O-$FeF_3{\cdot}0.33H_2O$ show low capacity attenuation (0.039% per cycle) even after 1000 cycles at 2C (1C = 273 mA h/g) (Fig. 3A, B).

Graphite-coated lithium manganese oxide (GLMO) was produced for various growth periods, e.g. the sample with 10 min growth period was termed as GLMO-10, similarly, the other versions were termed as GLMO-20, GLMO-30, GLMO-60. All GLMO electrodes demonstrated an improved cycling efficiency compared with the BLMO (Bare lithium manganese oxide) electrodes by retaining 96.2% after 200 cycles [33, 34].

Sui et al. [35] synthesized cathode material using impregnation of 4,8-dihydrobenzo[1,2-b:4,5-b′] dithiophene-4,8-dione (BDT) in 3D graphene structures (Fig. 4a). The BDT/3DGraphene composite 80% capacity retention for BDT (~100 mAhg) at 0.5 C and capacity of 210 mAh $g^{-1}$ at 4.0 C for 200 cycles) (Fig. 4b). The introduction of a 3D graphene as an electrode substrate reduces the dissolution of the BDT in electrolyte to a greater extent and also improved the electrical conductivity.

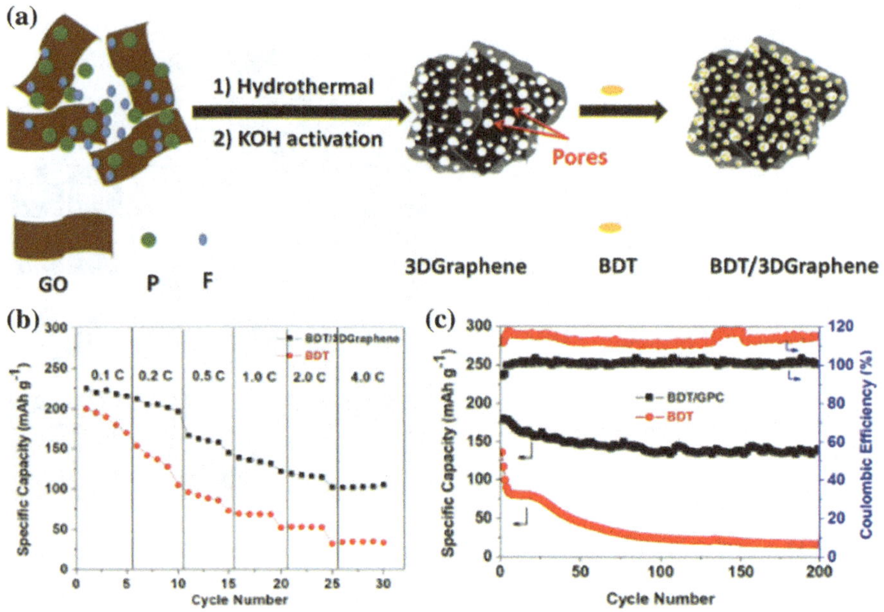


**Fig. 4** (**a**) Schematic of synthesis. (**b**) Plot of specific capacity versus cycle number. (**c**) Cyclic stability curve of BDT/3D graphene and BDT at 0.5 C rate (Carbon 157 (2020) 656–662)

Ghiyasiyan-Arani and Salavati-Niasari [36] fabricated 3D nanocomposite architecture of aluminum vanadate ($AlV_3O_9$) and carbon nanostructures in diverse dimensions. Graphene (GR), CNTs and graphene quantum dots (GQDs) are taken as 2D, 1D and 0D carbon nanostructures, respectively. Through a simple surfactant-assisted hydrothermal technique, morphology-controlled $AlV_3O_9$ was synthesized. Various nanocomposites that have been synthesized were abbreviated as $AlV_3O_9$ MWCNTs (AV-CNT), $AlV_3O_9$Graphene (AV-G) and $AlV_3O_9$/GQDs (AV-Q) (Fig. 5A, B). The specific capacity of pristine $AlV_3O_9$ spheres at 200 mA/g is 140 mAh/g. In contrast to pristine AV, the basic capability of AV-CNT increased to 235.63 mAh/g (at 200 mA/g). Other nanocomposites, AV-G and AV-Q, showed no major impact on the performance of the battery. The AV-CNT composite (1% MWCNTs), yielded optimal electrochemical properties among all composites tested, by delivering initial capacity of 228.15 mAh/g at 200 mA/g, and maintaining 198 mAh/g after 50 cycles (Fig. 6).

The techniques of nanostructure formation and nanoscale surface coating have continuously evolved over time and have been thoroughly researched to acquire control over thickness, and uniformity to significantly improve the results of battery cathode materials. In the next section, we will discuss and highlight the important work done in the field of electrolyte for improving battery performance, keeping in view the contribution of nanotechnology.

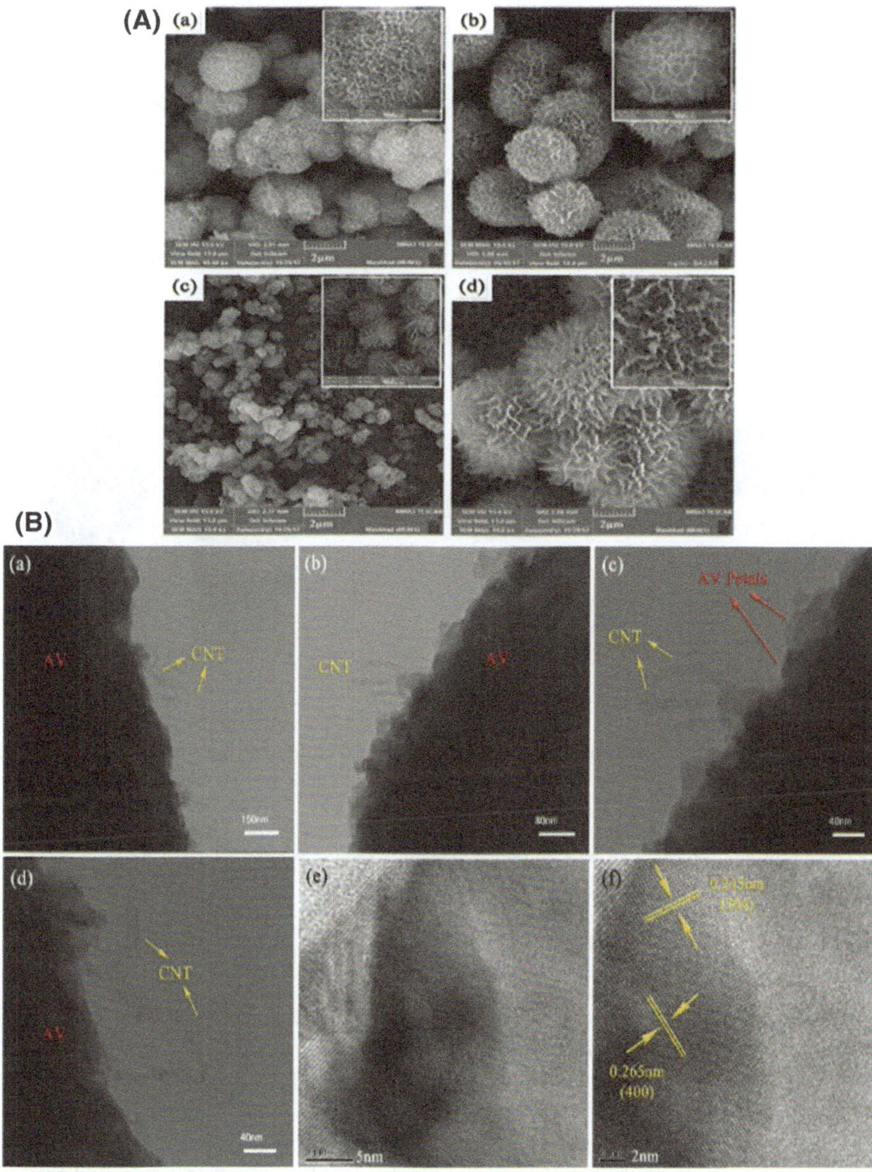


**Fig. 5** [**A**] FE-SEM images of AV microspheres at various synthesis conditions. (**a**) 160–6 h, (**b**) 180–6 h, (**c**) 200–6 h, (**d**) 160–10 h and [**B**] (**a**)–(**d**) TEM images of AV-CNT nanocomposites, (**e**) and (**f**) HR-TEM images of AV-CNT

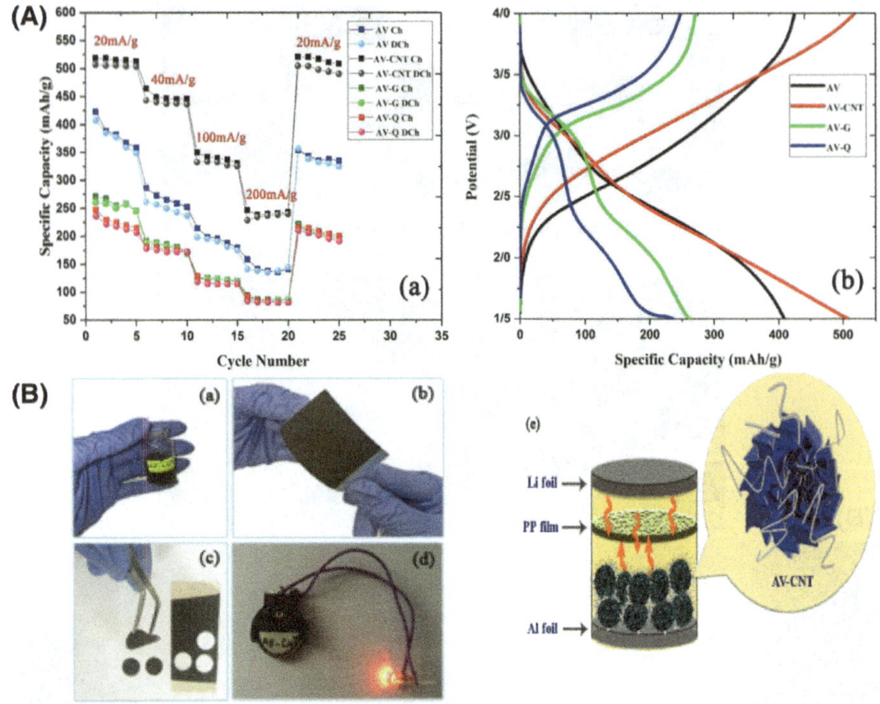


**Fig. 6** [**A**] (**a**) Cycling performance of as-synthesized electrodes at different rates. (**b**) Initial charge/discharge curves of as-synthesized different electrodes at 20 mA/g and [**B**] Schematic representation of (**a**) as-synthesized slurry of AV-CNT. (**b**) The picture of evenly coated Al foil. (**c**) Shows the flexibility of fabricated electrode. (**d**) The photo of the assembled coin cell in a working condition. (**e**) The representation of a LIB half-cell arrangement (Compos B Eng (2020), p. 183)

# 4 Nanotechnology in Electrolyte

In ideal context during battery operation, the electrolyte does not undergo chemical changes, thus is a sort of inert battery component that needs stability against both positive and negative surfaces. The electrochemical stability (ES) of the electrolyte is achieved in the actual system by passivation rather than thermodynamics [37]. Besides having ES, the electrolyte should also have: High ionic conductivity for enhanced motion of charges between the electrodes; low electronic conductivity to avoid self-discharge and short circuiting, wide electrochemical stability window to prevent its decomposition in the range of working potentials of cells [38]. Electrolyte also plays an important role so as to avoid the unwanted dendrite formation during cycling.

Most electrolyte solutions are organic, aprotic and solvent-based [38]. While the hot electrolyte vapor can pour out of a cell and lead to an explosion hazard. Low charge retention and problems at elevated temperatures are the other key challenges associated with liquid electrolytes [38]. To improve safety in liquid electrolytes, the researchers are looking into solid-state electrolytes (SSEs). Additionally, SSEs can provide other advantageous properties, such as higher stable operating voltages (>4.5 V), which are too high for typical organic solvents. Thus SSEs can have improved battery operation protection [38] and also offered improved chemistry of intercalation [37]. Keeping the above in view, we will confined ourselves to the discussion of SSEs in the upcoming section.

### *4.1 Solid Electrolytes*

Solid electrolytes are high ionic conductivity materials, either of cations or anions, but typically not both, or of insignificant electronic conductivity. It is the equivalent of solid state of molten salts or heavy liquid electrolytes [39]. Two major candidate groups are divided into solid electrolytes, solid polymer electrolytes (SPEs) and gel polymer electrolytes (SPEs) [40–42] (Dudney et al. 2000). However due to various disadvantages in these two groups of polymer electrolyte a new class of polymer discussed below are been researched now-a-days.

### *4.2 Hybrid Nanostructured Electrolyte*

Bae et al. [40] developed a 3D nanostructured hydrogel-derived $Li_{0.35}La_{0.55}TiO_3$ (LLTO) as a nanofiller for Li-ion electrolyte composite polymer. The systematic percolation analysis showed that at room temperature, the pre-percolating structure of the LLTO framework improved Li-ion conductivity to $8.8 \times 10^{-5}$ S/cm (Fig. 7). An increase in conductivity is mainly due to interconnected 3D nanostructured LLTO structure, which possibly forms a continuous interphase. This new 3D nanostructured hydrogel-derived percolating structure manufacturing technique leads to efficient composite polymer electrolytes (Fig. 7a).

Dai et al. [43] use a top-down approach to synthesize ion-conductive poly(ethylene oxide) (PEO) in the mesoporous, wood-templated aligned garnet nanostructure, resulting in a flexible composite electrolyte solid state. PEO improves the mechanical strength mesoporous garnet membrane and improves the motion of ion through the amorphous conductive and garnet polymer interfaces. This synergistic integration results in high ionic conductivity (at room temperature $1.8 \times 10^{-4}$ S/cm and $1.1 \times 10^{-4}$ S/cm at 95 °C) and strong mechanical stability (Fig. 8A, B). The scalable compressed wood model provides the special low tortuosity and mesostructure of the high surface area, which is crucial for improved electrochemical and mechanical stability. Figure 8B shows a cycling (for 180 h) at 0.1 mA/cm$^2$ for 30 min in each direction.

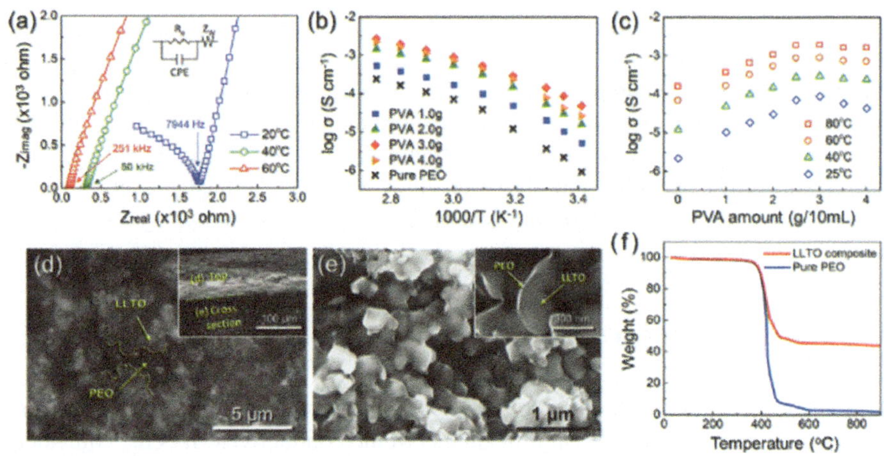


**Fig. 7** (**a**) Electrochemical impedance spectroscopy (EIS). (**b**) Arrhenius plot and (**c**) conductivity. (**d**) Top view (with low magnification inset) and (**e**) cross-section (with high magnification inset) SEM images of LLTO composite electrolytes. (**f**) TGA analysis of PEO-LiTFSI (pure PEO) and LLTO framework composite electrolyte (Angew. Chem. Int. Ed. 57, (2018) 2096.)

Wu and Guo [44] synthesized Uio-66 organic metal frameworks UIO/Li-IL SEs (MOFs) in UIO/LiIL SE nanostructured, by absorbing Li-IL in nanostructured Uio-66 MOF nanopores, and used in solid-state $LiFePO_4$/Li batteries (Fig. 9a). The UIO/Li-IL (15/16) SE's ionic conductivity was observed to be $3.2 \times 10^{-4}$ S/cm at 25 °C. With a high surface tension of nanostructured UIO/Li-IL SEs, the interaction between electrodes and SE is excellent, thus, Li/SE and $LiFePO_4$/SE interfacial resistances are around 44 and 206 Ω $cm^2$ at 60 °C, respectively. In addition, at the Li/SE interface, a stable solid conductive layer forms, making li plating/stripping stable (Fig. 3.5.3B). UIO/Li-IL SE solid-state batteries have high discharge capacities and excellent delays of retention (passing 100% of 0.2 C after 100 cycles; 119 mA h g $^{-1}$ maintaining 94% after 380 cycles at 1 C). This new form of UIO/Li-IL SE is very promising for solid-state batteries that will pave the way for stable long-life energy storage systems. This work can be popularized to other MOFs in addition to the Uio 66 in order to explore their applications in energy storage systems.

Electrolytes play a major role in efficiency and in cell aging, considering the high importance of active materials in an energy storage system. Fundamental research has shown that cell output is highly dependent on the electrolyte's chemical composition, resulting in wide variations in cycling power, rate capability and capacity retention at different temperatures. Reformulation of the electrolytes by using nanotechnology will be needed to create new cell chemistries or to improve the existing ones. In controlling the surface morphology of metallic lithium, solid-based electrolytes are also suggested to be sufficient and are far safer than liquid electrolyte. Further development of a solid-like electrolyte device for lithium metal battery assembly is the subject of ongoing research. Nanostructured solid-state or solid-like electrolytes may address key challenges in terms of improving interface stability and resistance.

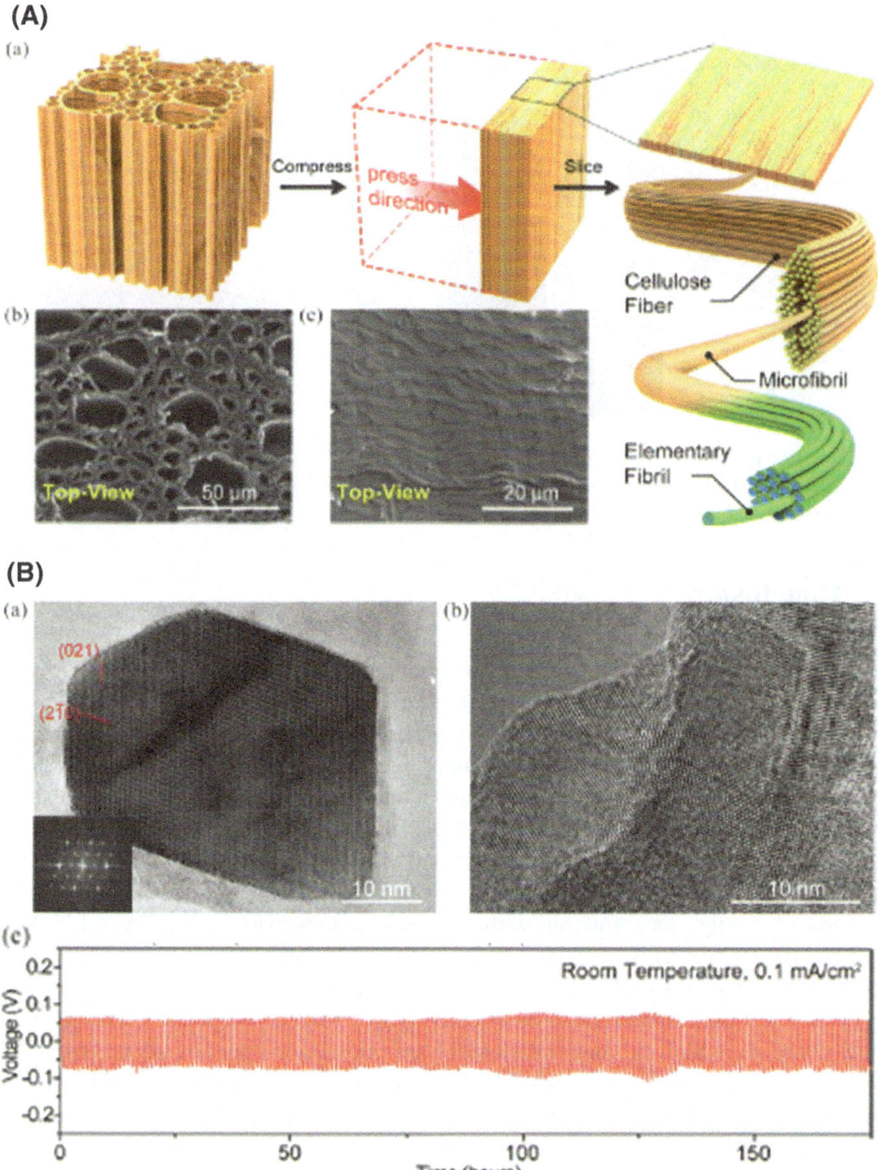


**Fig. 8** [**A**] (**a**) Fabrication schematic of the wood template using different physical techniques. (**b**) SEM image of pristine wood. (**c**) SEM image of compressed wood (ACS Materials Lett. 1, 3, (2019), 354–361). [**B**] HRTEM image (**b**) TEM image of a nanoparticle derived from the aligned garnet (**c**) voltage versus time curve of Li/garnet-wood/Li at RT and 0.1 $mA/cm_2$ current density (ACS Materials Lett. 1, 3, (2019), 354–361)

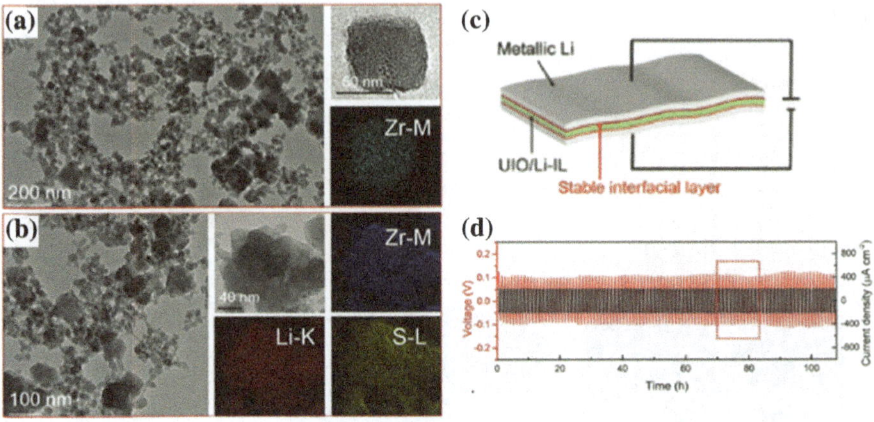


**Fig. 9** TEM micrographs of (**a**) UIO-66 and (**b**) UIO/Li-IL together with EELS mapping of Zr, Li and S. (**c**) Schematic of the structure of the symmetric cell. (**d**) Voltage versus time curve of UIO/Li-IL SE (Small 2019, 1804413)

## 5 Conclusion

While Li-ion batteries have seen drastic changes over the years, a lot of work has yet to be done to make batteries an alternative comparable to the existing fuels. Nanotechnology does not constitute a panacea. Even with its use, energy density and other qualities, lag well behind other fuels. Nanotechnology actually offers new ways of designing, synthesizing and manipulating cathode materials to solve power limitations and dramatically increase the efficiency of the battery. Undeniably, nanostructured materials have opened a new performance paradigm in the production of rechargeable battery cells. Principally, nanostructured materials offer various advantages as compared with their bulk counterpart: (1) facilitate Li+/Na+ and electron transport by providing shorter diffusion paths, (2) provide greater area, (3) provide spaces to allow expansion and thus preserve structural integrity and (4) enable different materials to be effectively hybridized/composited. The geometrical component and the material element are two major interdependent aspects of nanostructure materials. In terms of ion transport, stability and so on, 0D (such as nanoparticles), 1D (such as nanotubes, nanorods) and 2D (such as nanosheets) each have unique properties. Each of them alone cannot effectively fulfill all the requirements of robust battery materials for overall high efficiency. Nanostructuring offers new ways of designing and synthesizing materials to address power constraints and dramatically boost battery efficiency. Designing 3D hybrid structures can fulfill most of the requirements by combining low-dimensional structures. Among the solid-state battery technologies, hybrid dimensional nanostructured electrolytes also need to address key challenges in terms of improving interface stability and resistance at solid–solid interfaces with the electrodes when incorporated into a complete electrochemical system (e.g. solid-state battery with Li metal anode), as well as avoiding the growth of Li dendrites and improving rate capability. Huge research efforts are being currently dedicated to face these challenges.

Although there is still a long way to go until it is possible to incorporate nanostructured materials into a complete system, the basic knowledge produced is of great impact and will undoubtedly accelerate the realization of new battery technology with the help of nanotechnology. It is expected that nanomaterial-based materials will be commercialized in the very near future and meet the high requirements of modern applications by solving the existing problems.

## References

1. Abu-Lebdeh Y, Davidson I (eds) (2012) Nanotechnology for lithium-ion batteries. Springer Science & Business Media
2. Goutam S, Omar N, Van Den Bossche P, Van Mierlo J (2017) Chapter Two—Review of nanotechnology for anode materials in batteries. In: Rodriguez-Martinez LM, Omar N (eds) Emerging nanotechnologies in rechargeable energy storage systems, 1st edn. Elsevier, Inc., Amsterdam, pp 45–82
3. Liu C, Neale ZG, Cao G (2016) Understanding electrochemical potentials of cathode materials in rechargeable batteries. Mater Today 19:109–123. https://doi.org/10.1016/j.mattod.2015.10.009
4. Tarascon JM, Armand M (2001) Issues and challenges facing rechargeable lithium batteries. Nature 414:359–367. https://doi.org/10.1038/35104644
5. Wang LP, Yu L, Wang X, Srinivasan M, Xu ZJ (2015) Recent developments in electrode materials for sodiumion batteries. J Mater Chem A 3:9353–9378. https://doi.org/10.1039/C4TA06467D
6. Jiang C, Honma I, Kudo T, Zhou H (2007) Nanocrystalline rutile $TiO_2$ electrode for high-capacity and highrate lithium storage. Electrochem Solid State Lett 10:A127. https://doi.org/10.1149/1.2712041
7. Su X, Wu Q, Li J, Xiao X, Lott A, Lu W, Sheldon BW, Wu J (2014) Siliconbased nanomaterials for lithiumion batteries: a review. Adv Energy Mater 4:n/a–n/a10. https://doi.org/10.1002/aenm.201300882
8. Tang Y, Zhang Y, Deng J, Wei J, Le Tam H, Chandran BK, Dong Z, Chen Z, Chen X (2014) Mechanical forcedriven growth of elongated bending $TiO_2$based nanotubular materials for ultrafast rechargeable lithium ion batteries. Adv Mater 26:6111–6118. https://doi.org/10.1002/adma.201402000
9. Chan CK, Peng H, Liu G, McIlwrath K, Zhang XF, Huggins RA, Cui Y (2008) Highperformance lithium battery anodes using silicon nanowires. Nat Nanotechnol 3:31–35. https://doi.org/10.1038/nnano.2007.411
10. Chan CK, Peng H, Liu G, McIlwrath K, Zhang XF, Huggins RA, Cui Y (2008) High performance lithium battery anodes using silicon nanowires. Nat Nano 3(1):31–35
11. Kim TJ, Kim C, Son D, Choi M, Park B (2007) Novel $SnS_2$nanosheet anodes for lithiumion batteries. J Power Sources 167:529–535. https://doi.org/10.1016/j.jpowsour.2007.02.040
12. Liu S, Jia H, Han L, Wang J, Gao P, Xu D, Yang J, Che S (2012) Nanosheetconstructed porous $TiO_2B$ for advanced lithium ion batteries. Adv Mater 24:3201–3204. https://doi.org/10.1002/adma.201201036
13. Zhong Y, Yang M, Zhou X, Luo Y, Wei J, Zhou Z (2015) Orderly packed anodes for highpower lithium ion batteries with superlong cycle life: rational design of $MnCO_3$/largearea graphene composites. Adv Mater 27:806–812. https://doi.org/10.1002/adma.201404611

14. Zhong Y, Yang M, Zhou X, Zhou Z (2015) Structural design for anodes of lithiumion batteries: emerging horizons from materials to electrodes. Mater Horiz 2:553–566. https://doi.org/10.1039/C5MH00136F
15. Yu C, Li X, Ma T, Rong J, Zhang R, Shaffer J, An Y, Liu Q, Wei B, Jiang H (2012) Silicon thin films as anodes for highperformance lithiumion batteries with effective stress relaxation. Adv Energy Mater 2:68–73. https://doi.org/10.1002/aenm.201100634
16. Zhang B, Xia G, Sun D, Fang F, Yu X (2018) Magnesium hydride nanoparticles self-assembled on graphene as anode material for high-performance lithium-ion batteries. ACS Nano 12:3816–3824
17. Melot BC, Tarascon JM (2013) Design and preparation of materials for advanced electrochemical storage. Acc Chem Res 46:1226–1238. https://doi.org/10.1021/ar300088q
18. Chaikittisilp W, Ariga K, Yamauchi Y (2013) J. Mater Chem A 1:14–19
19. Fan HJ, Göele U, Zacharias M (2007) Small 3:1660–1671
20. Zhang Q, Wang WS, Goebl J, Yin YD (2009) Nano Today 4:494–507
21. Xiong SL, Zeng HC (2012) Angew Chem Int Ed 51:949–952
22. Zhang L, Wu HB, Madhavi S, Hng HH, Lou XW (2012) J Am Chem Soc 134:17388–17391
23. Sambandam B, Soundharrajan V, Mathew V, Song J, Kim S, Jo J, Tung DP, Kim S, Kim J (2016) J Mater Chem A 4:14605–14613
24. Soundharrajan V, Sambandam B, Song J, Kim S, Jo J, Kim S, Lee S, Mathew V, Kim J (2016) ACS Appl Mater Interfaces 8:8546–8553
25. Yang H, Zhang K, Wang Y, Yan C, Lin S (2018) J Phys Chem Solids 115:317–321
26. Miao X, Ge X, Wang P, Zhao D, Yin L (2020). Size-tunable $SnO_2/Co_2SnO_4$ nanoparticles loaded 3D reduced graphene oxide aerogel architecture as anodes for high performance lithium ion batteries. Electrochimica Acta 356:136769
27. Xie X, Su D, Sun B, Zhang J, Wang C, Wang G (2014) Synthesis of single-crystalline spinel $LiMn_2O_4$ nanorods for lithium-ion batteries with high rate capability and long cycle life. Chem A Eur J 20:17125–17131
28. Vu A, Stein A (2011) Multiconstituent synthesis of $LiFePO_4$/C composites with hierarchical porosity as cathode materials for lithium-ion batteries. Chem Mater 23:3237–3245
29. Wang M, Yang Y, Zhang Y (2011) Synthesis of micro-nano hierarchical structured $LiFePO_4$/C composite with both superior high-rate performance and high tap density. Nanoscale 3:4434
30. Zhang Y, Li Y, Niu X, Wang D, Zhou D, Wang X et al (2015) A peanut-like hierarchical micro/nano-Li1.2Mn0.54Ni0.18Co0.08$O_2$ cathode material for lithium-ion batteries with enhanced electrochemical performance. J Mater Chem A 3:14291–14297
31. Abou-Rjeily J, Bezza I, Laziz NA, Autret-Lambert C, Sougrati MT, Ghamouss F (2020) High-rate cyclability and stability of $LiMn_2O_4$ cathode materials for lithium-ion batteries from low-cost natural $\beta-MnO_2$. Energy Storage Mater 26:423–432
32. Cheng Q, Pan Y, Chen Y, Zeb A, Lin X, Yuan Z, Liu J (2020) Nanostructured iron fluoride derived from Fe-based metal-organic framework for lithium ion battery cathodes. Inorg Chem 59:12700–12710
33. Wang J, Zhang Q, Li X, Wang Z, Guo H, Xu D et al (2014) Sputtering graphite coating to improve the elevated-temperature cycling ability of the $LiMn_2O_4$ electrode. Phys Chem Chem Phys 16:16021–16029
34. Wang W, Ruiz I, Guo S, Favors Z, Bay HH, Ozkan M, Ozkan CS (2014) Hybrid carbon nanotube and graphene nanostructures for lithium ion battery anodes. Nano Energy 3:113–118. https://doi.org/10.1016/j.nanoen.2013.10.005
35. Sui D, Xu L, Zhang H, Sun Z, Kan B, Ma Y, Chen Y (2020) A 3D cross-linked graphene-based honeycomb carbon composite with excellent confinement effect of organic cathode material for lithium-ion batteries. Carbon 157:656–662
36. Ghiyasiyan-Arani M, Salavati-Niasari M (2020) Strategic design and electrochemical behaviors of Li-ion battery cathode nanocomposite materials based on $AlV_3O_9$ with carbon nanostructures. Compos B Eng 183
37. Scrosati B (2000) Recent advances in lithium ion battery materials. Electrochim Acta 45:2461–2466

38. Bekaert E, Buannic L, Lassi U, Llordés A, Salminen J (2017) Chapter One—Electrolytes for Li- and Na-ion batteries: concepts, candidates, and the role of nanotechnology. In: Rodriguez-Martinez LM, Omar N (eds) Emerging nanotechnologies in rechargeable energy storage systems. Micro and nano technologies. Elsevier, pp 1–43
39. Dudney NJ (2003) Lithium batteries. In: Nazri GA, Pistoia G (eds) Lithium batteries, science and technology. Springer Science + Business Media, LLC, New York, NY, pp 623–642
40. Bae J, Li Y, Zhang J, Zhou X, Zhao F, Shi Y, Goodenough JB, Yu G (2018) A 3D nanostructured hydrogel framework-derived high-performance composite polymer lithium-ion electrolyte. Angew Chem Int Ed 57:2096
41. Weston JE, Steele BCH (1982) Effects of inert fillers on the mechanical and electrochemical properties of lithium salt-poly(ethylene oxide) polymer electrolytes. Solid State Ion 7(1):75–79
42. Meyer WH (1998) Polymer electrolytes for lithium-ion batteries. Adv Mater 10(6):439–448
43. Dai J, Fu K, Gong Y, Song J, Chen C, Yao Y, Pastel G, Zhang L, Wachsman E, Hu L (2019) Flexible solid-state electrolyte with aligned nanostructures derived from wood. ACS Mater Lett 1(3):354–361
44. Wu J-F, Guo X (2019) Nanostructured metal–organic framework (MOF)-derived solid electrolytes realizing fast lithium ion transportation kinetics in solid-state batteries. Small 1804413
45. Balaya P, Li H, Kienle L, Maier J (2003) Fully reversible homogeneous and heterogeneous Li storage in $RuO_2$ with high capacity. Adv Funct Mater 13(8):621–625
46. Qin G, Xue S, Ma Q, Wang C (2014) The morphology controlled synthesis of 3D networking $LiFePO_4$ with multiwalled-carbon nanotubes for Li-ion batteries. CrystEngComm 16:260–269
47. Li H, Zhou H (2012) Enhancing the performances of Liion batteries by carboncoating: present and future. Chem Commun 48:1201. https://doi.org/10.1039/c1cc14764a
48. Tang J, Dysart AD, Pol VG (2015) Advancement in sodiumion rechargeable batteries. Curr Opin Chem Eng 9:34–41. https://doi.org/10.1016/j.coche.2015.08.007
49. Zhu J, Fiore J, Li D, Kinsinger NM, Wang Q, DiMasi E et al (2013) Solvothermal synthesis, development, and performance of $LiFePO_4$ nanostructures. Cryst Growth Des 13:4659–4666
50. Saravanan K, Balaya P, Reddy MV, Chowdari BVR, Vittal JJ (2010) Morphology controlled synthesis of $LiFePO_4/C$ nanoplates for Li-ion batteries. Energy Environ Sci 3:457
51. Xu K (2004) Nonaqueous liquid electrolytes for lithium-based rechargeable batteries. Chem Rev
52. Ji L, Lin Z, Alcoutlabi M, Zhang X (2011) Recent developments in nanostructured anode materials for rechargeable lithiumion batteries. Energy Environ Sci 4:2682. https://doi.org/10.1039/c0ee00699h
53. Ma L, Hendrickson KE, Wei S, Archer LA (2015) Nanomaterials: science and applications in the lithium–sulfur battery. Nano Today 10:315–338. https://doi.org/10.1016/j.nantod.2015.04.011
54. Li H, Richter G, Maier J (2003) Reversible formation and decomposition of LiF clusters using transition metal fluorides as precursors and their application in rechargeable Li batteries. Adv Mater 15(9):736–739
55. Sawicki M, Shaw LL (2015) Advances and challenges of sodium ion batteries as post lithium ion batteries. RSC Adv 5:53129–53154. https://doi.org/10.1039/C5RA08321D
56. Slater MD, Kim D, Lee E, Johnson CS (2013) Sodiumion batteries. Adv Funct Mater 23:947–958. https://doi.org/10.1002/adfm.201200691
57. Balogun MS, Luo Y, Qiu W, Liu P, Tong Y (2016) A review of carbon materials and their composites with alloy metals for sodium ion battery anodes. Carbon (NY) 98:162–178. https://doi.org/10.1016/j.carbon.2015.09.091.
58. Nitta N, Yushin G (2014) Highcapacity anode materials for lithiumion batteries: choice of elements and structures for active particles. Part Part Syst Charact 31:317–336. https://doi.org/10.1002/ppsc.201300231
59. Guo P, Song H, Chen X (2009) Electrochemical performance of graphene nanosheets as anode material for lithiumion batteries. Electrochem Commun 11:1320–1324. https://doi.org/10.1016/j.ele-com.2009.04.036

60. Poizot P, Laruelle S, Grugeon S, Dupont L, Tarascon JM (2000) Nanosized transitionmetal oxides as negativeelectrode materials for lithiumion batteries. Nature 407:496–499. https://doi.org/10.1038/35035045
61. Roy P, Srivastava SK (2015) Nanostructured anode materials for lithium ion batteries. J Mater Chem A 3:2454–2484. https://doi.org/10.1039/C4TA04980B
62. Bruce PG, Scrosati B, Tarascon JM (2008) Nanomaterials for rechargeable lithium batteries. Angew Chem Int Ed Engl 47:2930–2946. https://doi.org/10.1002/anie.200702505
63. Zhang Q, Uchaker E, Candelaria SL, Cao G (2013) Nanomaterials for energy conversion and storage. Chem Soc Rev 42:3127–3171. https://doi.org/10.1039/c3cs00009e
64. Cao R, Xu W, Lv D, Xiao J, Zhang JG (2015) Anodes for rechargeable lithiumsulfur batteries. Adv Energy Mater 5:n/a–n/a10. https://doi.org/10.1002/aenm.201402273
65. Goriparti S, Miele E, De Angelis F, Di Fabrizio E, Proietti Zaccaria R, Capiglia C (2014) Review on recent progress of nanostructured anode materials for Liion batteries. J Power Sources 257:421–443. https://doi.org/10.1016/j.jpowsour.2013.11.103
66. Maiti S, Pramanik A, Dhawa T, Sreemany M, Mahanty S (2018) Mater Sci Eng B Adv 229:27–36
67. Wu S, Ge R, Lu M, Xu R, Zhang Z (2015) Graphenebased nanomaterials for lithium–sulfur battery and sodiumion battery. Nano Energy 15:379–405. https://doi.org/10.1016/j.nanoen.2015.04.032
68. Choi SH, Kang YC (2013) Yolkshell, hollow, and singlecrystalline $ZnCo_2O_4$ powders: preparation using a simple onepot process and application in lithiumion batteries. Chemsuschem 6:2111–2116. https://doi.org/10.1002/cssc.201300300
69. Cho S-J, Uddin M-J, Alaboina P (2017) Chapter Three—Review of nanotechnology for cathode materials in batteries. In: Rodriguez-Martinez LM, Omar N (eds) Emerging nanotechnologies in rechargeable energy storage systems. Micro and nano technologies. Elsevier, pp 83–129
70. Osaka T, Ogumi Z (eds) (2014) Nanoscale technology for advanced lithium batteries. Springer, New York, NY. https://doi.org/10.1007/978-1-4614-8675-6
71. Palomares V, Serras P, Villaluenga I, Hueso KB, CarreteroGonzález J, Rojo T (2012) Naion batteries, recent advances and present challenges to become low cost energy storage systems. Energy Environ Sci 5:5884. https://doi.org/10.1039/c2ee02781j
72. Rui X, Tan H, Yan Q (2014) Nanostructured metal sulfides for energy storage. Nanoscale 6:9889–9924. https://doi.org/10.1039/c4nr03057e
73. Xiao X, Liu X, Wang L, Zhao H, Hu Z, He X et al (2012) LiCoO2 nanoplates with exposed (001) planes and high rate capability for lithium-ion batteries. Nano Res 5:395–401
74. Park Y, Roh KC, Shin W, Lee J (2013) Novel morphology-controlled synthesis of homogeneous $LiFePO_4$ for Li-ion batteries using an organic phosphate source. RSC Adv 3:14263
75. Wang Y, Li H, He P, Hosono E, Zhou H (2010) Nano active materials for lithiumion batteries. Nanoscale 2:1294. https://doi.org/10.1039/c0nr00068j
76. Zhong Y, Su L, Yang M, Wei J, Zhou Z (2013) Rambutanlike $FeCO_3$ hollow microspheres: facile preparation and superior lithium storage performances. ACS Appl Mater Interfaces 5:11212–11217. https://doi.org/10.1021/am403453r
77. Guo YG, Hu JS, Wan LJ (2008) Nanostructured materials for electrochemical energy conversion and storage devices. Adv Mater 20:2878–2887. https://doi.org/10.1002/adma.200800627

# Chapter 3
# Energy Storage Application

**Manoj Tripathi, Akanksha Verma, and Ashish Bhatnagar**

## 1 Introduction

As sustainability is cease to be asymptotic in today's world and everyday new challenges are overcoming by human race and such a solution is also needed in energy conservation and storage of energy through various methods of generation. Making of supercapacitor electrode by organic means counted in green energy generation columns and for making such electrodes we need a biomass feedstock which turns into bio char and other essential components like syngas, bio-oil, etc., by channelizing feedstock from biomass to activated carbon and syngas and bio-oil. Ratio of component we are getting depends upon various condition vary from physical properties to chemical properties mainly due to temperature or pressure and type of gas used in process, respectively.

Researchers are doing commendable work in this field from decades and many new technologies have been discovered through. Coming from agricultural waste as biomass to municipal sewage treatment and laboratory made graphene as a biomass results into effective and hefty amount of char for making of electrode. Such a continuation has been named by a new material called nanomaterials. First observed by the American physicist Richard Feynman in the mid-1900s and his statement at an American Physical Society meeting held at Caltech, "There's Plenty of Room at the Bottom," laid the foundation of nanotechnology. The chaos begins from here. Researchers gaining areas of interest in this field by the discovery of carbon nanotubes in the 1900s to expose the hidden dimension of this material and respective possible arears of interest for green energy alternate [29].

Nanoparticles (NPs) in terms of dimensions can be defined as the particle having dimension ranging from 1 to 100 nm but dimensions don't define the properties of

M. Tripathi (✉) · A. Verma · A. Bhatnagar
Department of Physics and Materials Science and Engineering, Jaypee Institute of Information Technology, A-10, Sector 62, Noida 201309, India
e-mail: manoj.tripathi@jiit.ac.in

 
M. Mujawar et al. (eds.), *Nanotechnology for Electronic Applications*, Materials Horizons: From Nature to Nanomaterials, https://doi.org/10.1007/978-981-16-6022-1_3

NPs as suggested by the Royal Society and the Royal Academy of Engineering. They can be natural or artificial based on their origin, inorganic, organic, and both based on their chemical behavior and single or hybrid based on composition [42].

According to ISO (the International Organization for Standardization) and they are categorized as 0D, 1D (one dimension), 2D (two dimension), or 3D (three dimension). Initially originated for organic fibers, 1D NPs are generated by electrospinning techniques [45]. The property of increase in efficiency of storage mechanism and getting electric energy as output comes under the intrinsic property of 1D NPs [38]. Some examples of 1D NPs are clay, nanoclays, graphite, silicate nanoplatelets nanolayers, nanosheets, and nanoflakes [34].

Graphene and graphene-based nanocomposites [59, 78], metalorganic frameworks [18, 37] and MXenes [22, 43], whiskers, CNT (carbon nanotubes), nanotubes, CNFs (carbon nanofibers), nanofibers, nanorods holds the category of 2D NPs. The vitality of nanomaterials is in the development of novel energy generation devices. 3D, also known as equiaxed NPs, can be referred as particles having 3D structure but at nano-scale. Some examples which are able to describe above context are fullerenes, nanogranules, dendrimers (concentric shells of polymer NPs), nanocrystals (nanogranules), and quantum dots [34].

Considering the structure of any raw NP, there is a complexity in the arrangement of atoms and surface morphology and, consequently, it rather form three layers, viz. surface layer, shell layer, and the core, than a single surface. All three layers shows different properties in a single material at nanolevel. Small entity molecules, polymeric chains, metal ions, and some surfactants constitutes to the surface layer of a NP. If we are talking about any nanomaterial and about its properties, then we are talking about the core layer of that material. The core level is itself is nanomaterial. Core layer and the shell layer are different from each other chemically [58].

Coming to very ground level of every-day life, we can encounter nanomaterials everywhere. Some of them are including sensing devices, car bumpers, in clothes which are wrinkle-constraints, bacteria-free clothes, toothpastes, tennis rackets, coating layers made up of polymers, UV-resistant creams [7, 54]. Nanotechnology and its uses are readily used by the automotive industry at global level. Some of the examples to validate this are automobile paints, their lubricants, electric and electronic devices and sensors, tires, suspension fluid (vibration damping), super-thin antireflective layers, and self-cleaning for windows and mirrors [3].

Materials having nanostructure or are made up of NPs have been exploring by researcher in many domains and its accountability for energy storage devices and in its mechanism is increasing exponentially technically and economically.

## 2 Fundamental of Energy Storage

Sustainable energy conversion and storage based on renewable energy sources solar energy, wind energy from power plants, hydropower, tidal energy, geothermal energy and nuclear energy, are some sustainable energy conversion storage sources and are renewable in nature due to fast growth in global energy consumptions, non-renewable resources are on verge of extinction as people in some part of the world practicing ancient methods of burning coal and fossil fuels and due to which these resources are exploiting and large emission of $CO_2$ in the environment. There are various methods of energy storing techniques from mechanical to electrochemical range. Mechanical energy storage signifies storage of mechanical energy by using mechanics of a particular body can be due to kinetic energy and sometimes due to potential energy of the body. Hydropower plant store pumped energy, suppressed air energy storage, and flywheel energy storage are some of the significant examples of mechanical storage. Basic principle underlying mechanical energy storage is work done by accelerating rotor (in flywheels). Electromagnetic energy storage system store energy which is in the form of both electric and magnetic field. Electrostatic capacitors conventional one, electrical double-layer capacitors (EDLCs), and superconducting magnetic energy storage systems (SMES) come under the category of former energy storage techniques [9, 60, 63]. According to the ancient data provided by the researchers, the invention of the first battery was before Christ. The first practically usable battery was invented by Alessandro Volta in 1800 [56]. In terms of thermal energy storage applications, its chief resources are industrial waste heat energy, fossil fuel power plants, geothermal energy, nuclear power plants, and biomass energy storage systems [15, 48]. There is an outgrowing demand for thermal energy presently because of use of various portable and compact electronic devices such as computers, medical appliances, cell-operated car keys, mobile phones, and laptops. We need a clean and green energy and to extract former we need to rely on renewable resources to operate machines to suppress the environmental disaster and extinction of this world. To the rescue, nanomaterials (NMs) plays a vital and efficient role in achieving such causes [8].

Some of the NMs such as activated carbon (AC), graphene, carbon nanotubes (CNTs) are most important NMs for electric double-layer capacitor (EDLC-type) electrodes in supercapacitor [50].

# 3 Nanomaterials Used for the Energy Storage Applications

## *3.1 Biomass Derived Char*

Biomass is an important basic building block for any material which requires high carbon content and a good surface area as their properties and some of them are coke, biochar, and activated carbon. Biochar accounts for some drawbacks of having limited surface area, while activated carbon have pore size very unmanageable, which hinder them and limit their applications.

Presently, owing to huge demand in these biomasses derived char that allow them to be called as nanomaterial, carbon nanomaterials (CNMs) are gaining top position due to their absolute and helpful properties like high conductivity, both electrical conductivity and thermal conductivity, we are also encounter stability in environmental and chemical approaches in these materials. Thus, initiatives are being taken by researchers in the making of CNMs from biomass, but to accomplish this, there are some drawback in achieving high performance for CNMs derived from biomass because of the complex structure of precursor, it also contains high oxygen content and strong chemical bond which make them really complex and difficult to mold them in CNMs.

Making CNMs from biomass is an important task for researcher as biomass holds good carbon content and also a substitute for renewable energy source. Bulk properties are always different from the properties at nano-level and so biomass-derived CNMs can contribute to renewable energy sources which is much more effective than other renewable energy sources like wind and solar energy [44, 65]. Biomass feedstock is readily available to us and thus this encourages researcher to do continuous efforts in the making of CNMs from former.

Nanocellulose are also biomass driven nanomaterials accomplished by the hydrolysis of cellulose to generate nano-sized crystalline material and we can refer them as nanowhiskers (CNW) [62], nanocrystalline cellulose (NCC), nanocrystals (CNC) [33], cellulose nanofibers (CNF), as nanocrystallites of cellulose (NCC), and bacterial nanocellulose (BNC).

## 3.2 Polymer Nanocomposites

For the formation of polymer nanocomposites (PNCs), a matrix is formed by the combination of a polymer and a corresponding co-polymer with the dispersion of NPs or nanofillers in it [14]. Thus, Polymer Nanocomposites (PNCs) belongs to a group of materials that are hybrid in nature and are using nanomaterials as its filler components [1]. Technologies using these composites are in a wide-range microwave absorption/electromagnetic interference shielding [25, 77], storage of highly dense information [23], bimodal imaging agent [27], environmental remediation [41, 52, 53, 76], drug delivery [26], and energy storage [66, 73].

Matrix component generally made up of polymers have the property of elasticity which is very high and due to its structure as matrix, it constitutes low bulk modulus and also high load capacity so that heavy material can be imposed on it. The composite component constitutes materials that are effective at nano-scale range. Both matrix and composite are merge together and they account some properties which are completely different from individual ones. The desired shape whether it is spherical or flat or platy of PNCs are fully commanded by composite material and not by the matrix element.

These PNCs are categorized into multiple phase system such as blends, composites, and foams are some different categorization of PNCs and they can also be used as biosensors which are made up of electrode for the detection of glucose [1]. Nanocomposites have wide research scope in every field of science and technology and this is due to their unique properties.

Demand for graphene and graphene-derived polymer composites are also increasing due to its unmatched behavior in composites, in the field of green energy, in defense community, aerospace technologies, electronics business, automobile sector, etc. Graphene is cheap as compared to carbon nanotubes (CNTs) as they belong to graphite which is readily available in huge amount and this makes it even specific for researcher to use as a composite [20].

## 3.3 Carbon–Metal Nanocomposites

Carbon nanotubes have some attention seeking qualities which are appropriate for the fabrication of supercapacitor electrode designing as they have pore structure which is very unique in their own way, high ends electrical properties, and better mechanical and thermal stability [6, 49].

For carbon–metal nanocomposites, an amalgamation of matrix and fillers is required. More specifically, nanostructured carbon acts as reinforcing agents and metal as a filler for matrix composites. Different shapes of allotropes of carbon, viz. spherical (fullerene), cylindrical (CNTs), and planar (graphene) and these shapes corresponds to different arrangement of molecules in carbon atom. Metals as fillers when combined with nanocarbon forms a metal matrix nanocomposite (MMNCs) and this arrangement of two has gained importance in the field of research due to unmatched mechanical properties and stabilized chemical activities [4, 35]. We can alter strength of MMNCs through transfer load mechanism. High-resolution transmission electron microscopy (HRTEM) images and electron energy loss spectroscopy (EELS) analysis tools can be used for observation of the carbon and metal matrix nanocomposites, mainly for Ti and Al [57].

Being a matrix and a filler combination for carbon nanocomposites metal and carbon amalgamation have also a property of catalysis and this directly due to their size at nanoscale and physical and chemical properties that are purely shape-dependent. They gained an interest in research field mainly due to its catalytic property in many fields like material sciences, medical science, in agro-based firm, organic fields.

For preparing carbon metal nanocomposites, noncovalent functionalization is a very generalize method and it is basically a physical absorption with the involvement of very weak interaction also called π-interactions [21]. Noble metals are also acting as a promoter, whereas CNTs and graphite analogs such as graphene or graphene oxides acts as supporter for them to intact these noble metal nanoparticles. These materials as promoters have significantly gained attention due to stabilized thermal, chemical, and mechanical phenomenon, regular pattern order, and a steeply high surface area [2, 61].

### 3.4 *Metal and Metal Oxides Nanomaterials*

If a gold solution having blue, purple, and ruby colors on reducing the gold salt then the main reason behind the different colors of gold is due to its different dimension in different nano-units and this was initially observed by Michael Faraday in 1857. From then metal NPs, viz. Ag, Au, Au–Ag, Au–Cu, etc., were explored by researcher for advancement in technology and other world issues [17]. Metal having their specific properties but, when we measure them at nanoscale the properties of former likely to be change but to a better one and due to such extraordinary properties of metal at nanoscale. Thus, researchers are doing exceptionally good work to explore them a bit more.

It has also been found that the corresponding color for a particular metal due to some physical properties like shape, size, and composition. Au and Ag NPs may take a color like blue, cyan, indigo, brown, yellow, mustard, and many more depending upon the arrangement of NPs, their size and in molding shape in 1D, 2D, and 3D.

Optical properties of metal NPs are also sublime to encounter as we can actually alter and tune them practically and thus this works great in sensors [46].

Among the MNPs, silver (Ag) and gold (Au) have been the specific focus for plant-based synthesis via the biogenic reduction of metal salts due to their potential applications in a number of fields. Various plant extracts have been reported for the synthesis of $CeO_2$, $Co_3O_4$, CuO, $Fe_3O_4$, MgO, and ZnO [47].

## 4 Applications of Nanomaterials

### 4.1 *Electrode Material for Batteries*

Specific area should be large for rapid Faradic reaction, reduced mean free path for mass and electric charge mobilization and no enlargement in volume during reaction are some properties that should be adhere by NMs to be used as an electrode fabrication for lithium ion batteries. It has also been found that carbon coating results in boosting up electrical conductivity of a particular electrode material with the stimulation of interfacial Faradaic reactions and phase transitions. In spite of having various advantages, they have some lowkey which includes low thermal stabilization because of surface energy which is generally high [72].

Transition metal oxide nanoparticles (TMONPs) electrode have specific capacity of 700 mAh/g by retaining it fully for almost up to 100 charge/discharge cycles [51]. They also show that rate of recharging currents in charges/discharged cycle have a electrochemical potential window of 3.5–0.01 V (versus $Li^+$/Li). TMONPs shows high Li storage capacity at low potential [30, 79]. Electrodes when examine at nanoscales, shows tremendous incomparable properties of tweaking electrochemical properties by reaction mechanism specifically build up specific energy storage capacity, high current charge/discharge rate, and cycle stabilization, compare to their bulk properties. Short diffusion length and high contact area between the active materials and electrolyte are due to the properties described former [40, 75].

## 4.2 Nanomaterials in Energy Conversion Techniques

Nanomaterials are good path for the conversion of energy in various energy storage devices due the involvement of physical or chemical interaction. The surface impacts such as surface energy, area of that surface and chemical reaction involved to that surface are not solely depended upon the movement of particular ion or molecule in particular reaction but on some other factors also, such as the knowledge of heterogeneous reaction occurring at the junction and involvement of phase transition. Nanoscale is an important factor to define entity and it has been shown practically that smaller is the dimension of NMs, larger is the probability of favoring heat, mass and electric charge transfer. For example, larger the specific surface area, it will offer large number of sites for recombination of charge in PVs [10, 55]. Small size of pores of an electrolyte hinders the movement of ions in supercapacitors [19, 64]. Inorganic PVs called for larger specific surface area for the absorption of quantum dots and dye molecules into them. Dye-sensitized solar cells (DSSCs) and quantum dot sensitized solar cells (QDSSCs) are examples of inorganic PVs.

Hollow nanomaterials (HNMs) which includes hollow spheres and nanotubes is a current topic to be explore by researchers. These HNMs shows unmatchable morphology characteristics [11] to enunciate monumental electrochemical performance. HNMs found to have excellent immense surface area, low densities and build up surface permeability. These materials are the backbone of drug delivery machine, optical and electronic sensors, catalytic properties and Li-ion batteries [39].

Quantum dots such as CdTe and PbSe are of urgent importance and recent studies are going through that core but quantum dots as an entity is an ancient technology with organic dyes and photo luminescent, comparatively, C-dots are a better alternative to compensate smooth processing, powerful chemical inertness, high aqueous solubility, photo bleaching at resistance level, less toxic and better biocompatibility [31].

## 4.3 Applications in Supercapacitor

The design and construction of NMs has done in such a way so that it provides (i) a larger an embedded surface area between electrode and electrolyte, (ii) the reduction in mean free path of ion diffusion and (iii) blocking expansion of volume when electrochemical device is operating.

Among NMs, carbon nanomaterials are the most general one to be used as electrode for better performance of the supercapacitors. Supercapacitor shows compatible properties as a energy storing devices from technical an economical growth viz., efficient performance device with compatibility of high power and energy density.

Nanotechnology is the backbone for newer researcher in genre of energy storing device with flexible properties on altering NMs at our ease. High performance supercapacitors are embedded with carbon nanomaterials but more precisely, graphite carbon nanomaterials hold position which includes, CNTs, graphene sheets etc. [12] and this is due to their high pore size in mesopores range, good electrical and thermal stabilities [13].

Iron oxides are presently a very state-of-the-art topic for young researchers for producing electrodes for supercapacitors [67], which include $Fe_2O_3$, $Fe_3O_4$, $Fe_xO_y$, and FeOOH as iron oxides [24]. Recent studies show that CNTs still holds good in comparison to the iron oxide ($Fe_2O_3$). In composite form, CNTs suppress the pitfall of $Fe_2O_3$ in composite form. Carbon cloth, carbon fiber, CNTs, graphene, porous, carbon are some examples which can be used with $Fe_2O_3$ to form a composite. Shape of NMs is an overriding factor, if graphene shell is used a matrix element and we fold $Fe_2O_3$ transparent film on graphene shell, this configuration can be used to fabricate a supercapacitor with properties like flexibility and transparency [28]. Iron oxide layer enclosing the graphene layer of shell to store more and more charge and to increase electron movement. $Fe_2O_3$–conducting polymer composites shows flexibility and stability with good electric conductivity, these properties show the alternative for green energy a consequently degradation in hazardous environmental issues which are increasing exponentially due the thermal decomposition for blending carbon coating. Polypyrrole (PPy), poly(3,4-ethyl-enedioxythiophene) (PEDOT), and polyaniline (PANI) are some practically found iron oxide polymer-composites for the enhanced recital of iron oxide based super capacitors [68].

### *4.4 Nanomaterials as LHTES*

Phase change materials (PCMs) are materials that can absorb and release large amount of thermal energy during phase change and thus they are thermal storage materials. These are the key points why PCMs are becoming researchers' vital component in many applications such as battery for energy storage application, thermal stabilization, heating and cooling mechanism. Latent heat thermal energy storage (LHTES) system uses (PCMs) to hold and emit thermal energy to suppress energy consumption of a system.

(LHTES) technique is a streamlined and reliable means, due to its superlative properties of storing large amount of heat and also phase change behavior at thermal equilibrium. CNTs play a vital role of increasing thermal conductivity up to 5.0 wt%, exceeding this proportion can alter the graph of thermal conductivity due to low weight density of CNTs [70].

Bio-based and paraffin with xGnP are some PCMs that shows higher thermal conductivity, paraffin is the most general to be studied under thermal energy storage applications. Paraffin waxes exhibit high latent heat and melting point. Polymers are also used as PCMs sometimes. Mainly, polyethylene glycol (PEG), water, salts, salt hydrates and metals are come under the category of inorganic PCMs. Inorganic PCMs works over a spectrum of different temperature range. Inorganic PCMS have higher density than organic PCMs and thus they have large phase change enthalpy.

## *4.5 Nanomaterials in Fuel Cells*

The underlying principle behind fuel cell is the conversion of chemical energy into electrical energy by hindering emission of gases that are toxic in nature [16]. Thus, fuel cell is scrutinized as the purest form of technique with coherent energy, integrated circuits (ICs) are lacking in former. Proton exchange membrane fuel cell (PEMFC) is labelled as a most general and readily used fuel cell in the genre of energy storage mechanism and hold a dominant role to apply results. PEMFC uses hydrogen as a fuel which results it certain privilege of low operating temperature comparatively, rapid fluctuations of on an off, huge power density, low emission, stiffness, silent operations, a low operating temperature range [71]. Fuel cells covers a spectrum of applications in various domains such as in transportation e.g., in vehicles, compact devices, and stationary power systems. Chief properties of fuel cells that make them to use in power generation technology are (i) low to negligible emission rate, (ii) lofty energy density and (iii) condescending efficiency. Role of nanomaterials as an electro catalyst is directly depends upon fuel cell amplification for generation of power.

To make fuel cell as efficient energy storage device at commercial ground, there is a need of proper strategy of preparing electro catalyst that are inexpensive and are active in nature and robust as a material. Neglecting such key points can lead to energy loss. For example, 33% of total energy at gross level is wasted due to worst catalytic activity of PEMFCs and DMFCs at anode and cathode respectively [74]. Co-precipitation or impregnation [5] is the fundamental preparation method of fuel cells and also an ancient one. Thus, there are several challenges to be encounter in respective field. Range from optimize size of, to composition of different constituents of multi-metallic catalysts, researcher is finding ways to make fuel cell more commercialized. Bulk analogues and NP properties are two completely different aspects to study and to understand bulk properties with the knowledge of alteration in properties at nanoscale is important for coaction [32].

### *4.6 Nanomaterials in Solar Cell Applications*

The distinctive and peculiar properties that we can confront in case of NPs are depends upon the type of property we're looking for, in terms of knack of NPs (i) recombining probability of charge carrier get reduced significantly in favor of light generated particle, (ii) we can reshape NPs and this results into energy bandgap flexibility and (iii) optical path can be increased by providing numerous reflections, and Photo Voltic (PV) applications are starving for such properties mention above. Thus, both act as a key-lock system to one another. Nanomaterials (NMs) acts as a limber body for PV fillers to convert solar energy to practical applications. By using NMs as a matrix, PV efficiency and its ability to capture sunlight and create photo-generated current has increased significantly. The basic underlying path which helps these PVs for better performance are mainly, due to involvement of new PVs mechanism e.g., MEG in QDs, and (ii) providing a huge surface area, e.g., nanocrystalline films in dye-sensitized solar cells. PV made of inorganic NMs known as quantum dots is acquiring a better place in the practical world for its clean and renewable energy motto. Also, inorganic solar cells offer several advantages over organic solar cells in terms of stability, absorption properties, lifetime, and efficiency with less dependence on manufacturing equipment.

## 5 Challenges and Future Scope

Nanotechnology is expected to have a significant impact on the world's economy in the near future. The market volumes are a good indication of its significance. The market forecasts for nanotechnology began in 2000, with a time horizon up to 2015. The National Science Foundation (NSF) published the best-known data for a future nanotechnology forecast in 2001. The NSF data showed the market forecast for nanotechnology and its products at around US$1 trillion. In 2004 Lux Research showed the market forecast for nanotechnology to be $2.6 trillion in 2014.

Recent extensive studies on nanostructures of conventional semiconducting materials, especially 3D structures in the form of nanowires, nanocones, etc., demonstrated enhanced photon absorption and improved photo-carrier collection efficiencies [69]. Extensive work has been reported on organic molecules, inorganic molecules, and their hybrid structures to achieve better selectivity and improved light ECE [36].

Metal NPs are toxic to living cells. So, synthesis and use of metal NPs in powder form is considered to be hazardous. Further, recovery of precious metal NPs after catalysis or from solutions is difficult. Potential research should be directed towards expansion of the laboratory-scale synthetic procedures and applications of metal NPs in hosts to industrial scale [46].

## 6 Conclusions

NMs are one of the few functional materials which have proved their presence in almost all the fields of science and engineering. NMs due to their excellent properties are suitable for energy storage applications also. The NM-based energy storage materials devices are capable of storing a comparatively large amount of charge and thus can be highly useful. These NMs can store energy in different forms like chemical energy, thermal energy, in the form of latent heat, etc. These materials can be used to store the charge in batteries, cells, or supercapacitors. Different organo-nanometals have shown their usefulness in solar energy-related applications. Bio-based PCM composites with carbon nanomaterials like xGnP and CNT can be considered as effective energy-saving materials. Furthermore, in order to utilize PCMs in various fields, they need to be prevented from instability in liquid state by using shape stabilization, micro-encapsulation, and incorporation techniques. Biomass-derived char materials which are highly porous in nature and exhibit a large BET surface area are also used extensively for energy storage applications. The bio-based NMs can be used in supercapacitors to store the charge.

## References

1. Abdulkadir A, Sarker T, He Q, Guo Z, Wei S (2016) Chapter 13—Mössbauer spectroscopy of polymer nanocomposites. In: Thomas S, Rouxel D, Ponnamma D (eds) Spectroscopy of polymer nanocomposites. William Andrew Publishing, pp 393–409
2. Allen MJ, Tung VC, Kaner RB (2010) Honeycomb carbon: a review of graphene. Chem Rev 110(1):132–145
3. Asmatulu R, Nguyen P, Asmatulu E (2013) Chapter 5—Nanotechnology safety in the automotive industry. In: Asmatulu R (ed) Nanotechnology safety. Elsevier, Amsterdam, pp 57–72
4. Barrera EV, Sims J, Callahan DL, Provenzano V, Milliken J, Holtz RL (1994) Processing of fullerene-reinforced composites. J Mater Res 9(10):2662–2669
5. Bond GC, Thompson DT (1999) Catalysis by gold. Catal Rev 41(3–4):319–388
6. Bordjiba T, Mohamedi M, Dao LH (2008) New class of carbon-nanotube aerogel electrodes for electrochemical power sources 20(4):815–819
7. Brayner R, Ferrari-Iliou R, Brivois N, Djediat S, Benedetti MF, Fiévet F (2006) Toxicological impact studies based on Escherichia coli bacteria in ultrafine ZnO nanoparticles colloidal medium. Nano Lett 6(4):866–870
8. Bruce PG, Scrosati B, Tarascon J-M (2008) Nanomaterials for rechargeable lithium batteries 47(16):2930–2946
9. Burt R, Birkett G, Zhao XS (2014) A review of molecular modelling of electric double layer capacitors. Phys Chem Chem Phys 16(14):6519–6538
10. Cahen D, Hodes G, Grätzel M, Guillemoles JF, Riess I (2000) Nature of photovoltaic action in dye-sensitized solar cells. J Phys Chem B 104(9):2053–2059
11. Caruso F, Caruso RA, Möhwald H (1998) Nanoengineering of inorganic and hybrid hollow spheres by colloidal templating 282(5391):1111–1114
12. Chen T, Dai L (2013) Carbon nanomaterials for high-performance supercapacitors. Mater Today 16(7):272–280
13. Dai L, Chang DW, Baek J-B, Lu W (2012) Carbon nanomaterials for advanced energy conversion and storage 8(8):1130–1166

14. Das R, Pattanayak AJ, Swain SK (2018) 7—Polymer nanocomposites for sensor devices. In: Jawaid M, Khan MM (eds) Polymer-based nanocomposites for energy and environmental applications. Woodhead Publishing, pp 205–218
15. Denholm P, King JC, Kutcher CF, Wilson PPH (2012) Decarbonizing the electric sector: combining renewable and nuclear energy using thermal storage. Energy Policy 301–311
16. Dyer CK (2002) Fuel cells for portable applications 106:31
17. Faraday M (1857) X. The Bakerian Lecture.—Experimental relations of gold (and other metals) to light 147:145–181
18. Farha OK, Yazaydın A, Eryazici I, Malliakas CD, Hauser BG, Kanatzidis MG, Nguyen ST, Snurr RQ, Hupp JT (2010) De novo synthesis of a metal-organic framework material featuring ultrahigh surface area and gas storage capacities. Nat Chem 2(11):944–948
19. Fernández JA, Tennison SR, Kozynchenko O, Rubiera F, Stoeckli F, Centeno T (2009) Effect of mesoporosity on specific capacitance of carbons. Carbon 47:1598–1604
20. Gazal U, Khan I, Usmani MA, Bhat AH (2018) 11—Modification of polymer nanocomposites and significance of ionic liquid for supercapacitor application. In: Jawaid M, Khan MM (eds) Polymer-based nanocomposites for energy and environmental applications. Woodhead Publishing, pp 315–332
21. Georgakilas V, Otyepka M, Bourlinos AB, Chandra V, Kim N, Kemp KC, Hobza P, Zboril R, Kim KS (2012) Functionalization of graphene: covalent and non-covalent approaches, derivatives and applications. Chem Rev 112(11):6156–6214
22. Ghidiu M, Lukatskaya MR, Zhao MQ, Gogotsi Y, Barsoum MW (2014) Conductive two-dimensional titanium carbide 'clay' with high volumetric capacitance. Nature 516(7529):78–81
23. Goesmann H, Feldmann C (2010) Nanoparticulate functional materials 49(8):1362–1395
24. Gong X, Li S, Lee PS (2017) A fiber asymmetric supercapacitor based on FeOOH/PPy on carbon fibers as an anode electrode with high volumetric energy density for wearable applications. Nanoscale 9(30):10794–10801
25. He Q, Yuan T, Zhang X, Yan X, Guo J, Ding D, Khan MA, Young DP, Khasanov A, Luo Z, Liu J, Shen TD, Liu X, Wei S, Guo Z (2014) Electromagnetic field absorbing polypropylene nanocomposites with tuned permittivity and permeability by nanoiron and carbon nanotubes. J Phys Chem C 118(42):24784–24796
26. Hoare T, Timko BP, Santamaria J, Goya GF, Irusta S, Lau S, Stefanescu CF, Lin D, Langer R, Kohane DS (2011) Magnetically triggered nanocomposite membranes: a versatile platform for triggered drug release. Nano Lett 11(3):1395–1400
27. Howes P, Green M, Bowers A, Parker D, Varma G, Kallumadil M, Hughes M, Warley A, Brain A, Botnar R (2010) Magnetic conjugated polymer nanoparticles as bimodal imaging agents. J Am Chem Soc 132(28):9833–9842
28. Huang X, Zhang H, Li N (2017) Symmetric transparent and flexible supercapacitor based on bio-inspired graphene-wrapped $Fe_2O_3$ nanowire networks. Nanotechnology 28(7):075402
29. Iijima S (1991) Helical microtubules of graphitic carbon. Nature 354(6348):56–58
30. Jamnik J, Maier J (2003) Nanocrystallinity effects in lithium battery materials: aspects of nano-ionics. Part IV. Phys Chem Chem Phys 5(23):5215–5220
31. Kalyanasundaram K, Grätzel M (2012) Themed issue: nanomaterials for energy conversion and storage. J Mater Chem 22(46):24190–24194
32. Kariuki NN, Luo J, Maye MM, Hassan SA, Menard T, Naslund HR, Lin Y, Wang C, Engelhard MH, Zhong C-J (2004) Composition-controlled synthesis of bimetallic gold-silver nanoparticles. Langmuir: ACS J Surf Colloids 20(25):11240–11246
33. Kaushik M, Moores A (2016) Review: nanocelluloses as versatile supports for metal nanoparticles and their applications in catalysis. Green Chem 18(3):622–637
34. Khan WS, Asmatulu R (2013) Chapter 1—Nanotechnology emerging trends, markets, and concerns. In: Asmatulu R (ed) Nanotechnology safety. Elsevier, Amsterdam, pp 1–16
35. Kondoh K, Threrujirapapong T, Imai H, Umeda J, Fugetsu B (2009) Characteristics of powder metallurgy pure titanium matrix composite reinforced with multi-wall carbon nanotubes. Compos Sci Technol 69(7):1077–1081

36. Kumar S, Nehra M, Deep A, Kedia D, Dilbaghi N, Kim K-H (2017) Quantum-sized nanomaterials for solar cell applications. Renew Sustain Energy Rev 73(C):821–839
37. Li J-R, Kuppler RJ, Zhou H-C (2009) Selective gas adsorption and separation in metal–organic frameworks. Chem Soc Rev 38(5):1477–1504
38. Lieber CM (2011) Semiconductor nanowires: a platform for nanoscience and nanotechnology. MRS Bull 36(12):1052–1063
39. Liu J, Xue D (2010) Cation-induced coiling of vanadium pentoxide nanobelts. Nanoscale Res Lett 5(10):1619
40. Liu XY, Sawant PD (2002) Mechanism of the formation of self-organized microstructures in soft functional materials 14(6):421–426
41. Liu Z, Chen L, Zhang L, Poyraz S, Guo Z, Zhang X, Zhu J (2014) Ultrafast Cr(VI) removal from polluted water by microwave synthesized iron oxide submicron wires. Chem Commun 50(59):8036–8039
42. Lucena R, Simonet BM, Cárdenas S, Valcárcel M (2011) Potential of nanoparticles in sample preparation. J Chromatogr A 1218(4):620–637
43. Lukatskaya MR, Mashtalir O, Ren CE, Dall'Agnese Y, Rozier P, Taberna PL, Naguib M, Simon P, Barsoum MW, Gogotsi Y (2013) Cation intercalation and high volumetric capacitance of two-dimensional titanium carbide. Science 341(6153):1502–1505
44. Ma P-C, Zhang Y (2014) Perspectives of carbon nanotubes/polymer nanocomposites for wind blade materials. Renew Sustain Energy Rev 30:651–660
45. Mathur S, von Hagen R, Fischer T, Mueller R, Mettenboerger A, Raccis R, Lepcha A (2013) One dimensional (1D) inorganic nanomaterials for energy storage and production. In: International photonics and optoelectronics meetings (POEM), Wuhan, Optical Society of America
46. Mitra A, De G (2016) Chapter 6—Sol-gel synthesis of metal nanoparticle incorporated oxide films on glass 145–163
47. Nasrollahzadeh M, Atarod M, Sajjadi M, Sajadi SM, Issaabadi Z (2019) Chapter 6—Plant-mediated green synthesis of nanostructures: mechanisms, characterization, and applications. In: Nasrollahzadeh M, Sajadi SM, Sajjadi M, Issaabadi Z, Atarod M (eds) Interface science and technology, vol 28. Elsevier, pp 199–322
48. Pantaleo A, Camporeale S, Adio M, Russo V, Mugnozza G, Markides C, Shah N (2017) Thermo-economic assessment of an externally fired hybrid CSP/biomass gas turbine and organic Rankine combined cycle. Energy Procedia 105:174–181
49. Park SK, Mahmood Q, Park HS (2013) Surface functional groups of carbon nanotubes to manipulate capacitive behaviors. Nanoscale 5(24):12304–12309
50. Pereira C, Pereira AM (2016) Functional carbon-based nanomaterials for energy storage: towards smart textile supercapacitors
51. Poizot P, Laruelle S, Grugeon S, Dupont L, Tarascon JM (2000) Nano-sized transition-metal oxides as negative-electrode materials for lithium-ion batteries. Nature 407(6803):496–499
52. Qiu B, Gu H, Yan X, Guo J, Wang Y, Sun D, Wang Q, Khan M, Zhang X, Weeks BL, Young DP, Guo Z, Wei S (2014) Cellulose derived magnetic mesoporous carbon nanocomposites with enhanced hexavalent chromium removal. J Mater Chem A 2(41):17454–17462
53. Qiu B, Guo J, Zhang X, Sun D, Gu H, Wang Q, Wang H, Wang X, Zhang X, Weeks BL, Guo Z, Wei S (2014) Polyethylenimine facilitated ethyl cellulose for hexavalent chromium removal with a wide pH range. ACS Appl Mater Interfaces 6(22):19816–19824
54. Ray PC, Yu H, Fu PP (2009) Toxicity and environmental risks of nanomaterials: challenges and future needs. J Environ Sci Health Part C 27(1):1–35
55. Robertson N (2006) Optimizing dyes for dye-sensitized solar cells 45(15):2338–2345
56. Scrosati B (2011) History of lithium batteries. J Solid State Electrochem 15(7):1623–1630
57. Shin SE, Choi HJ, Hwang JY, Bae DH (2015) Strengthening behavior of carbon/metal nanocomposites. Sci Rep 5(1):16114
58. Shin WK, Cho J, Kannan AG, Lee YS, Kim DW (2016) Cross-linked composite gel polymer electrolyte using mesoporous methacrylate-functionalized $SiO_2$ nanoparticles for lithium-ion polymer batteries. Sci Rep 6:26332

59. Stoller MD, Park S, Zhu Y, An J, Ruoff RS (2008) Graphene-based ultracapacitors. Nano Lett 8(10):3498–3502
60. Tixador P (2008) Superconducting magnetic energy storage: status and perspective
61. Vairavapandian D, Vichchulada P, Lay MD (2008) Preparation and modification of carbon nanotubes: review of recent advances and applications in catalysis and sensing. Anal Chim Acta 626(2):119–129
62. Visakh PM, Thomas S (2010) Preparation of bionanomaterials and their polymer nanocomposites from waste and biomass. Waste Biomass Valoriz 1(1):121–134
63. Vulusala G VS, Madichetty S (2018) Application of superconducting magnetic energy storage in electrical power and energy systems: a review 42(2):358–368
64. Wang G, Zhang L, Zhang J (2012) A review of electrode materials for electrochemical supercapacitors. Chem Soc Rev 41(2):797–828
65. Wang Z, Shen D, Wu C, Gu S (2018) State-of-the-art on the production and application of carbon nanomaterials from biomass. Green Chem 20(22):5031–5057
66. Wei H, Wang Y, Guo J, Yan X, O'Connor R, Zhang X, Shen NZ, Weeks BL, Huang X, Wei S, Guo Z (2015) Electropolymerized polypyrrole nanocoatings on carbon paper for electrochemical energy storage 2(1):119–126
67. Xia Q, Xu M, Xia H, Xie J (2016) Nanostructured iron oxide/hydroxide-based electrode materials for supercapacitors 2(7):588–600
68. Xu B, Zheng M, Tang H, Chen Z, Chi Y, Wang L, Zhang L, Chen Y, Pang H (2019) Iron oxide-based nanomaterials for supercapacitors. Nanotechnology 30(20):204002
69. Yu R, Lin Q, Leung S-F, Fan Z (2012) Nanomaterials and nanostructures for efficient light absorption and photovoltaics. Nano Energy 1(1):57–72
70. Yu S, Jeong S-G, Chung O, Kim S (2014) Bio-based PCM/carbon nanomaterials composites with enhanced thermal conductivity. Sol Energy Mater Sol Cells 120:549–554
71. Zhang J, Xia Z, Dai L (2015) Carbon-based electrocatalysts for advanced energy conversion and storage. Sci Adv 1(7):e1500564
72. Zhang Q, Uchaker E, Candelaria SL, Cao G (2013) Nanomaterials for energy conversion and storage. Chem Soc Rev 42(7):3127–3171
73. Zhao Y, Wei H, Arowo M, Yan X, Wu W, Chen J, Wang Y, Guo Z (2015) Electrochemical energy storage by polyaniline nanofibers: high gravity assisted oxidative polymerization vs. rapid mixing chemical oxidative polymerization. Phys Chem Chem Phys 17(2):1498–1502
74. Zhong C-J, Luo J, Njoki PN, Mott D, Wanjala B, Loukrakpam R, Lim S, Wang L, Fang B, Xu Z (2008) Fuel cell technology: nano-engineered multimetallic catalysts. Energy Environ Sci 1(4):454–466
75. Zhou H, Li D, Hibino M, Honma I (2005) A self-ordered, crystalline–glass, mesoporous nanocomposite for use as a lithium-based storage device with both high power and high energy densities 44(5):797–802
76. Zhu J, Gu H, Rapole SB, Luo Z, Pallavkar S, Haldolaarachchige N, Benson TJ, Ho TC, Hopper J, Young DP, Wei S, Guo Z (2012) Looped carbon capturing and environmental remediation: case study of magnetic polypropylene nanocomposites. RSC Adv 2(11):4844–4856
77. Zhu J, Wei S, Ryu J, Sun L, Luo Z, Guo Z (2010) Magnetic epoxy resin nanocomposites reinforced with core−shell structured Fe@FeO nanoparticles: fabrication and property analysis. ACS Appl Mater Interfaces 2(7):2100–2107
78. Zhu Y, Murali S, Stoller MD, Ganesh KJ, Cai W, Ferreira PJ, Pirkle A, Wallace RM, Cychosz KA, Thommes M, Su D, Stach EA, Ruoff RS (2011) Carbon-based supercapacitors produced by activation of graphene. Science 332(6037):1537–1541
79. Zhukovskii YF, Balaya P, Kotomin EA, Maier J (2006) Evidence for interfacial-storage anomaly in nanocomposites for lithium batteries from first-principles simulations. Phys Rev Lett 96(5):058302

# Chapter 4
# Nanotechnology and Nanomaterials for Medical Applications

**Bridgid L. F. Chin, Filbert H. Juwono, and Kelvin S. C. Yong**

## 1 Introduction

Triggered by the talk of Richard Feynman titled "There's Plenty of Room at the Bottom" at the American Physical Society meeting in 1959, the development of nanotechnology has been showing significant results in many fields, including medical applications. It is worth mentioning that nanotechnology is a multidisciplinary area that comprises materials science, biology, physics, chemistry, and engineering [112]. In particular, the blend of nanotechnology and biotechnology studies creates the nanobiotechnology study, which underpins the technology beyond the molecular level of biological systems. This includes viruses, small molecules (proteins, hormones), and other nanosized biological systems. Therefore, it is obvious that nanobiotechnology is fundamental for medical applications which involve the molecular level of disease treatment and detection. Nanobiotechnology is expected to pave the way for new breakthroughs in essential disease prevention and cure, such as cancer and HIV [112].

Nanobiotechnology is the science that has lately assisted in the progress of medical research. For instance, it can be used for in vitro diagnostics (IVD) which is able to detect pregnancy, glucose level, and cancer biomarkers [135]. Another example is

B. L. F. Chin
Department of Chemical and Energy Engineering, Faculty of Engineering and Science, Curtin University Malaysia, 98009 Miri, Sarawak, Malaysia

F. H. Juwono (✉)
Department of Electrical and Computer Engineering, Faculty of Engineering and Science, Curtin University Malaysia, 98009 Miri, Sarawak, Malaysia
e-mail: filbert@ieee.org

K. S. C. Yong
Faculty of Engineering, Computing and Science, Swinburne University of Technology Sarawak Campus, Kuching, Malaysia

 
M. Mujawar et al. (eds.), *Nanotechnology for Electronic Applications*, Materials Horizons: From Nature to Nanomaterials, https://doi.org/10.1007/978-981-16-6022-1_4

nanoimaging which has been developed to assist real-time tracking during surgery [94].

In general, nanobiotechnology study for medical applications can be categorized into two [107]: tissue and implant engineering, and nanomedicine. Tissue and implant engineering is related to the process of replacing or reconstructing damaged tissue or organs by using nanoparticles, such as stem cell-based therapeutic. This concept is sometimes known as regenerative medicine. On the other hand, the nanomedicine field can be divided into two applications [39]: diagnosis and treatment. Nanomedicine can be considered as the field related to monitoring, repair, construction, and control of human biological systems at the molecular level [104]. Diagnosis applications include diseases detection and quantum dots imaging. Treatment-based applications include drug delivery, immune system triggering, and nanosurgery. Nanomedicine applications use nanoscale materials to perform the tasks, which may include biocompatible nanoparticles and nanorobots. The development of these technologies cannot be achieved without the development of nanomaterials and nanobiosensors.

In this chapter, we will thoroughly discuss the applications of nanobiotechnology in medical applications as well as the recent developments and challenges in this area.

## 2 Nanomaterials and Nanotools

Nanomaterials can be categorized into five types, which are nanoparticles, carbon-based, dendrimers, quantum dots (QDs), and conjugates, as shown in Fig. 1 [79]. Examples of metallic nanoparticles are gold (Au) and silver (Ag). The unique electronic and optical properties of gold nanoparticles, such as mediator-less electrochemical sensing, enhanced Raman scattering, fluorescence properties coupled with the chemical inertness, lack of toxicity, and good biocompatibility, make gold one of the popular nanomaterials. Silver nanoparticles have physicochemical properties,

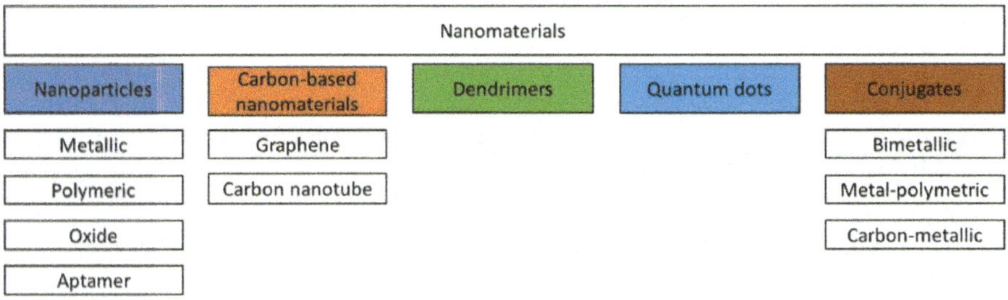


**Fig. 1** The categories of nanomaterials

which make them commonly being used for medicinal purposes. This nanomaterial has the benefit of good thermal and electrical conductivity, chemical inertness, antibacterial properties, catalytic activity, and optical properties [48, 126].

Polymeric-based nanoparticles are attractive for the fabrication of molecularly imprinted polymers (MIR) [129]. Such materials have the advantages of low cost, high sensitivity and selectivity, low limit of detection (LOD), robust against environmental conditions, and facile preparation [73]. $TiO_2$ is one example of metal oxide nanoparticles. It has the benefit of chemical inertness, biocompatibility, and optical properties [38]. Another example is porous silica ($SiO_2$) which shares the same strength of $TiO_2$ of having biocompatibility combined with other benefits of wide surface area, controllable particle size, and pore volume [82].

Aptamer-based nanoparticles use DNA/RNA oligonucleotide sequences to create unique 3D conformation for the detection and interaction with a wide array of complementary target molecules' sequences [54]. This material has the advantages of being small in size, high sensitivity, selectivity and affinity, and efficient immobilization. An example of macromolecules of dendrimers is poly (amidoamine) dendrimers (PD). In the fabrication of nanobiosensors, this material has the advantages of good chemical and mechanical stability, hydrophilicity, controllable structure and size, and mono-dispersity [66]. These advantages lead to good sensitivity, stability, specificity, reusability, and detection of the biosensors [123].

Graphene is of interest in the fabrication of electrochemical nanobiosensors as it has a wide surface area, good electrical conductivity, and mechanical performance [2]. Another carbon-based nanomaterial is the carbon nanotube [10]. Its attractive properties of high biocompatibility and tensile strength, antibacterial and antifungal properties, chemical stability, lightweight, wide surface area, support protein immobilization, and fast electron transfer kinetics suited for biosensor applications [15]. The use of quantum dots in biosensors demonstrated the enhancement of sensitivity, LOD, and response time of biosensor's detection [120]. This material is commonly being used as fluorophores [11] in biosensing applications [62].

Nanomaterials for biomedical applications which will be discussed in detail in this chapter include gold nanoparticles, QDs, carbon nanotubes, and chitin and chitosan. Meanwhile, one example of nanotools is nanowires.

## *2.1 Gold Nanoparticles*

A gold nanoparticle is a small gold particle with a diameter of 1 to 100 nm, which is known as colloidal gold when dispersed in water. Gold nanoparticles were discovered more than 100 years ago and yet, they still have attracted much attention in many research areas due to their unique characteristics such as biocompatibility, low short-term toxicity, high atomic number, and high X-ray absorption coefficient [88]. Gold nanoparticles are synthesized by reducing the gold salts, which can lead to the nucleation of gold ions. A stabilizing agent is needed to absorb or to create a chemical bond to the gold surface. Trisodium citrate or sodium borohydride acts as a reducing agent to generate gold nanoparticles in the aqueous medium. Meanwhile, controlling citrate concentration gives the formation of small and uniform gold nanoparticles [41, 113].

Gold nanoparticles are commonly used as a contrast agent for X-ray computerized tomography (CT). The X-ray contrast agents are used to increase the contrast between normal and, for example, cancerous tissue. Note that X-ray CT visualizes the body structures and tissues based on the X-ray absorption differences of the human tissues [128]. Further, gold nanoparticles can be used as colorimetric biosensors for diagnostics purposes [4, 21].

## 2.2 *Chitin and Chitosan*

Chitosan is derived from chitin through partial deacetylation under alkaline conditions. Meanwhile, chitin comes from natural sources of marine crustaceans' shells such as shrimps, lobsters, crabs, and the fungi cell walls. The shells of the same size and species are grouped to be cleaned, dried, and ground into small shell pieces [130]. Chitin is extracted by adding acid treatment to dissolve the calcium carbonate and the alkaline solution to dissolve the proteins. Next, decolorization is performed to remove the pigments and obtain colorless pure chitin. The chitin source must get to the treatments for producing a high-quality chitin followed by chitosan [33]. Chitosan is a semi-synthetic polyaminosaccharide obtained by N-deacetylation of chitin. Chitosan is the second most important polymer in the world because it is a natural polymer that possesses biodegradable, biocompatible, bioactivity, non-toxicity, and polycationic properties [103]. The unique properties have made chitosan be used widely for different biological and biomedical applications such as in water treatment, wound-healing materials, drug carrier, obesity treatment, and also scaffold for tissue engineering [8, 25, 37, 85, 91].

One of the chitosan applications is for drug delivery via non-injection to mucosal sites. The biggest challenge of this application is the absorption of the drugs at these sites since the drug delivery system must have mucoadhesive particles and release the drug over time. The positive charge of chitosan can bind with the negatively charged mucus, where chitosan acts like an excellent carrier for the mucoadhesive drugs [83, 106].

## 2.3 *Quantum Dots (QDs)*

QDs are fluorescent semiconductor nanocrystals composed of atoms from groups II–VI or III–V elements of the periodic table [20]. QDs have unique optical and electrical properties because of their quantum and size effects. They have the properties of biocompatibilty, strong fluorescence intensity, larger stokes shift, higher stability, long fluorescence lifetime, and negligible photobleaching [133]. These properties aid

QDs in the medical imaging application particularly targeting the specific cells like labeling neoplastic cells, DNA, and cell membrane receptors after bioactive moieties. For the synthesis of QDs, trictylphosphine TOP-selenide is added to the hot solution of cadmium oleate and hexadecylamine under inert atmospheric conditions [55, 108]. Further, the solution of diethyl zinc and hexamethyldisilathiane are added to the mixture for improving the photoluminescence efficiency [61]. The pure metalloid crystalline core nanocrystals can be recovered through the process of purification using liquid–liquid extraction [101].

One of the QD applications is as fluorophores for in vivo fluorescence imaging. This fluorescence imaging gives several advantages such as good sensitivity, noninvasive, readily available and inexpensive [125]. Recent research has developed a QD-based contrast agent where the agent is based on surface modification of QDs and functioned with wheat germ agglutinin for brain imaging. Using the means of intranasal administration to deliver the QD-based imaging agent, we can accumulate QDs in the brain for more than four hours and clear them in 8 h after administration [136].

## *2.4 Carbon Nanotubes*

Carbon nanotubes (CNTs) have received much attention since 1991 due to their unique optical, electrical, mechanical, and thermodynamic properties [98]. CNTs are a cylinder fabricated of rolled up monolayered graphite sheet. It can be single-walled carbon nanotubes (SWCNTs) or multi-walled carbon nanotubes (MWCNTs). High electrical and thermal conductivity and stronger tensile strength are the characteristics of the CNTs used in the field emission devices [99]. The three methods used to produce CNTs are the chemical vapor decomposition (CVD) technique, the laser-ablation technique, and the carbon arc-discharge technique [47, 90, 105]. The CVD method has been widely used for CNT production compared with the other two methods. In the CVD technique, the hydrocarbon is heated with a transition metal catalyst to aid the tub's growth. The catalyst is deposited on the substrate followed by carbon source is placed in a gas phase reaction chamber [57]. The laser ablation technique is limited to the lab scale but it is able to produce a pure 90% SWCNTs. The laser is used to vaporize a graphite in an oven with inert gas which can give a high yield of SWCNTs with narrow size distribution [1]. The arch-discharge technique is the easiest and common way to produce CNTs by arc evaporation of two carbon rods end-to-end in an inert gas environment.

CNTs can be applied in tissue engineering for several purposes, such as to substitute damaged or diseased tissues with biologic alternates that can repair and preserve the normal original function, and to give the full potential for measuring the biodistribution and modified with radiotracers for gamma scintigraphy [35].

### 2.5 Nanowires

Nanowires have a very small diameter, i.e., 3–100 nm. Nanowires have a much longer length compared with their diameter. Silicon nanowires (SiNWs) are mostly adopted for biomedical and clinical applications [117]. The applications of SiNWs include neural interface, tissue engineering, and force sensing. In addition, SiNWs can be used as the material for electrode arrays in the neuro-prosthetic devices. The electrode arrays are used to interface the nervous system with the device. In tissue engineering, SiNWs are used to create scaffolds for tissue regeneration and as a sensor network for regeneration monitoring. Finally, in force sensing, SiNWs are used as the force sensor placed at the guidewire for minimally invasive surgery (MIS).

## 3 Diagnosis and Treatment Applications

Many diseases can be detected through the existence of their biomarkers in the cardio-circulatory system [84]. A biomarker can be used as an indication for a certain disease state. Biomarkers can be found in the forms of proteins, nucleic acids, antibodies, and peptides. For example, a protein called the cancer antigen 19-9 (CA19-9) is used to mark pancreatic cancer [58]. One can test one's blood to observe the concentration level of the biomarkers. From the nanotechnology point of view, to detect the biomarkers, nanoparticles and nanotools are required.

### 3.1 Diagnosis Applications

Nanoparticles usually work as a contrast agent in biomedical imaging diagnoses. Biomedical imaging is popularly known as radiology. There are two types of imaging: optical (or fluorescence) imaging and magnetic resonance imaging (MRI). Optical imaging uses near-infrared (NIR) light due to the deeper penetration of the tissue.

Two well-known optical imaging applications are endoscopy and optical coherence tomography (OCT). On the other hand, MRI uses a strong magnetic field which forces the protons in the body to align with the field. Radiofrequency (RF) waves are then released for a few microseconds to disrupt the protons. After the RF is stopped, the protons will be realigned with the magnetic field and release energy. The energy is captured by the sensors. To enhance certain tissues or body parts, contrast agents may be injected intravenously. Based on the types of imaging, the contrast agents can be grouped into optical contrast agents and MRI contrast agents. QDs and gold nanoparticles belong to optical contrast agents while gadolinium and iron oxide belong to MRI contrast agents.

Further, a deep-tissue optical imaging with the help of QDs has been developed [81]. It has been reported that CdTe and HgTe, which belong to group II–VI QDs,

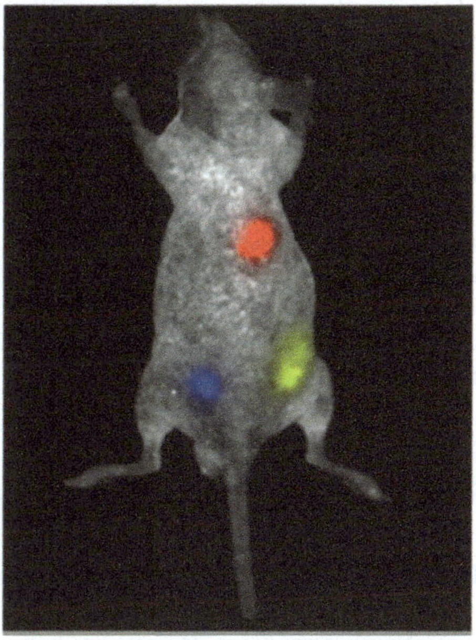

**Fig. 2** Imaging results with CdTe/CdSe QDs with three wavelengths [60]

are commonly used in the NIR region for optical imaging [26]. Figure 2 shows the results of CdTe/CdSe QDs imaging with three wavelengths (700, 750, and 800 nm) of a mouse [60].

## *3.2 Drug Delivery Applications and Concepts*

As mentioned previously, one of the nanomedicine applications is for treatment. The most popular nanomedicine treatment is drug delivery. In principle, the drug substances (molecules) are released to the blood vessel where they can be delivered through the bloodstream to the target. The targeted drug delivery has many applications such as: (1) tumor/cancer therapy, (2) high activity antiretroviral therapy (HAART) for HIV patients, (3) therapeutic strategy for genetic diseases, (4) therapy for gastrointestinal diseases, and (5) therapy for neurological diseases. As the drug is delivered directly to the target (e.g., tumor/cancer), this method is useful to reduce the side effects of the chemotherapy treatment. The following is the concept of drug delivery. The transport mechanism of drug delivery can be passive (e.g., diffusion) or active (e.g., molecular motors). An advanced active transport mechanism is to employ nanorobots. Nanorobots can be in the form of flagellated bacteria. To detect tumors, flagellated bacteria are attracted to proteins secreted by tumors that cause them to be drawn to the tumors [23].

The drug delivery mechanism is determined by three functions [39]: delivery, monitoring, and control. The delivery function is done by nanoactuators where they release a sufficient amount of drug to the target. The monitoring function ensures the release rate can be sufficient to use. The control function gives feedback to the nanoactuator to regulate the delivery rate.

In principle, a drug nanocarrier is needed for bringing the drug. The drug may be absorbed or attached to the surface or encapsulated to the nanocarrier [127]. More importantly, nanocarriers need to be biocompatible and harmless. The first nanocarrier was a liposome. A liposome is 80–300 nm size range. A drug is encapsulated in liposomes and released based on the liposome composition, pH, osmotic gradient, and surrounding environment [127]. Other nanocarriers are solid lipid nanoparticles, nanostructured lipid carriers, lipid drug conjugates, polymeric nanoparticles, dendrimer nanocarriers, silica materials, carbon nanomaterials, and magnetic nanoparticles.

### *3.3 Recent Developments and Challenges*

The challenges of the drug delivery system are about the efficacy and the accuracy of the drug delivery [95]. There are a few emerging strategies for drug delivery [75]: stimuli-responsive strategy, co-delivery strategy and theragnostic, biomimetic delivery strategy, and ligand-modified target drug delivery strategy. The stimuli-responsive strategy is the most popular one, especially for tumor/cancer-related drug delivery. The stimuli are needed to release the drug at the target site. In co-delivery strategy, combination therapy with various drugs has been developed, e.g., co-delivery of photosensitizer and chemotherapeutic agent. Nanoparticles have been used as the co-delivery strategy by combining therapeutic effects with diagnostic and prognostic. Biomimetic is a novel strategy for nanosystems by mimicking the structures, functions, and biosynthetic pathways of biological systems. Lastly, a ligand-modified strategy can be used to improve various characteristics of the delivery, such as the penetration at the target site. Further, there have been some emerging techniques developed for drug delivery, such as 3D printing, microneedle, nanocrystals, and prodrug nanomedicines [75].

## 4 Tissue and Implant Engineering

Nanotechnology is an emerging technology for both tissue and implant engineering. The material formulated for nanotechnology is said to influence the performance of the tissue and implant engineering applications. Higher yield and improved tissue implant mechanical interlocking could be achieved by increasing the surface area and reducing the surface roughness [107]. Furthermore, strong wear resistance implants can be improved by enhancing both the hardness and strength of the nanomaterials [107]. Hence, the surface modifications involved in nanotechnology need to be conducted comprehensively to achieve the optimum performance of the nanotechnology in this particular area. In this section, recent developments of tissue and implant engineering for the applications of (i) tissue repair, (ii) nanobiomaterial in craniofacial tissue engineering, and (iii) dental filler and antimicrobial nanomaterial are reviewed. Furthermore, the challenges faced by various applications in the tissue and implant engineering with the implementation of nanotechnology are also discussed.

## *4.1 Tissue Repair Applications*

Gold and titanium nanoparticles are frequently used to improve cell proliferation rates for both bone and cardiac tissue regeneration, respectively [56]. From the literature, it has been evidently shown that gold nanoparticles possess a good biocompatibility and exterior modification capability [34, 50]. Gold nanoparticles were first discovered to stimulate osteogenic differentiation of an osteoblast precursor cell line known as MC3T3-EI [19, 24]. Furthermore, these nanoparticles have the ability to act like a bone-resorbing cell, which could stimulate the formation of osteoclast from the hematopoietic cell, which permits mitochondrial dysfunction in osteoblastic cell to be protected [72, 116]. Additionally, these nanoparticles can also aid in both osteogenic and adipogenic differentiation of mesenchymal stem cells [56]. This usually takes place during the formation of the p38 mitogen-activated protein kinase pathway [24] and the size of the gold nanoparticles can influence the osteogenic differentiation.

In Ko et al. [72], it was found out that particles with size between 30 and 50 nm are suitable for human adipose-derived stem cell function. Meanwhile, Liu et al. [78] mentioned that gold nanoparticle sizes of 20–40 nm are most preferred for MC3T3-E1 osteoblast-like cell function. Although gold nanoparticle is suitable for bone tissue regeneration, there are several disadvantages associated when using these nanoparticles for tissue repair. Few disadvantages of utilizing gold nanoparticles for tissue engineering application are cost-ineffectiveness (expensive), the tendency on forming unnecessary bone, and possible reactions associated with local inflammatory [16]. Hence, it is significantly important to conduct a continuous research to find a sustainable and effective material for bone growth in which gold nanoparticles could be a potential option for this purpose.

## *4.2 Nanobiomaterials in Craniofacial Tissue Engineering*

The presence of nanobiomaterials in tissue engineering permits the improvement of mechanical strength, encourages cell growth development, and provides aid in bone-forming [118]. A unique scaffold can be specially formulated by involving any key properties based on the specific requirement of the tissue engineering applications. Table 1 displays the summary of different main types of nanobiomaterials that can be

**Table 1** Summary of different main types of nanomaterials for tissue engineering with its advantages and disadvantages

| Nanomaterial types | Advantages and disadvantages |
|---|---|
| Metallic (e.g., gold, silver, titanium) | • It possesses high mechanical strength [32, 122]<br>• It promotes osseointegration, osteoconductivity, mineralization, proliferation, osteoblast, and chondrocyte cell adhesion<br>• It enhances collagen production, alkaline phosphatase activity, and calcium deposition [32, 51, 122]<br>• It accumulates in organs due to small size, it can attribute to oxidative stress and inflammatory reactions in human body [76] |
| Polymeric (e.g., carbon, hydrogen, oxygen, nitrogen) | • It has wide applications such as drug delivery, surface coatings, and filtration [42]<br>• It has ability to encapsulate numerous molecules which includes vaccine antigens, proteins, drugs, various growth cells and factors which aid in providing a useful transportation and delivery tool [53]<br>• Polymeric nanomaterial which possesses nanofiber and human extracellular matrix (ECM) resemble-like could provide benefit in scaffold nurturing and tissue regeneration [44] |
| Carbon (e.g., graphene and carbon nanotubes) | • It has excellent mechanical strength, electrical conductivity, and thermal stability which are displayed closely to biological ECM behavior [12] |
| Ceramic (e.g., bioactive glass, zirconia ceramic alumina ceramics, titania-based ceramics, nanohydroxyapatite (nHAp), and calcium phosphate (CaP), among many other oxides, phosphates, nitrides, carbides) | • It possesses the characteristics of both metal and nonmetal elements [12]<br>• It possesses physical properties such as hard, brittle, electrical insulation, high tolerance towards chemical environment [12]<br>• A similar inorganic component that can be found in human bone [12] |

(continued)

**Table 1** (continued)

| Nanomaterial types | Advantages and disadvantages |
|---|---|
| Composite (e.g., silica-calcium phosphate (SCPC), titania-hydroxyapatite ($TiO_2$-HAp), hydroxyapatite-multi-walled carbon nanotubes (HAp-MWCNT)) | • This nanomaterial is composed by merging two or more nanomaterials which consist of matrix materials and nanosized particles [28]<br>• Factors such as biocompatibility, biofunctionality, physical properties (modulus of elasticity, surface hardness, polymerization shrinkage degree, strength of compressive and tensile, and material flow) are considered for the development of this nanomaterial [3, 28, 124]<br>• This composite-based nanomaterial is commonly used in dentistry by combining both ceramic and polymer together [28]<br>• The major drawback of synthetic implants is low succeeding rate for the local tissue adaption toward the human cells. Hence, this mechanism can be resolved via intimate tissue in-growth [28] |

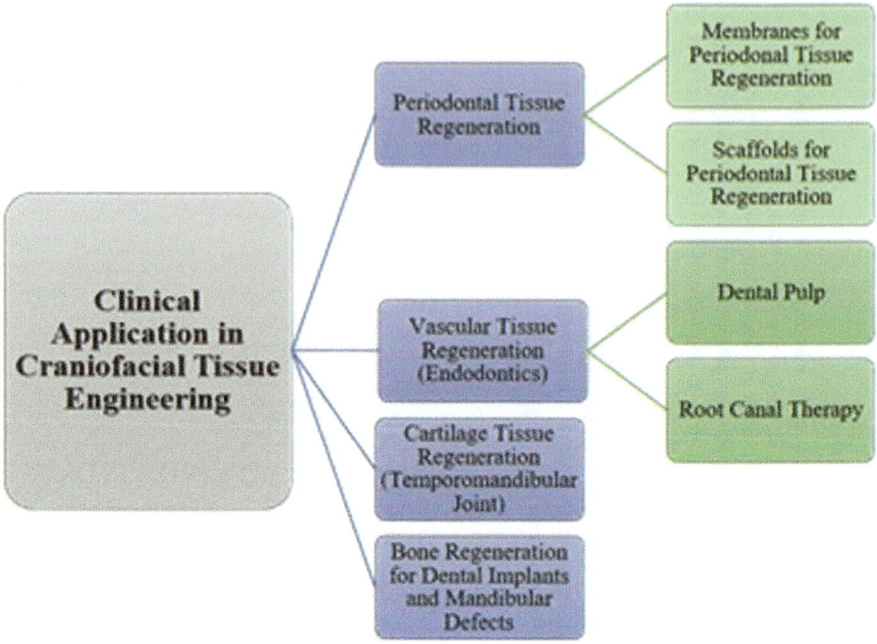


**Fig. 3** The clinical applications in craniofacial tissue engineering

applied in craniofacial tissue engineering with their advantages and disadvantages. It can be divided into five main groups, namely metallic, polymeric, carbon-based, ceramic, and composite nanomaterials.

The craniofacial tissue engineering for clinical applications can be divided into (i) periodontal tissue regeneration, (ii) vascular tissue regeneration (endodontics), and (iii) cartilage tissue regeneration (temporomandibular joint), and (iv) bone regeneration for dental implants and mandibular defects as summarized in Fig. 3.

For periodontal tissue regeneration via using membrane, the membrane must be able to fulfil several aspects consisting of (i) biocompatibility, (ii) space maintenance, (iii) tissue integration, and (iv) clinical manageability [121]. A synthetic membrane, such as polycaprolactone (PCL) membrane added with anti-inflammatory drugs and growth factors, is commonly used for periodontal regeneration [14]. Besides, membrane from biomimetic fish collagen, bioactive glass, or chitosan composite nanofiber is another alternative that possesses characteristic of good hydrophilicity, thereby enhancing interaction between cells and matric interaction on the surface area, great porosity, sufficient tensile strength, and low antibacterial properties [134]. Furthermore, poly-DL-lactic acid (PDLLA) polymers with antibiotics and growth factors could assist in the healing and regenerative process and at the same time reduce the infections from bacteria [59].

For vascular tissue regeneration, also known as endodontics, the implementation of nanomaterials can assist in dental pulp and root canal therapy. Galler et al. [44]

fabricated a biomaterial scaffold consisted of personalized peptide nanofibers and encapsulate dental pulp stem cells (DPSCs) for pulp tissue regeneration. From this investigation, it was observed that the encapsulation of the growth factor was seen over a period of time without proper monitoring required and the authors had successfully induced in vivo angiogenesis for the implanted artificial teeth experimented on the mice in the peptide nanofibers. Besides, there are also other researchers who applied a similar approach as Galler et al. [44] for encapsulating purpose with the aim of facilitating tissue engineering [74, 77]. The obturating material is an alternative method to improve root canal therapy [121]. Composite materials such as nHa and synthetic polymers (polylactic (PLA), poly lactic-co-glycolic acid (PLGA), polyamide, chitosan, and PCL) are used to assist and enhance the biding and retention of nHA to periodontal defects in order to reduce the dissemination and phagocytosis [45]. Regenerative effects (formation of acellular cementum and proliferation stimulation, periodontal fibroblasts differentiation, and osteoblasts differentiation) [7, 97] can be improved with the presence of biologics such as enamel matrix derivative (EMD).

Nanocomposite-based scaffolds (nanosurfaces, nanofibers, and nanocomposites) and nanofibrous-type scaffolds (collagen, elastin) are a few approaches for cartilage tissue regeneration (temporomandibular joint) [43]. An example of nanocomposite-based scaffolds like PLA with hydroxyapatite nanoparticles has been proven to enhance the chondrogenic differentiation of mesenchymal stem cells. Furthermore, the combination of gold nanoparticles with porcine collagen hydrogels is said to improve the mechanical properties of the present hydrogen and also associated a rise in the chondrocyte proliferation [27]. A similar observation was also seen in the study by Kay et al. [71], where the PLGA/titanium dioxide composite scaffolds had transformed the surfaces via using a range of sodium hydroxide (NaOH) concentration which led to an incremental of osteoblast and chondrocyte adhesion. Meanwhile, for nanofibrous-type scaffold, it is evidently shown that human-adipose-derived stromal cells could experience selectively differentiation for the osteochondral tissue to regenerate [36]. Overall, it is said that the cartilage tissue regeneration specifically for the temporomandibular joint is considered new and more research should be encouraged in this research area [43]. Most research is said to mainly focus on the cartilage of the knee joint rather than cartilage tissue regeneration particularly on the temporomandibular joint [76].

For bone regeneration of dental implants and mandibular defects, it is said that osteoconductive scaffold using macropores size (>200 $\mu$m) can increase the success rate of bone regeneration. This is further explained that such geometry provides sufficient flow and biofactors distribution throughout the healing process [9]. Furthermore, the deposition of new bone can improve these scaffolds with the presence of certain bacteria which could aid in post-infection treatment [67]. For an in-depth understanding of the various nanomaterial investigated for dental implants and mandibular bone defects, Tao et al. [121] conducted a critical literature review on the different types of nanomaterial for dental implants and mandibular bone defects in the aspects of nHA composite scaffolds, nanomodified mineral trioxide aggregate, and graphene.

## 4.3 Dental Filler and Antimicrobial Nanomaterials

The applications of nanotechnology in dentistry cover a broad range from dental diagnostics, preventive dentistry, dental materials (prosthodontics, endodontics, conservative and aesthetic dentistry, periodontics, implantology, and regenerative dentistry) to nanoproducts [6]. Jandt and Watts [65] conducted a comprehensive review on the scientific knowledge, principles, and applications related to the commercially available dental nanomaterials. From their studies, it was highlighted that fillers in nanocomposites are one of the main applications of nanoparticles in dentistry. By applying the simple rule-of-mixtures, this aids in predicting Young's modulus or the strength of the material, which provides insights information when designing a novel particle-based composite material. This compulsory rule refers to the composite strength which is defined as the volume-weighted average of the strengths of the filler and the matrix [22]. An interesting finding from the rule of mixtures is that despite the parameter particle size is not involved, a boundary condition exists, however, that is only applicable with large filler particles in the composites and there are no rules for the determination of properties of nanocomposites in the mixtures [65]. Note that the elasticity theory is considered in complicated situations. The actual size ratio of a nanoparticle to a larger particle is often not revealed by the producers. Furthermore, it is also challenging to conduct reverse engineering to quantify the formulations through techniques that are sufficient with filler of bigger particle's size.

For the development of the dental adhesive formulations, the primary dental materials need to be exclusively reinforced with the nanoparticles. In an ideal state, the nanocomposites are supposedly to possess nanohybrids that consist of greater volume-fractions of micron-sized particles. The visualization of the packing structure in either dental composites or dental ceramics is illustrated in Fig. 4. The packing structure helps to increase both the density and toughness of the dental composites or dental ceramics with minimal deficiencies. High surface-to-volume ratio and shorter inter-particle separation are the two benefits of nanocomposites compared to other composite materials [89]. This can be further explained that a high surface-to-volume ratio permits filler with smaller size and shorter inter-particle separation improved the properties of mechanical, optical, thermal, and scratch proof [69]. However, there are some drawbacks in nanocomposites which include decreasing in both toughness and impact performance, lack of basic understanding between the relationship of the material formulation and structure-property, and complicated control mechanism occurring during the particle dispersion stage [89].

To date, many researchers are still trying to prove the mechanical properties between dental nanocomposites and conventional dental nanocomposites [65]. Beun et al. [17] conducted comparative studies of the inorganic fraction and the mechanical properties of three nano-filled composites involving four universal hybrid and two micro-filled composites. In addition, halogen and LED units are used to study the conversion degrees of the photopolymerized materials. It is reported that the investigated nano-filled resin composites demonstrated an elasticity modulus higher compared to universal and micro-filled composites. Furthermore, the micro-filled

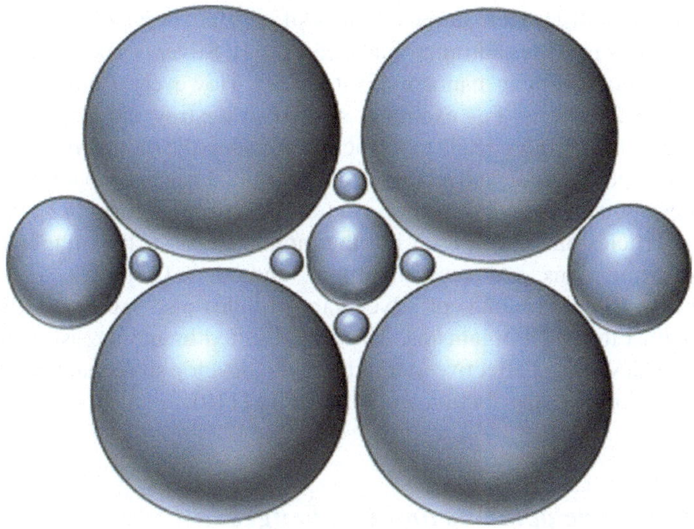

**Fig. 4** The packing structure of particles exists in the dental composites and dental ceramics whereby the nanoparticles fill the spaces between the larger particles to allow the enhancement in the density of the particle packing

composites displayed weaker mechanical properties. In addition, the mechanical properties of nano-filled resin composites are said to be comparable to the universal hybrids which results in providing similar clinical signals and anterior renewal owing to its good aesthetic properties. However, there are also some studies that do not validate the nanocomposite's mechanical properties [68, 102]. This is because to distinguish between nanocomposites and conventional composites is not a straightforward process as several analyses on the filler content, filler geometry, and composition are involved. In Karabela and Sideridou [70], the dental resin composites were investigated using various average particles size of nanosilica particles ranging from 40 to 7 nm. It was found that the silica filler differs in the prepared composites, however, with the presence of fixed amount of silanized silica and organic matrix, it leads to similar flexural strength and flexural modulus. In an exceptional case when a minimal particle size is used in the composite, a reduced flexural modulus is achieved. In other words, a minimum threshold particle size could influence the mechanical properties of a composite. In order to validate this hypothesis, more studies related to this area are required.

Nanoparticles in composites have the ability to provide both therapeutic and preventive effects. It has been evidently shown in the studies conducted by Zhang et al. [132]. The authors formed amorphous calcium phosphate (CaP) nanoparticles filled composites, which aided in maintaining everlasting cavities inhibition. In particular, the released and recharged activities of the Ca and P ions in the nanocomposite permit cavities-inhibiting restorations, which provide opportunities in broad applications such as dental composites, adhesives, cements, and sealants.

The presence of different materials such as silver, zinc, or copper nanoparticles in dental composites provides antimicrobial effects [64]. Silver is said to be the earliest used antimicrobial agent in nanocomposites [29]. However, it attributes harmful effects to the living cells. For example, silver can destroy the cell wall and cell membrane, and furthermore leads to a reaction between the biomacromolecules and cells [29]. On a positive note, silver-based antimicrobial has the capability to combat various types of microorganisms such as bacteria, fungi, and viruses [87]. In addition, silver nanoparticles have been used in other applications such as implant coating, anti-caries formulations, oral cancer treatment, and others [87].

The surface-to-volume ratio of silver nanoparticles also influences the dosage of silver against the microorganisms. This had resulted in an increasing interest by researchers to develop an effective methodology to develop silver nanoparticles with minimal deviation size distribution and uniformity [119].

## 4.4 Recent Developments and Challenges

For tissue engineering applications, it is observed that nanoparticles play a promising role in this area. Since toxicity, carcinogenicity, and teratogenicity are the important aspects in the practicality use of nanoparticles in tissue engineering application, the involved tools and methodologies used for conducting assessment for safe handling and risk management need to be relevant and effective. This is because the dosage and exposure of nanoparticles toward living cells are very much dependent on three aspects, namely toxicity, carcinogenicity, and tetratogenicity. Besides, the harmful effects on human health could lead to possible negative implications if bioaccumulation exists for a longer period. Hence, an international standard for the safe handling of nanomaterial is essential and urgently required in handling these nanoparticles.

The clinical settings in terms of integration, application, and translation have imposed challenges for both tissue and implant engineering. Cell viability, immunogenic properties, and innervation of the engineering tissue through nanomaterials are said to be the key challenges as mentioned by Tao et al. [121]. Although there are many types of nanopolymer that can be utilized for hydrogel to support the duration for both the biofabrication and fabrication, they can directly affect the cell viability [92]. Hence, continuous innovative scaffold techniques specifically for craniofacial tissue engineering are necessary and currently it is still at its infancy stage of research [92]. Furthermore, the immune response in living things should also be emphasized by researchers and scientists working in this area instead of just primarily focusing on the nanomaterial development to illustrate a similar structure to the native tissue. Additionally, an in-depth understanding of the biomechanisms involved in tissues engineered via nanomaterials should be well-understood before implementing these bionanomaterials in a clinical setting for the applications of tissue and implant engineering.

For the dental microbial nanomaterials, there is still so much room for improvement and further development. It would be suggested to develop a novel and improved quality dental material in the presence of nanotechnology associated with other new areas of its application dentistry to be discovered. This will be great benefits for both the patient and also continuous improvement in the dental treatment if nanotechnology is considered as a disruptive technology by replacing the older technologies particularly on the huge particles in dental composites.

# 5 Nanobiosensors

Nanobiosensors are biosensors that are fabricated using nanomaterials. A biosensor consists of a bio-receptor and a signal transducer [110]. The bio-receptor acts as the sensing element of the biosensors while the signal transducer is tasked to convert the biological or chemical signal to physical signals, being it as optical, electrical, mass, or thermal. There are many types of transducers, namely electrochemical, optical, piezoelectric, thermometric, and semiconductor [86]. The signal generated by the transducer then undergoes processing using an amplifier and processor before producing a result for reading. The various components of a biosensor are summarized in Fig. 5. The use of nanomaterials in the fabrication of biosensors especially in medical applications is of great interest due to the many benefits and advantages of nanomaterials which will be the focus of this subchapter.

## *5.1 Nanobiosensor Design and Fabrication*

There are several types of signal transducer in the nanobiosensors, namely electrochemical, optical, and mass [31, 49, 86, 93]. The most common type of transducer of nanobiosensors is the electrochemical approach. This is due to the good

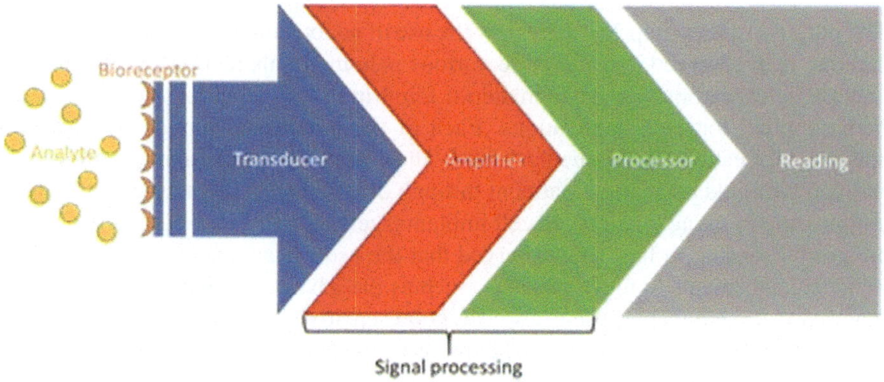


**Fig. 5** The components of a biosensor

sensitivity and selectivity, fast response, and cost-effectiveness of the approach [63]. This approach consists of various methods such as amperometric, voltametric, and photoelectrochemical [49], where the electrochemical signal is generated through the interaction between the biomarker and biorecognition molecules [30, 52, 100]. The optical approach uses light interaction with target molecules such as surface plasmon resonance imaging (SPRi), fluorescence, absorption, luminescence, total internal reflection, electrochemiluminescence (ECL), and surface plasmon resonance (SPR) [40, 46, 49, 114, 131]. Such an approach provides the advantage of label-free detection [96].

The mass-based transducer detects the changes in mass during the interaction between the biomarker and biorecognition molecule [80, 111, 115]. An example of a mass-based transducer is the piezoelectric transducer where it is then categorized into two approaches: quartz crystal microbalance (QCM) and surface acoustic wave (SAW) [109]. Both approaches involve the measurement of voltage changes caused by a change in mass. Such a transducer is commonly used as immunosensors.

### *5.2 Recent Development and Challenges*

The challenges for nanobiosensor design are related to the characteristics: sensitivity, response time, selectivity, and linearity [13]. An ideal nanobiosensor should have high sensitivity, high selectivity, high linearity, and fast response time. In the coronavirus disease 2019 (COVID-19) pandemic time, nanobiosensor can be employed to detect the virus [5]. The sensor can be integrated with internet-of-things (IoT) platform to be more practicable [18].

## 6 Conclusion

The purpose of this chapter is to review the use of nanotechnology for medical applications. It is observed that there are various nanomaterials and nanotools, namely nanoparticles, carbon-based nanomaterial, dendrimer, quantum dots, and conjugates being applied for such applications. Each of these materials has its strength and specific applications in the medical field. These specific applications are diagnosis through biomarker sensing, treatment through drug delivery, and tissue and implant engineering in tissue repair, stem cell implants, gene therapy, and dental filler. In the area of biosensing, it has been observed that the use of nanomaterials brings about many enhancements to the biosensors, such as the advantages and suitability of the materials for medical purposes, low cost, good sensitivity, and selectivity along with specific benefits of each material. The trends show that the study of nanotechnology for medical purposes is very much focused on identifying and developing nanomaterials that are complied with the acceptable toxicity and required safety while having a simpler approach. Also, the use of nanotechnology for medical applications is moving toward commercialization, with the focus on the ease and practicality of the use of medical tools based on this technology.

**Acknowledgements** We thank Angie Liong who helped us in collecting references. We also thank our colleagues at Curtin University Malaysia and Swinburne University of Technology Sarawak Campus for their support.

## References

1. Abdalla S, Al-Marzouki F, Al-Ghamdi AA, Abdel-Daiem A (2015) Different technical applications of carbon nanotubes. Nanoscale Res Lett 10. https://doi.org/10.1186/s11671-015-1056-3
2. Aghili Z, Nasirizadeh N, Divsalar A, Shoeibi S, Yaghmaei P (2018) A highly sensitive miR-195 nanobiosensor for early detection of Parkinson's disease. Artif Cells Nanomed Biotechnol 46:32–40. https://doi.org/10.1080/21691401.2017.1411930
3. Ahmed W, Elhissi A, Subramani K (2013) Chapter 1—Introduction to nanotechnology. Nanobiomaterials in clinical dentistry. William Andrew Publishing, Norwish, NY, USA, pp 3–16
4. Aldewachi H, Chalati T, Woodroofe MN, Bricklebank N, Sharrack B, Gardiner P (2017) Gold nanoparticle-based colorimetric biosensors. Nanoscale 10:18–33. https://doi.org/10.1039/c7nr06367a
5. Alhalaili B, Popescu IN, Kamoun O, Alzubi F, Alawadhia S, Vidu R (2020) Nanobiosensors for the detection of novel Coronavirus 2019-nCoV and other pandemic/epidemic respiratory viruses: a review. Sensors 20:6591
6. AlKahtani RN (2018) The implications and applications of nanotechnology in dentistry: a review. Saudi Dent J 30:107–116
7. Al Machot E, Hoffmann T, Lorenz K, Khalili I, Noack B (2014) Clinical outcomes after treatment of periodontal intrabony defects with nanocrystalline hydroxyapatite (Ostim) or enamel matrix derivatives (Emdogain): a randomized controlled clinical trial. BioMed Res Int 2014:786353
8. Al-Manhel AJ, Al-Hilphy ARS, Niamah AK (2018) Extraction of chitosan, characterisation and its use for water purification. J Saudi Soc Agric Sci 17:186–190. https://doi.org/10.1016/j.jssas.2016.04.001
9. Amini AR, Adams DJ, Laurencin CT, Nukavarapu SP (2012) Compatible scaffolds for large-area bone regeneration. Tissue Eng Part A 18:1376–1388
10. Anzar N, Hasan R, Tyagi M, Yadav N, Narang J (2020) Carbon nanotube—a review on synthesis, properties and plethora of applications in the field of biomedical science. Sens Int 1:100003. https://doi.org/10.1016/j.sintl.2020.100003
11. Baker SN, Baker GA (2010) Luminescent carbon nanodots: emergent nanolights. Angew Chem Int Ed 49:6726–6744
12. Balasubramanian S, Gurumurthy B, Balasubramanian A (2017) Biomedical applications of ceramic nanomaterials: a review. Int J Pharm Sci Res 8:4950–4959
13. Banigo AT, Azeez TO, Ejeta KO, Lateef A, Ajuogu E (2020) Nanobiosensors: applications in biomedical technology. IOP Conf Ser: Mater Sci Eng 805:12028
14. Batool F, Strub M, Petit C, Bugueno I, Bornert F, Clauss F, Huck O, Kuchler-Bopp S, Benkirane-Jessel N (2018) Periodontal tissues, maxillary jaw bone, and tooth regeneration approaches: from animal models analyses to clinical applications. Nanomaterials 8:337. https://doi.org/10.3390/nano8050337
15. Baughman RH, Zakhidov AA, de Heer WA (2002) Carbon nanotubes—the route toward applications. Science 297:787–792

16. Benglis D, Wang MY, Levi AD (2008) A comprehensive review of the safety profile of bone morphogenetic protein in spine surgery. Neurosurgery 62:ONS423–ONS431. https://doi.org/10.1227/01.neu.0000326030.24220.d8
17. Beun S, Glorieux T, Devaux J, Vreven J, Leloup G (2007) Characterization of nanofilled compared to universal and microfilled composites. Dent Mater 23:51–59
18. Bhalla N, Pan Y, Yang Z, Payam AF (2020) Opportunities and challenges for biosensors and nanoscale analytical tools for pandemics: COVID-19. ACS Nano 14:7783–7807
19. Boisselier E, Astruc D (2009) Gold nanoparticles in nanomedicine: preparations, imaging, diagnostics, therapies and toxicity. Chem Soc Rev 38:1759–1782
20. Bonilla CAM, Kouznetsov VV (2016) "Green" quantum dots: basics, green synthesis, and nanotechnological applications. IntechOpen
21. Chang C-C, Chen C-P, Wu T-H, Yang C-H, Lin C-W, Chen C-Y (2019) Gold nanoparticle-based colorimetric strategies for chemical and biological sensing applications. Nanomaterials 9:1–24. https://doi.org/10.3390/nano9060861
22. Chawla KK (2012) Composite materials
23. Chen Y, Kosmas P, Anwar PS, Huang L (2015) A touch-communication framework for drug delivery based on a transient microbot system. IEEE Trans Nanobiosci 14:397–408. https://doi.org/10.1109/TNB.2015.2395539
24. Chen Y, Li N, Yang Y, Liu Y (2015) A dual targeting cyclodextrin/gold nanoparticle conjugate as a scaffold for solubilization and delivery of paclitaxel. RSC Adv 5:8938–8941. https://doi.org/10.1039/C4RA13135E
25. Cheung RCF, Ng TB, Wong JH, Chan WY (2015) Chitosan: an update on potential biomedical and pharmaceutical applications. Mar Drugs 13:5156–5186. https://doi.org/10.3390/md13085156
26. Chinnathambi S, Shirahata N (2019) Recent advances on fluorescent biomarkers of near-infrared quantum dots for in vitro and in vivo imaging. Sci Technol Adv Mater 20:337–355. https://doi.org/10.1080/14686996.2019.1590731
27. Chiu L, Waldman S (2013) Nanomaterials for cartilage tissue engineering. IAPC Publishing, Zagreb, Croatia
28. Choi AH, Ben-Nissan B, Matinlinna JP, Conway RC (2013) Current perspectives: calcium phosphate nanocoatings and nanocomposite coatings in dentistry. J Dent Res 92:853–859
29. Clement JL, Jarrett PS (1994) Antibacterial silver. Met Based Drugs 1:467–482. https://doi.org/10.1155/MBD.1994.467
30. da Silva ETSG, Souto DEP, Barragan JTC, de Fátima Giarola J, de Moraes ACM, Kubota LT (2017) Electrochemical biosensors in point-of-care devices: recent advances and future trends. ChemElectroChem 4:778–794
31. Damborský P, Švitel J, Katrlík J (2016) Optical biosensors. Essays Biochem 60:91–100. https://doi.org/10.1042/EBC20150010
32. Dhivya S, Ajita J, Selvamurugan N (2015) Metallic nanomaterials for bone tissue engineering. J Biomed Nanotechnol 11:1675–1700
33. Divya K, Jisha MS (2018) Chitosan nanoparticles preparation and applications. Environ Chem Lett 16:101–112
34. Dykman L, Khlebtsov N (2012) Gold nanoparticles in biomedical applications: recent advances and perspectives. Chem Soc Rev 41:2256–2282
35. Eatemadi A, Daraee H, Karimkhanloo H, Kouhi M, Zarghami N, Akbarzadeh A, Abasi M, Hanifehpour Y, Joo SW (2014) Carbon nanotubes: properties, synthesis, purification, and medical applications. Nanoscale Res Lett 9
36. Erisken C, Kalyon DM, Wang H, Örnek-Ballanco C, Xu J (2011) Osteochondral tissue formation through adipose-derived stromal cell differentiation on biomimetic polycaprolactone nanofibrous scaffolds with graded insulin and Beta-glycerophosphate concentrations. Tissue Eng Part A 17:1239–1252
37. Esmaeili F, Heuking S, Junginger HE, Borchard G (2009) Progress in chitosan-based vaccine deliver systems. J Drug Del Sci Technol 20:53–61. https://doi.org/10.1016/S1773-2247(10)50006-6

38. Fei Yin Z, Wu L, Gui Yang H, Hua SuY, Yin ZF, Wu L, Yang HG, Su YH (2013) Recent progress in biomedical applications of titanium dioxide. Phys Chem Chem Phys 15:4844–4858. https://doi.org/10.1039/c3cp43938k
39. Felicetti L, Femminella M, Reali G, Liò P (2016) Applications of molecular communications to medicine: a survey. Nano Commun Netw 7:27–45
40. Fernández Gavela A, Grajales García D, Ramirez J, Lechuga L, Gavela AF, García DG, Ramirez J, Lechuga L (2016) Last advances in silicon-based optical biosensors. Sensors 16:285. https://doi.org/10.3390/s16030285
41. Freitas LF, Varca GHC, dos Santos Batista JG, Lugao AB (2018) An overview of synthesis of gold nanoparticles using radiation technologies. Nanomaterials (Basel) 8:939. https://doi.org/10.3390/nano8110939
42. Fu R, Fu G-D (2011) Polymeric nanomaterials from combined click chemistry and controlled radical polymerization. Polym Chem 2:465–475
43. Gaharwar AK, Sant S, Hancock MJ, Hacking SA (2013) Nanomaterials in tissue engineering: fabrication and applications. Elsevier, Amsterdam, The Netherlands
44. Galler KM, Hartgerink JD, Cavender AC, Schmalz G, D'Souza RN (2012) A customized self-assembling peptide hydrogel for dental pulp tissue engineering. Tissue Eng Part A 18:176–184
45. Gamal AY, Iacono VJ (2013) Mixed nano/micro-sized calcium phosphate composite and EDTA root surface etching improve availability of graft material in intrabony defects: an in vivo scanning electron microscopy evaluation. J Periodontol 84:1730–1739
46. Gamal R, Ismail Y, Swillam MA (2015) Optical biosensor based on a silicon nanowire ridge waveguide for lab on chip applications. J Opt (United Kingdom) 17:045802. https://doi.org/10.1088/2040-8978/17/4/045802
47. Gately RD, Panhuis M (2015) Filling of carbon nanotubes and nanofibres. Beilstein J Nanotechnol 6:508–516. https://doi.org/10.3762/bjnano.6.53
48. Ge L, Li Q, Wang M, Ouyang J, Li X, Xing MMQQ (2014) Nanosilver particles in medical applications: synthesis, performance, and toxicity. Int J Nanomed 9:2399–2407. https://doi.org/10.2147/IJN.S55015
49. Ghorbani F, Abbaszadeh H, Mehdizadeh A, Ebrahimi-Warkiani M, Rashidi M-R, Yousefi M (2019) Biosensors and nanobiosensors for rapid detection of autoimmune diseases: a review. Springer 186. https://doi.org/10.1007/s00604-019-3844-4
50. Giljohann DA, Seferos DS, Daniel WL, Massich MD, Patel PC, Mirkin CA (2014) Gold nanoparticles for biology and medicine. Angew Chem (int Ed Engl) 49:3280–3294. https://doi.org/10.1002/anie.200904359
51. Griffin MF, Kalaskar DM, Seifalian A, Butler PE (2016) An update on the application of nanotechnology in bone tissue engineering. Open Orthop J 10:836–848
52. Hammond JL, Formisano N, Estrela P, Carrara S, Tkac J (2016) Electrochemical biosensors and nanobiosensors. Essays Biochem 60:69–80. https://doi.org/10.1042/EBC20150008
53. Han J, Zhao D, Li D, Wang X, Jin Z, Zhao K (2018) Polymer-based nanomaterials and applications for vaccines and drugs. Polymers 10:31
54. Hanif A, Farooq R, Rehman MU, Khan R, Majid S, Ganaie MA (2019) Aptamer based nanobiosensors: promising healthcare devices. Saudi Pharm J 27:312–319
55. Harris DK, Allen PM, Han H-S, Walker BJ, Lee J, Bawendi MG (2011) Synthesis of cadmium arsenide quantum dots luminescent in the infrared. J Am Chem Soc 133:4676–4679. https://doi.org/10.1021/ja1101932
56. Hasan A, Morshed M, Memic A, Hassan S, Webster TJ, Marei HE (2018) Nanoparticles in tissue engineering: applications, challenges and prospects. Int J Nanomed 13:5637–5655. https://doi.org/10.2147/IJN.S153758
57. Hedman D, Barzegar HR, Rosen A, Wagberg T, Larsson JA (2015) On the stability and abundance of single walled carbon nanotubes. Sci Rep 5. https://doi.org/10.1038/srep16850
58. Henry NL, Hayes DF (2012) Cancer biomarkers. Mol Oncol 6:140–146
59. Ho MH, Chang H, Chang Y, Claudia J, Lin TC, Chang PC (2017) PDGF-metronidazole-encapsulated nanofibrous functional layers on collagen membrane promote alveolar ridge regeneration. Int J Nanomed 12

60. Hu D, Zhang P, Gong P, Lian S, Lu Y, Gao D, Cai L (2011) A fast synthesis of near-infrared emitting CdTe/CdSe quantum dots with small hydrodynamic diameter for in vivo imaging probes. Nanoscale 3:4724–4732. https://doi.org/10.1039/c1nr10933b
61. Hu MZ, Zhu T (2015) Semiconductor nanocrystal quantum dot synthesis approaches towards large-scale industrial production for energy applications. Nanoscale Res Lett 10.https://doi.org/10.1186/s11671-015-1
62. Hu SW, Qiao S, Xu BY, Peng X, Xu JJ, Chen HY (2017) Dual-functional carbon dots pattern on paper chips for $Fe^{3+}$ and ferritin analysis in whole blood. Anal Chem 89:2131–2137. https://doi.org/10.1021/acs.analchem.6b04891
63. Huang Y, Xu J, Liu J, Wang X, Chen B (2017) Disease-related detection with electrochemical biosensors: a review. mdpi.com. https://doi.org/10.3390/s17102375
64. Jandt KD, Al-Jasser AMO, Al-Ateeq K, Vowles RW, Allen GC (2002) Mechanical properties and radiopacity of experimental glass-silica-metal hybrid composites. Dent Mater 18:429–435
65. Jandt KD, Watts DC (2020) Nanotechnology in dentistry: present and future perspectives on dental nanomaterials. Dent Mater 36:1365–1378
66. Jayakumar K, Camarada MB, Rajesh R, Venkatesan R, Ju H, Dharuman V, Wen Y (2018) Layer-by-layer assembled gold nanoparticles/lower-generation (Gn $\leq$3) polyamidoamine dendrimers-grafted reduced graphene oxide nanohybrids with 3D fractal architecture for fast, ultra-trace, and label-free electrochemical gene nanobiosensors. Biosens Bioelectron 120:55–63. https://doi.org/10.1016/j.bios.2018.08.032
67. Johnson CT, García AJ (2015) Scaffold-based anti-infection strategies in bone repair. Ann Biomed Eng 43:515–528
68. Junior SAR, Ferracane JL, della Bona Á (2008) Flexural strength and Weibull analysis of a microhybrid and a nanofill composite evaluated by 3- and 4-point bending tests. Dent Mater 24:426–431
69. Kaizer MR, de Oliveira-Ogliari A, Cenci MS, Opdam NJM, Moraes RR (2014) Do nanofill or submicron composites show improved smoothness and gloss? A systematic review of in vitro studies. Dent Mater 30:e41–e78
70. Karabela MM, Sideridou ID (2011) Synthesis and study of properties of dental resin composites with different nanosilica particles size. Dent Mater 27:825–835
71. Kay S, Thapa A, Haberstroh KM, Webster TJ (2004) Nanostructured polymer/nanophase ceramic composites enhance osteoblast and chondrocyte adhesion. Tissue Eng 8:753–761
72. Ko W-K, Heo DN, Moon H-J, Lee SJ, Bae MS, Lee JB, Sun I-C, Jeon HB, Park HK, Kwon IK (2015) The effect of gold nanoparticle size on osteogenic differentiation of adipose-derived stem cells. J Colloid Interface Sci 438:68–76
73. Kryscio DR, Fleming MQ, Peppas NA (2012) Protein conformational studies for macromolecularly imprinted polymers. Macromol Biosci 12:1137–1144. https://doi.org/10.1002/mabi.201200068
74. Kuang R, Zhang Z, Jin X, Hu J, Shi S, Ni L, Ma PX (2016) Nanofibrous spongy microspheres for the delivery of hypoxia-primed human dental pulp stem cells to regenerate vascularized dental pulp. Acta Biomater 33:225–234
75. Li C, Wang J, Wang Y, Gao H, Wei G, Huang Y, Yu H, Gan Y, Wang Y, Mei L, Chen H, Hu H, Zhang Z, Jin Y (2019) Recent progress in drug delivery. Acta Pharm Sin B 9:1145–1162
76. Li G, Zhou T, Lin S, Shi S, Lin Y (2017) Nanomaterials for craniofacial and dental tissue engineering. J Dent Res 96:725–732
77. Li X, Ma C, Xie X, Sun H, Liu X (2016) Pulp regeneration in a full-length human tooth root using a hierarchical nanofibrous microsphere system. Acta Biomater 35:57–67
78. Liu D, Zhang J, Yi C, Yang M (2010) The effects of gold nanoparticles on the proliferation, differentiation, and mineralization function of MC3T3-E1 cells in vitro. Chin Sci Bull 55:1013–1019. https://doi.org/10.1007/s11434-010-0046-1
79. Mahato K, Baranwal A, Srivastava A, Maurya PK, Chandra P (2018) Smart materials for biosensing applications. In: Techno-Societal 2016. Springer International Publishing, pp 421–431

80. Marrazza G (2014) Piezoelectric biosensors for organophosphate and carbamate pesticides: a review. Biosensors 4:301–317. https://doi.org/10.3390/bios4030301
81. Matea CT, Mocan T, Tabaran F, Pop T, Mosteanu O, Puia C, Iancu C, Mocan L (2017) Quantum dots in imaging, drug delivery and sensor applications. Int J Nanomed 12:5421–5431
82. McNamara K, Tofail SAMM (2017) Nanoparticles in biomedical applications. Adv Phys: X 2:54–88. https://doi.org/10.1080/23746149.2016.1254570
83. Mohammed MA, Syeda JTM, Wasan KM, Wasan EK (2017) An overview of chitosan nanoparticles and its application in non-parenteral drug delivery. Pharmaceutics 9. https://doi.org/10.3390/pharmaceutics9040053
84. Nie S (2009) Biomedical nanotechnology for molecular imaging, diagnostics, and targeted therapy. In: 2009 annual international conference of the IEEE engineering in medicine and biology society, pp 4578–4579
85. No HK, Meyers SP (1995) Preparation and characterization of chitin and chitosan—a review. J Aquat Food Prod Technol 4. https://doi.org/10.1300/J030v04n02_03
86. Noah NM, Ndangili PM (2019) Current trends of nanobiosensors for point-of-care diagnostics. J Anal Methods Chem 2019
87. Noronha VT, Paula AJ, Durán G, Galembeck A, Cogo-Müller K, Franz-Montan M, Durán N (2017) Silver nanoparticles in dentistry. Dent Mater 33:1110–1126
88. Nune SK, Gunda P, Thallapally PK, Lin Y-Y, Forrest ML, Berkland CJ (2009) Nanoparticles for biomedical imaging. Expert Opin Drug Deliv 6:1175–1194. https://doi.org/10.1517/17425240903229031
89. Omanovic-Miklicanin E, Badnjevic A, Kazlagic A, Hajlovac M (2020) Nanocomposites: a brief review. Health Technol 10:51–59. https://doi.org/10.1007/s12553-019-00380-x
90. Ong YH, Ahmad AL, Zein SHS, Tan SH (2010) A review on carbon nanotubes in an environmental protection and green engineering perspective. Braz J Chem Eng 27. https://doi.org/10.1590/S0104-66322010000200002
91. Onsoyen E, Skaugrud O (1990) Metal recovery using chitosan. J Chem Technol Biotechnol 49. https://doi.org/10.1300/J030v04n02_03
92. Orciani M, Fini M, di Primio R, Mattioli-Belmonte M (2017) Biofabrication and bone tissue regeneration: cell source, approaches, and challenges. Front Bioeng Biotechnol 5:17
93. Ozsoz M (2012) Electrochemical DNA biosensors
94. Parveen A, Deshpande R (2018) Nanobiotechnology in the health care: The game and the goal. In: Phytotoxicity of nanoparticles. Springer International Publishing, pp 395–407
95. Pasut G (2019) Grand challenges in nano-based drug delivery. Front Med Technol 1:1
96. Perumal V, Hashim U (2014) Advances in biosensors: principle, architecture and applications. J Appl Biomed 12:1–15
97. Pilloni A, Saccucci M, di Carlo G, Zeza B, Ambrosca M, Paolantonio M, Sammartino G, Mongardini C, Polimeni A (2014) Clinical evaluation of the regenerative potential of EMD and NanoHA in periodontal infrabony defects: a 2-year follow-up. Biomed Res Int 2014:9
98. Popov VN (2004) Carbon nanotubes: properties and application. Mat Sci Eng: R 43:61–102. https://doi.org/10.1016/j.mser.2003.10.001
99. Qian D, Wagner GJ, Liu WL, Yu MF, Ruoff RS (2002) Mechanics of carbon nanotubes. Appl Mech 55:495–533
100. Rackus DG, Shamsi MH, Wheeler AR (2015) Electrochemistry, biosensors and microfluidics: a convergence of fields. Chem Soc Rev 44:5320–5340. https://doi.org/10.1039/c4cs00369a
101. Ratnesh RK, Mehata MS (2015) Controlled synthesis and optical properties of tunable CdSe quantum dots and effect of pH. AIP Adv 5.https://doi.org/10.1063/1.4930586
102. Rodrigues SA, Scherrer SS, Ferracane JL, della Bona Á (2008) Microstructural characterization and fracture behavior of a microhybrid and a nanofill composite. Dent Mater 24:1281–1288
103. Saeed RM, Dmour I, Taha MO (2020) Stable chitosan-based nanoparticles using polyphosphoric acid or hexametaphosphate for tandem ionotropic/covalent crosslinking and subsequent investigation as novel vehicles for drug delivery. Front Bioeng Biotechnol 8:4. https://doi.org/10.3389/fbioe.2020.00004

104. Sahoo SK, Parveen S, Panda JJ (2007) The present and future of nanotechnology in human health care. Nanomedicine 3:20–31. https://doi.org/10.1016/j.nano.2006.11.008
105. Saifuddin N, Raziah AZ, Junizah AR (2013) Carbon nanotubes: a review on structure and their interaction with proteins. J Chem
106. Sailaja AK, Amareshwar P, Chakravarty P (2010) Chitosan nanoparticles as a drug delivery system. J Pharm Biol Chem Sci 1:474–484
107. Saji VS, Choe HC, Yeung KWK (2010) Nanotechnology in biomedical applications: a review. Int J Nano Biomater 3:119–139. https://doi.org/10.1504/IJNBM.2010.037801
108. Salaheldin AM, Walter J, Herre P, Levchuk I, Jabbari Y, Kolle JM, Brabec CJ, Peukert W, Segets D (2017) Automated synthesis of quantum dot nanocrystals by hot injection: mixing induced self-focusing. Chem Eng J 320:232–243. https://doi.org/10.1016/j.cej.2017.02.154
109. Seo SE, Tabei F, Park SJ, Askarian B, Kim KH, Moallem G, Chong JW, Kwon OS (2019) Smartphone with optical, physical, and electrochemical nanobiosensors. J Ind Eng Chem 77:1–11. https://doi.org/10.1016/j.jiec.2019.04.037
110. Sharifi M, Avadi MR, Attar F, Dashtestani F, Ghorchian H, Rezayat SM, Saboury AA, Falahati M (2019) Cancer diagnosis using nanomaterials based electrochemical nanobiosensors. Biosens Bioelectron 126:773–784
111. Skládal P (2016) Piezoelectric biosensors. TrAC Trends Anal Chem 79:127–133. https://doi.org/10.1016/j.trac.2015.12.009
112. Solaiman SM, Yamauchi Y, Kim JH, Horvat J, Dou SX, Alici G, Ooi L, Martinac B, Shiddiky MJA, Gopalan V, Hossain MSA (2017) Nanotechnology and its medical applications: Revisiting public policies from a regulatory perspective in Australia. Nanotechnol Rev 6:255–269
113. Souza CDD, Nogueira BR, Rostelato MECM (2019) Review of the methodologies used in the synthesis gold nanoparticles by chemical reduction. J Alloys Compd 798:714–740. https://doi.org/10.1016/j.jallcom.2019.05.153
114. Špačková B, Wrobel P, Bocková M, Homola J (2016) Optical biosensors based on plasmonic nanostructures: a review. Proc IEEE 104:2380–2408. https://doi.org/10.1109/JPROC.2016.2624340
115. Su L, Fong CC, Cheung PY, Yang M (2017) Development of novel piezoelectric biosensor using PZT ceramic resonator for detection of cancer markers. In: Methods in molecular biology. Humana Press Inc., pp 277–291
116. Suh KS, Lee YS, Seo SH, Kim YS, Choi EM (2010) Gold nanoparticles attenuates antimycin A-induced mitochondrial dysfunction in MC3T3-E1 osteoblastic cells. Biol Trace Elem Res 153:428–436. https://doi.org/10.1007/s12011-013-9679-7
117. Sun T, Tsang WM (2018) 4—Nanowires for biomedical applications. In: Narayan R (ed) Nanobiomaterials. Woodhead Publishing, pp 95–111
118. Sun T, Tsang WM (2018) Chapter 15—Nanomaterials for dental and craniofacial tissue engineering. In: Nanomaterials in tissue engineering. Woodhead Publishing Series in Biomaterials, pp 95–111
119. Sun Y (2013) Controlled synthesis of colloidal silver nanoparticles in organic solutions: empirical rules for nucleation engineering. Chem Soc Rev 42:2497–2511
120. Taguchi M, Ptitsyn A, McLamore ES, Claussen JC (2014) Nanomaterial-mediated biosensors for monitoring glucose. J Diabetes Sci Technol 8:403–411. https://doi.org/10.1177/1932296814522799
121. Tao O, Wu D, Pham H, Pandey N, Tran S (2019) Nanomaterials in craniofacial tissue regeneration: a review. Appl Sci 9:317. https://doi.org/10.3390/app9020317
122. Tevlin R, McArdle A, Atashroo D, Walmsley GG, Senarath-Yapa K, Zielins ER, Paik KJ, Longaker MT, Wan DC (2014) Biomaterials for craniofacial bone engineering. J Dent Res 93:1187–1195
123. Trinchi A, Muster TH (2007) A review of surface functionalized amine terminated dendrimers for application in biological and molecular sensing. Supramol Chem 19:431–445
124. Wang W, Liao S, Zhu Y, Liu M, Zhao Q, Fu Y (2015) Recent applications of nanomaterials in prosthodontics. J Nanomater 2015:1–11

125. Wegner KD, Hildebrandt N (2015) Quantum dots: bright and versatile in vitro and in vivo fluorescence imaging biosensors. Chem Soc Rev 44:4792–4834. https://doi.org/10.1039/C4CS00532E
126. Wei L, Lu J, Xu H, Patel A, Chen ZS, Chen G (2015) Silver nanoparticles: synthesis, properties, and therapeutic applications. Drug Discov Today 20:595–601
127. Wilczewska AZ, Niemirowicz K, Markiewicz KH, Car H (2012) Nanoparticles as drug delivery systems. Pharmacol Rep 64:1020–1037
128. Xi D, Dong S, Meng X, Lu Q, Meng L, Ye J (2012) Gold nanoparticles as computerized tomography (CT) contrast agents. RSC Adv 2:12515–12524
129. Yazdani Z, Yadegari H, Heli H (2019) A molecularly imprinted electrochemical nanobiosensor for prostate specific antigen determination. Anal Biochem 566:116–125. https://doi.org/10.1016/j.ab.2018.11.020
130. Younes I, Rinaudo M (2015) Chitin and chitosan preparation from marine sources. Structure, properties and applications. Mar Drugs 13:1133–1174. https://doi.org/10.1300/J030v04n02_03
131. Yuan LL, Herman PR (2016) Laser scanning holographic lithography for flexible 3D fabrication of multi-scale integrated nano-structures and optical biosensors. Sci Rep 6:1–15. https://doi.org/10.1038/srep22294
132. Zhang L, Weir MD, Chow LC, Antonucci JM, Chen J, Xu HHK (2016) Novel rechargeable calcium phosphate dental nanocomposite. Dent Mater 32:285–293
133. Zhao M-X, Zeng E-Z (2015) Application of functional quantum dot nanoparticles as fluorescence probes in cell labeling and tumor diagnostic imaging. Nanoscale Res Lett 10.https://doi.org/10.1186/s11671-015-0873-8
134. Zhou T, Liu X, Sui B, Liu C, Mo X, Sun J (2017) Development of fish collagen/bioactive glass/chitosan composite nanofibers as a GTR/GBR membrane for inducing periodontal tissue regeneration. Biomed Mater 12:55004
135. Zhou W, Gao X, Liu D, Chen X (2015) Gold nanoparticles for in vitro diagnostics. Chem Rev 115:10575–10636
136. Zhu Y, Hong H, Xu ZP, Cai W (2013) Quantum dot-based nanoprobes for in vivo targeted imaging. Curr Mol Med 13:1549–1567

# Chapter 5
# Application of Nanotechnology in Enhanced Oil Recovery

**Hisham Ben Mahmud, Walid Mohamed Mahmud, Mian Umer Shafiq, Mansur Ermila, Ziad Bennour, and Saber Elmabrouk**

## 1 Introduction

Enhanced Oil Recovery (EOR) is a process used when the primary and secondary recovery does not improve oil production. EOR is also known as tertiary recovery technique can extract 30–60% or more of oil, while primary and secondary recovery can yield 20–40% [12]. According to the International Energy Agency [51], there are approximately 374 EOR projects in operation worldwide in 2017, producing more than 2 MMbbl/d, 166 $CO_2$-EOR projects, 120 thermal-EOR projects, 45 hydrocarbon-gas EOR projects, 35 chemical-EOR projects, and 8 additional EOR projects. Since introducing EOR, EOR projects have been largely implemented in North America, while other countries have begun to use these technologies in recent years. Malaysia has initiated offshore EOR, while India, Ecuador, Colombia, and some Middle East countries (i.e., Kuwait, Kingdom of Saudi Arabia, and the United Arab Emirates) have only developed EOR pilot projects. Furthermore, Libya and

H. B. Mahmud (✉) · Z. Bennour
Department of Petroleum Engineering, Faculty of Engineering and Science, Curtin University, 98009 Miri, Sarawak, Malaysia
e-mail: hisham@curtin.edu.my

W. M. Mahmud
Department of Petroleum Engineering, Faculty of Engineering, University of Tripoli, Tripoli, Libya

M. U. Shafiq
Petroleum and Gas Engineering, NFC IET Multan, Multan, Pakistan

M. Ermila
Petroleum Engineering Department, Colorado School of Mines, Golden, USA

S. Elmabrouk
Chemical and Petroleum Engineering Department, Libyan Academy for Postgraduate Studies, Janzour, Libya



M. Mujawar et al. (eds.), *Nanotechnology for Electronic Applications*, Materials Horizons: From Nature to Nanomaterials, https://doi.org/10.1007/978-981-16-6022-1_5

Oman have significantly increased EOR projects [90]. The current development and application of EOR technologies in Libya is limited to hydrocarbon-miscible floods.

Obviously, the world oil recovery still does not heavily rely on EOR, but it is clear that EOR techniques have an essential role in improving oil production, especially in combination with nanomaterials (nano-EOR). Nanotechnology is seen as an alternative method used as a new EOR technique to solve residual oil problems in oil reservoirs. However, nanotechnology is science, engineering, and technology on the nanoscale. It has been a known field of research since 1960 when Richard Feynman introduced nanotechnology in his well-known lecture in 1959: "There is plenty of space at the bottom" [35]. Since then, various revolutionary developments have taken place in this field, such as chemistry, biology, physics, materials science, engineering, and many sectors of the economy, including consumer goods, health care, transportation, energy, and agriculture. Nanotechnology is the study, manipulation, and fabrication of ultra-small structures and machines that consist of only one molecule. In the International System of Units, a nanometer is one-billionth of a meter ($10^{-9}$ m), which means it is about a hundred thousand times smaller than the diameter of human hair, a thousand times smaller than a red blood cell, or about half the diameter of DNA. In principle, ISO/TS 80004:2015 defines a nanomaterial as "a material that has any external dimension at the nanoscale, or has an internal structure or surface structure at the nanoscale". These include both nano-objects, which are separate pieces of material, and nanostructured materials that have an internal or surface structure at the nanoscale; a nanomaterial can be a member of these two categories [16]. Besides, in 2007, the British Standards Institution proposed a terminology for nanomaterials of particular relevance to nanotechnology [109], such as nanoscale, nanoscience, nanotechnology, nanomaterials, nano-objects, nanoparticles, nanorods, nanoplates, nanofibers, nanocomposites, and nanostructures. Although this terminology seeks to exclude terms that are used as defined in the Oxford English Dictionary and terms that already have a well-established meaning and for which the addition of the prefix "nano" means only those changes to which they are applied but do not change their meaning. Recently, however, some of the new nanotechnology terms have been traded among petroleum engineers, such as nanofluid, nano-polymer flooding, nano-suspensions, nano-smart waters, nanoparticle-flooding, nano-EOR, nano-based drilling fluid.

Of note, nanoparticles consist of three layers: (1) the surface layer, which is functionalized with a number of small molecules, metal ions, surfactants, and polymers. (2) The shell layer, which is chemically different from the core in all respects, and (3) the core, which is essentially the central part of the nanoparticle and generally refers to the nanoparticle itself [106]. Thus, nanomaterials have unique physicochemical, biological, optical, electronic, and mechanical properties compared to their larger counterparts because the surface area of the nanoparticles is relatively larger [50, 94]. This means that the ratio of surface area to volume increases as the radius of the sphere decreases [28, 38, 54]. Note that particle size and surface area determine how the system reacts, distributes, and eliminates the materials [95]. However, reducing the size of the materials obviously leads to an exponential increase in the volume of the surface, thereby making the surface of the nanomaterial more reactive toward itself and the adjacent medium [38].

Due to their specific size, shape, chemical composition, surface structure, charge, solubility, and agglomeration, these materials have aroused tremendous interest among multidisciplinary researchers. Different nanofluids may be designed by adding nanoparticles to different base fluids. The stability of the dispersion of nanoparticles in solutions depends on the functionality or surface activity of the nanoparticles. As a result, there are changes in the physical properties of such mixtures, such as viscosity, density, and thermal conductivity, relative to the base fluid [1, 17, 32, 40, 67, 68, 112, 125].

Nanotechnology in the petroleum industry has made tremendous interest in recent years, where nanotechnology has proven useful at all stages of the petroleum industry, from exploration, drilling, well extraction, manufacturing to EOR. However, nanoparticles had a significant positive effect on the rheological and filtration characteristics and thermal stability of drilling fluids [9, 10, 14, 39, 43, 73]. In addition, the nanoparticles showed a significant improvement in the rheological properties of the fracturing fluids [24, 87]. In contrast, there are very few field trial applications reported in the literature, suggesting the need for further field studies. The field trial is to inhibit and resolve remedy damage to Colombian formation [36], to evaluate the stability of nanoparticles under harsh formation conditions as a tracer of water injection in Saudi Arabia based on recovery percentage [61, 63, 70], increasing the mobility of heavy oils in Colombia [124] and stabilizing slate formation in Brazil with nano-water-base mud [13]. In addition, a significant increase in oil recovery was observed when nanoparticles were applied on a laboratory scale to apply EOR by altering wettability and reducing oil viscosity [2, 12, 26, 33, 48, 90].

The main challenge with the use of nanotechnology is the production of nanomaterials. Large-scale production of nanomaterials for industrial use by conventional synthetic methods is expensive [8, 22, 27]. Nevertheless, many experts predict that the application of nanotechnologies will ultimately permeate all areas of life and allow for dramatic advances in most areas of communication, health care, manufacturing, materials and knowledge-based technologies. Even if this is only partially true, there is a clear need to provide industry and research with the right tools to support the development, application, and communication of technologies. This chapter reviews the state-of-the-art of research findings on the application of nano-EOR. Particularly, this review focuses on the current attempts to analyze the influence of different nanoparticle additives on EOR methods and their mechanisms.

## 2 Applications of Nanotechnology in EOR

As the nanotechnology terminology indicates, it involves nanoparticles. This, relatively new technology that is somewhat inexpensive and environmentally harmless, has become widely popular in enhanced oil recovery (EOR) operations at nanoparticles in range size between 1 and 100 nm. The extremely small size of nanoparticles results in elevated concentration of atoms on the surface, thus, various properties compared to the same larger scale molecules. For example, particularly at elevated temperatures, the large surface area of nanoparticles creates a great diffusion driving force. Among many efforts that nanoparticles are believed to be EOR by decreasing interfacial tension and oil viscosity, modifying reservoir wettability toward water-wet, increasing mobility ratio, and altering reservoir permeability [89].

## 2.1 Reservoir Characterization

Nano-characterization technology plays a significant role in mineral composition analysis, rock physical properties, and micro pore structure of unconventional reservoirs. Nanomaterial is the basis of nanotechnology and nano-characterization technology assists in developing new nanotechnologies. Reservoir rocks are composed of porous material originated from different mineral particles with different levels of homogeneity/heterogeneity and significant presence of pores, organic clusters, and micro-nano mineral particles. There are several aspects of nano-characterization technology that analyze nanoscale objects, material components, structure, and property. Mineral composition is the first aspect that represents the crystalline component and structure of reservoir mineral particles. Micro-pore structure is the second aspect that characterizes micro-pore structures of reservoirs including unconventional reservoirs. Analysis of organic components is the third aspect that deals with the solid organic matter in the reservoir such as kerogen and recoverable liquid organic matter such as crude oil and bitumen. Another aspect is the petrophysical property that focuses on analyzing and characterizing chemical and physical rock properties. And finally, the in situ characterization aspect deals with petroleum accumulation and migration in reservoirs down to the microscopic, molecule and atom levels at different pressure and temperature conditions [78].

Bera and Belhaj [18] suggested investing in the high-risk nanosensors for formation evaluation in geochemical exploration and seismic characterization and interpretation in order to establish feasible technical and economical approaches for exploration. As reservoir fluids are flooded by nanoparticles, the grain surface adsorbs some of the nanoparticles while others flow in the porous material (pores and throats) of the rock along with the native reservoir fluids. Thus, adsorbed nanoparticles can be utilized as sensors to analyze and obtain formation distribution and rock properties because of the evident alteration in optical, electrical, and magnetic properties. Utilizing hyperpolarized silicon nanoparticles as petroleum exploration nanosensors was also suggested [108]. Moreover, using microbes was also suggested to indirectly estimate reservoir temperature, salinity, pressure, and other properties utilizing nano-optical fibers of Resonance Raman Spectroscopy (RRS) [53]. Utilizing magnetic nanosensors can also be used to identify magnetic nanoparticles in oil–water interfaces and flood front [6]. Superparamagnetic nanoparticles were also utilized to detect flood front through cross-well magnetic sensing [97]. Reservoir nano-robots have also been introduced to petroleum exploration, recovery, and well logging [78, 96]. However, the challenge remains in how to prevent nanosensors from deterioration while reaching the targeted porous media and fluids [3].

## 2.2 Characterization and Properties of Hydrocarbon Fluid

It is important to identify the chemical nature of hydrocarbon fluid before any potential EOR operation especially those involving nanotechnology. Hydrocarbon fluids based on their chemical composition can be paraffinic, naphthenic, or aromatic. Moreover, molecule structure is also used to classify hydrocarbon fluids. For instance, straight or open chains merged by single bonds are the characteristics of paraffinic hydrocarbon fluid compounds and their isomers that contain branched chains such as methane, ethane, propane, and decane.

At room temperature and pressure, starting from member one to four of the paraffinic series are found in the gaseous phase and compounds starting from heptadecane ($C_{17}H_{36}$) down to pentane ($C_5H_{12}$) are liquids. Heavier compounds, such as hentriacontane ($C_{31}H_{64}$), are wax-like solids and colorless. Naphthene hydrocarbon fluid molecules are ringed (cycloparaffins) and saturated and stable. Aromatic hydrocarbon fluids are also stable and cyclic, however, benzene derivatives having double-bonded rings.

In reality, hydrocarbon crude oils, produced around the world every day, are complicated combinations of all previously mentioned hydrocarbon fluids. Paraffin-based or paraffinic are crude oils that mainly contain paraffin hydrocarbon fluids. Petroleum is a mixture of these complex hydrocarbons. Paraffin-based or paraffinic are oils mainly composed of paraffin hydrocarbons. On the other hand, crude oils containing a significant amount of heavy cycloparaffins components are described as naphthenic-based hydrocarbon fluids. The least common hydrocarbon fluids are the highly aromatic oils. Therefore, it can be concluded that most hydrocarbon fluid mixtures produced are composed of paraffins and naphthenes while less common hydrocarbon fluid mixtures are composed of paraffins-naphthenes-aromatics.

As mentioned above, petroleum crude contains a mixture of mainly volatile hydrocarbon liquids that are carbon and hydrogen compounds with smaller quantities of oxygen, sulfur, and nitrogen. Therefore, hydrocarbon elements form complex miscellaneous molecular structures that are sometimes unidentifiable. However, almost entirely petroleum compounds contain 12–15% hydrogen and 82–87% carbon by weight. Overall, the hydrocarbon compound type that is dominant characterizes the hydrocarbon crude being aromatics, naphthenes, and paraffins that are most prevailing and valuable hydrocarbons. Naphthenes form a significant portion of refinery products and residues of the refinery processes that are heavy asphalt like. Benzene is the most popular aromatic in petroleum crudes. The physical properties of hydrocarbon crudes widely differ because of their varying proportions and constituents. Appearance might be black, colorless, or anything in between. However, specific gravity is the most crucial physical property.

The gravity scale known as the American Petroleum Institute (API) is the standard scale of the petroleum industry. The API method arbitrarily assigns pure water an API gravity of 10°. Petroleum crude oils, which are lighter than water, have API gravities greater than 10. Thus, petroleum crude oils can be classified based on their API gravity as light, medium, and heavy. Petroleum crude oils that have API gravity above 25° are light, 20–25° are medium, and 10–20° are heavy.

The level of sulfur, elemental or compounds such as hydrogen sulfide, in the petroleum crude oils, is also used to further categorize oils that can be sour or sweet. Sulfur contents by weight of 1 percent or more in the crude oils make it sour and sulfur contents of 0.5% or less in the crude oils make it sweet. Thus, the greater the sulfur contents in the crude oils the heavier they become.

## *2.3 Type of Nanoparticles*

Size, physical and chemical properties and morphology are the basis to classify different types of nanoparticles. Moreover, nanoparticles might be carbon-based, semiconductor nanoparticles, metal nanoparticles, ceramic nanoparticles, polymeric nanoparticles, and lipid-based nanoparticles. In EOR, the nanoparticles used are aluminium oxide, iron oxide, magnesium oxide, hydrophobic silicon oxide, zinc oxide, silicon oxide treated with saline, tin oxide, zirconium oxide and nickel oxide. However, the effectiveness of any type of nanoparticles, in an EOR operation, relies on the nature and conditions of the reservoir [60]. For example, mixing $SiO_2$ and $Al_2O_3$ nanoparticles yielded the highest oil recovery when flooded into heavy oil sandstones that is much higher than injecting water or individual nanoparticles of $Al_2O_3$, $TiO_2$, $SiO_2$, or NiO [7]. A glass micromodel was used to also flood heavy oil using $SiO_2$, NiO, and $Fe_3O_4$ nanoparticles that led to an additional oil recovery factor of 22.6%, 14.6%, and 8.1%, respectively, compared to just water flooding [66]. When $SiO_2$ and $CaCO_3$ nanoparticles are mixed with the base fluid and injected to alter wettability favorably towards higher oil recovery which indeed increased it by a factor of 4 $CaCO_3$ and a factor of 6 for $SiO_2$ [88]. Modified $SiO_2$ Nanoparticles were injected into intermediate and light core samples leading to an additional oil recovery of 25 and 14%, respectively, higher than the recovery by just water flooding [101].

It is important to mention that injecting nanoparticles might unfavorably affect reservoir parameters and lead to a decrease in oil recovery [8]. For instance, injecting zinc oxide leads to the formation of particles that block and settle in the pore space hindering the flooding process and therefrom less oil recovery [48].

### 2.3.1 Type of Nanoparticles

Metal oxides are hydrophilic or polar molecules with electronegativity differences of the bonded elements of $0.5 \times 10^{1.7}$. Nanoparticles are useful for applications in electrical and magnetic fields because of their unique magnetic properties. Some of the metal oxides nanoparticles that have been experimented as EOR include iron oxide ($Fe_2O_3/Fe_3O_4$), nickel oxide ($Ni_2O_3$), aluminum oxide ($Al_2O_3$), copper (II) oxide (CuO), magnesium oxide (MgO), tin oxide ($TiO_2$), zinc oxide (ZnO), and zirconium oxide ($ZrO_2$). For example, $Al_2O_3$ nanofluid was shown to decrease the interfacial tension between oil and brine and also to reduce oil viscosity that both lead to additional oil recovery [90]. Moreover, $TiO_2$ nanoparticles were shown to modify sandstone formation wettability from oil to water-wet by particles deposition on the internal pore surfaces of the rock leading to additional oil recovery [29]. Brine (NaCl 0.3 wt%) was flooded at 26 °C into an intermediate limestone sample yielding an oil recovery of 47.3% [15]. However, as three metal oxides nanoparticles or nanofluids ($Al_2O_3$, $TiO_2$, and $SiO_2$) were flooded at identical flooding conditions, oil recoveries increased to 52.6%, 50.9%, and 48.7%, respectively. $Al_2O_3$ nanofluid highly decreased oil capillary force yielded higher oil recovery.

### 2.3.2 Magnetic Nanoparticles

Magnetic Nanoparticles

Nanoparticles are ultrafine particles measured in nanometers and are in between 0.1 and 100 nm. They are used in a vast range of applications with different sizes and surface properties to meet the needs of research and industrial development. Among these nanoparticles are magnetic nanoparticles (MNPs). Magnetic nanoparticles are described as particles of electrons, protons, and holds negative and positive ions. Ferrite oxide-magnetite ($Fe_3O_4$) is the most naturally taking place magnetic mineral on the earth and is broadly utilized in the form of superparamagnetic nanoparticles for all applications categories.

Several common methods are reported in regards to the synthesis of MNPs. The uses of it offer numerous merits due to their distinctive size and physicochemical properties. As a result of the vast applications of MNPs in different fields such as material science, biotechnology, and biomedical, engineering and environmental areas, much thought and effort have been enforced to the synthesis of various types of MNPs. An important factor to note is that the application of MNPs requires obtaining different properties. This can be seen in data storage applications in which the particles must be in a stable, switchable magnetic state to present fragments of data that are not influenced by temperature variations. As for biomedical practices, the application of particles that show superparamagnetic behavior at 25 °C is preferred [19, 34, 85]. Moreover, applications in biology, therapy, and medical diagnosis need the magnetic molecules to be stable in water and in a physiological environment at a pH of 7. An additional application for MNPs is the environmental exercises in which particles can be used as versatile tools for the remediation of various kinds of contaminants on experimental and field scales within groundwater, soil, and air [113].

The application of nanotechnology has been growing promptly in the past two decades within the petroleum industry. MNPs were initially introduced as "smart-nano fluids" in conjunction with surfactants in an EOR process to decrease the interfacial tension [71]. When dealing with water-wet rock formations, oil bubbles occasionally become locked in the middle of small pores. Surfactant application also coats the ferromagnetic nanoparticles to prevent agglomeration.

When using ferro nanofluids these bubbles collapse. Additionally, due to the existence of dipole moment, the reservoir formation oil molecules align causing a reduction in the resistance to the flow and thereby increasing oil recovery. The ultra-fine size of nanoparticles allows them to pass through the pore throats; however, their delivery at the target points where there is the existence of oil within the rock requires additional emphasis [71]. This is a clear indication that nanoparticles support the enhanced heavy oil recovery process. In another experiment, cobalt ferrite nanoparticles were added to nanofluid and fed into a core sample while being exposed to an electromagnetic field for twenty-four hours under a temperature of 55 °C. This resulted in 72% oil recovery [120]. To further factualize this result, more research and investigation are required using these nanoparticles and core-flood to verify this result.

### Organic Nanoparticles

Organic nanoparticles (ONPs) are described as very small particles (10 nm to 1 μm) but consist of organic compounds as lipids or polymeric. ONPs are one of the most important points of interest in material and life sciences. ONPs can be found in nature and can be manufactured as well. The chemical composition and the shape of a nanoparticle also influence its specific properties. One example of ONPs is carbon nanoparticles (CNP). CNPs are made from carbon and are described as a spherical-shaped nanoparticle. The surface of these solids can be amended with polymers chemically bound to the particles' surfaces or organic molecules [123]. CNPs were used in the Berea sandstone and dolomite core flood experiment in the existence of salt ions. The aim of investigation was to analyze the impact of nanomaterials on the breakthrough time. The results concluded that the salt ions negatively affected the breakthrough time and retention throughout these cores. It also concluded that the effect of the salt ions can be reduced by coating the surface of nanoparticles [123].

Due to the difference in the surface charges between the dolomite core (+) and CNP (−), the retention for the dolomite core is higher; therefore, it is highly recommended to modify the surface of CNP when it is used in dolomite reservoirs. In carbonate reservoirs with temperatures higher than 100 °C and the formation water salinity greater than 120,000 ppm, CNP can be used normally [61, 63]. A-dots CNP is a new type of particles used in similar reservoir conditions. These particles are carbon-based fluorescent nanoparticles and their implementation in core samples from carbonate formation is proven to improve the recovery factor up to 96% [61, 63].

Another unique nanoparticle made from carbon is called carbon nanotubes (CNT) [21]. CNTs can be described as cylindrical molecules made of rolled-up sheets of single-layer carbon atoms (graphene). They are available as single walled (SWCNT) with a diameter <1 nm or multi-walled (MWCNT), including several concentrically interlinked nanotubes, with diameters >100 nm [21]. The length of them can get to several μm or even mm. There are not many studies that have been performed using CNT as an EOR agent. Chandran [21], used multi-walled (MWCNT) as an EOR agent in a core test representing high pressure, high-temperature reservoir. The test was conducted in two different methods. The first test was without applying the electromagnetic waves and the oil recovery from the first test was about 36%, while in the second test, the electromagnetic waves were applied, and the result of oil recovery was almost doubled.

Inorganic Nanoparticles

Inorganic nanoparticles refer to particles without carbon atoms in their molecular structure. The inorganic nanoparticles include metal and metal oxides such as sliver, iron oxide, magnesium oxide, copper oxide, titanium oxide, zinc oxide, nitrite oxide, and aluminum oxide. Inorganic nanoparticles have attracted a large amount of attention in the petroleum industry including in drilling muds, cementing, well stimulation, and EOR [90]. An estimate of 60–70% of oil in place remains in the pore space after the primary and secondary recovery process, and in order to recover that portion of oil, several enhanced recovery processes such as miscible gas injection, chemical injection, and thermal recovery need to be conducted. Even though the previous recovery methods were successful in achieving a small amount of oil recovery, low sweep efficiency, expensive and potential reduction of formation permeability, still obstruct the new applications of these EOR methods [90]. Up-to-date investigations on NPs are potential solutions to most of the challenges related to traditional EOR methods.

Hydrophobic and lipophilic (HLP) nanoparticles have been used as an injection fluid in the water-wet sandstone core flood experiment at 25 °C [42]. Researchers have revealed that the HLP nanoparticles have reduced the interfacial tension (IFT). Alumina-coated silica nanoparticles with modified surfaces resulted in a stable foam and the ability to improve the oil production from sandstone core samples compared to using only NPs or surfactant flooding. Iron oxide, copper oxide, and nickle oxide are used in EOR in tight carbonate cores. It is found that using NPs has increased oil recovery by 22% [42].

A nanoparticle known for its cost-effectiveness and abundancy is silica dioxide ($Sio_2$) also known as silica. When silica dioxide nanoparticles were used in a water-wet sandstone reservoir as an EOR agent with ethanol as a dispersing agent, it lowered the IFT and changed the wettability to mixed-wet. It could also recover up to 80% of oil in place [114]. Wang et al. [114] proved that $SiO_2$ has good thermal stability which means that the surface area of $SiO_2$ powders does not alter even at high temperatures (up to 650 °C).

Synergistic blends of $SiO_2$ nanoparticles and surfactants were also implemented in sandstone reservoirs to evaluate the EOR process at high temperatures (>100 °C) and high salinity (120,000 ppm). The results showed that a considerable potential for EOR applications because of their adsorption resistance onto the rock surface, thermostability, and also IFT reduction to vary low values [114].

By affecting viscosity, interfacial tension, and wettability, nanoparticle fluids can be used to modify the following fundamental flow-determining properties of a rock-multiphase fluid system such as relative permeability and capillary pressure of water and oil phase by targeting trapped oil and modify flow path diversion by targeting bypassed oil. Inorganic nanoparticles have many characteristics that grant them to be excellent candidates for the EOR process such as ultrafine sizes which have the ability to be modified to reservoirs, the ability to reduce the interfacial tension between water–oil, and the ability to target specific locations. Nanoparticles are used in the EOR process to help wettability alteration, decrease oil viscosity and interfacial tension, and stabilize foam or emulsion [105].

#### Non-silica Nanoparticles

Nanostructured zeolite can be used in absorbing cations such as the sodium ion, potassium ion, calcium ion, and magnesium ion from the formation which has a high water salinity (>120,000 ppm) due to their challenging form in the EOR process. One type of non-silica nanoparticles is nanosized Colloidal Dispersion Gel (CDG). They are applied as sweep improvement agents in EOR operation. Several studies reported that using nanosized colloidal dispersion gel in a sandstone core flood experiment resulted in an improvement in the oil recovery up to 40% [107].

One type of non-silica nanoparticles are Polyacrylamide Micro-gel nano-spheres particles which are employed in combination with sodium hydroxide to reduce the interfacial tension and increase the oil recovery by over 20% from sandstone reservoirs that consist of heavy crude oil viscosity (238 cp). These are also used in the EOR process to achieve the same benefits as mentioned at the end of section "Non-silica Nanoparticles". Table 1 shows some of the commonly applied nanoparticles in the field of the EOR process and their mechanisms lead to improve oil recovery.

**Table 1** Common NPs in EOR process and their effect on the recovery process

| Viscosity reduction | IFT reduction | Wettability alteration | Sweep and displacement efficiency |
|---|---|---|---|
| Aluminum oxide | Silica oxide NP | Tin oxide | Polymer-coated NP |
| Copper oxide | Polymer-coated NP | Silica oxide NP | Polymer NP |
| Iron oxide | Ferrofluid | Aluminum coated NP | Nanosized CDG |
| Nickle oxide | HLP nanoparticles | Hydrophobic $Sio_2$ | |
| Magnesium oxide | | Polymer-coated NP | |
| Polymer-coated NP | | | |

# 3 NP's-EOR Mechanism

## *3.1 Wettability Variation, Interfacial Tension Reduction, and Mobility Control*

The wettability of reservoir rock strongly influences the location, distribution, and flowing of oil and water through the formation during production. In some special cases, as in decreasing connate water saturation by rising the imposed capillary pressure, change the wettability from water- to oil-wet also improves recovery [72, 103]. Changes in wettability can occur at any stage in the life of the reservoir, e.g., during drilling, the mud and some of its used components can change the rock minerals, resulting in a change in wettability. Wetting can also change during production if asphaltenes precipitate during the drop-in reservoir pressure or gas injection [5, 11, 82, 111, 121, 122]. Another important role of the surfactant in chemical flooding is varying the reservoir rock wettability [104, 127]. However, one of these mechanisms can explain the change in wettability, the coating mechanism in which the surfactant particles are adsorbed on the solid surface, or the cleaning mechanism in which the surfactant molecules separate the oil molecules adsorbed on the rock surface [41].

The use of nanoparticles has increased the efficiency of the change in wettability controlling mobility due to two main properties of nanoparticles; very small size and ability to manipulate their behavior [56–58, 64, 65, 79, 91]. The nanoparticle size (diameter from 1 to 100 nm) explains a large specific surface area that can flow through container formations with a pore size of one μm or less. The bigger surface area causes an increase in surface energy. Adsorption of the material on a solid surface can completely change the wettability and surface energy. By exploiting the properties of nanoparticles, a "smart fluid" can be created that simply involves adding fractions of nanosized particles to a conventional fluid to improve or enhance some of the properties of the fluid [84, 128]. The impact of nanofluids on the change in wettability can be measured by assessing the contact angles before and after treatment, or assessed by imbibition tests.

Experimental studies on changing wettability showed that the oil production rate after water flooding was enhanced by adding nanoparticles to the injection fluid [25,

44, 46, 76]. This is because nanoparticles generally attempt to form a layer between the water–oil interface. This layer creates a lower interfacial tension between two liquid phases depending on the concentration of the nanoparticles spread out in the nanofluids [49, 98].

The decrease in IFT between fluids in the reservoir is one of the mechanisms of EOR that affects capillary forces, relative permeability, and flow patterns in the formations. The nanoparticles form a layer at the water–oil interface. This layer forms a lower interfacial tension between two immiscible liquid phases depending on the concentration and dispersed of nanoparticles in the nanofluids [23, 52, 116]. Some nanoparticles are recognized agents for the reduction of IFT, such as silica ($SiO_2$), alumina ($Al_2O_3$), and titanium dioxide ($TiO_2$). Table 2 shows a summary of nanoparticles types and effects on IFT.

### *3.2 Disjoining Pressure*

A disjoining pressure is an additional mechanism for changing the wettability with nanofluids. Many studies have been conducted to evaluate whether a nanofluid can remove the oil from the rock surface and become water-wet due to the disjoining pressure [117]. Because of the entropy enhancement in the entire nanofluids, the nanoparticles were ordered in a limited wedge between the oil drop and the grains. These ordered nanoparticles exert an overpressure that separates the two surfaces that enclose the nanofluids. This overpressure is defined as a structural disjoining pressure [61, 63]. The nanoparticles that find in the contact area can establish a wedge-like structure between the discontinuous phase and the substrate [74, 77].

The forces behind the disjoining pressure comprise; Brownian motion, van der Waals, and electrostatic repulsion. Due to nanoparticle size, the electrostatic repulsion among the nanoparticles is higher than the van der Waals and Brownian motion forces and it produces a higher structural disjoining pressure. The quantity of this separation pressure, which results from the nanoparticle dispersions, relies on the formation characteristics, salinity and temperature, particle size, volume fraction, and stability [99].

### *3.3 Improving Heat Transfer in Heavy Oil Reservoir*

Various thermal techniques used to improve the viscosity/density of the heavy oil reservoirs to increase their mobility involved steam flooding, cyclic steam stimulation, and steam gravity drainage. These methods include large processes such as heat generation, injection, and recycling, which add to operating costs. Nanotechnology in heat generation processes is the alternative to the production and recovery of heavy oil, which can result in small energy consumption, less pollution, and a high oil recovery [36, 37]. However, there is little data in the technical literature on the

**Table 2** Summary of type and effects of nanoparticles on IFT [43]

| NPs | NPs size (nm) | NPs cone | Dispersion media | Porous media | IFT (mN/m) | |
|---|---|---|---|---|---|---|
| | | | | | Clean | With NPs |
| FNP | 7–16 | 0.05 wt% | DIW | Sandstone cores | 16.41 | 12.61 |
| CNP | 8–75 | 0.05 wt% | DIW | Sandstone cores | 16.41 | 12.15 |
| HLP | N/A | 0.05 wt% | Surfactant | Quartz plate | 18.4 | 5.4 |
| $SiO_2$ | 7–12 | 1.0 wt% | SDS | Sandstone cores | 20 | 1.87 |
| $SiO_2$ | 20–30 | 5 wt% | Surfactant | Sandstone cores | 35 | 10.9 |
| $SiO_2$ | 40 | 0.05 wt% | Brine (3 wt% NaCl) | Quartz plates | 19.2 | 17.5 |
| $TiO_2$ | 21 | 0.05 wt% | Brine (3 wt% NaCl) | Quartz plates | 19.2 | n.a. |
| $Al_2O_3$ | 17 | 0.05 wt% | Brine (3 wt% NaCl) | Quartz plates | 19.2 | 12.8 |
| $SiO_2$ | 10–15 | 10 g/200 ml | Ethanol | Glass micromodel | 37.5 | 22.1 |
| FSPNs | 10–15 | 10 g/200 ml | Ethanol | Glass micromodel | 37.5 | 13 |
| $Al_2O_3$ | 20 | 0.05 wt% | CTAB | Carbonate dolomite | 8.46 | 1.65 |
| $Al_2O_3$ | 20 | 0.05 wt% | SDS | Carbonate dolomite | 9.88 | 2.75 |
| $Al_2O_3$ | 20 | 0.05 wt% | TX-100 | Carbonate dolomite | 9.13 | 2.55 |
| $ZrO_2$ | 40 | 0.05 wt% | CTAB | Carbonate dolomite | 8.46 | 1.85 |
| $ZrO_2$ | 40 | 0.05 wt% | SDS | Carbonate dolomite | 9.88 | 2.78 |
| $ZrO_2$ | 40 | 0.05 wt% | TX-100 | Carbonate dolomite | 9.13 | 2.64 |
| $Al_2O_3$ | 40 | 50 mg/L | DIW (26 C) | Limestone rocks | 26.5 | 18 |
| $TiO_2$ | 10–30 | 50 mg/L | DIW (26 C) | Limestone rocks | 26.5 | 17.5 |
| $SiO_2$ | 20 | 50 mg/L | DIW (26 C) | Limestone rocks | 26.5 | 17 |
| $Al_2O_3$ | 40 | 50 mg/L | DIW (60 C) | Limestone rocks | 21.1 | 13.2 |

(continued)

**Table 2** (continued)

| NPs | NPs size (nm) | NPs cone | Dispersion media | Porous media | IFT (mN/m) | |
|---|---|---|---|---|---|---|
| | | | | | Clean | With NPs |
| $TiO_2$ | 10–30 | 50 mg/L | DIW (60 C) | Limestone rocks | 21.1 | 12.4 |
| $SiO_2$ | 20 | 50 mg/L | DIW (60 C) | Limestone rocks | 21.1 | 11.2 |
| HLP | 10–40 | 4 g/L | Ethanol | Sandstone rocks | 26.3 | 1.75 |
| NWP | 10–20 | 4 g/L | Ethanol | Sandstone rocks | 26.3 | 2.55 |
| $Al_2O_3$ | ~60 | 0.5–3 g/L | Propanol | Sandstone cores | 38.5 | 2.25 |
| $Fe_2O_3$ | 40–60 | 0.5–3 g/L | Propanol | Sandstone cores | 38.5 | 2.75 |
| $SiO_2$ | 10–30 | 0.5–3 g/L | Propanol | Sandstone cores | 38.5 | 1.45 |
| $SiO_2$ | 12 | 1–4 g/L | Brine (5 wt% NaCl) | Sandstone cores | 26.5 | 1.95 |
| $ZrO_2$ | 5–15 | 10–500 mg/L | Surfactant | Bidentate carbonates | 48 | 10 |

effect of nanomaterials on enhanced thermal recovery techniques. On the other hand, the first real application of nano-EOR in oil fields was carried out in Liaohe oilfield in northeastern China, where cyclic steam simulation pilot tests using nano-nickel as catalyst were performed [75, 115, 118]. Despite the many benefits of using nanoparticles in thermal oil recovery processes, large-scale research is needed to improve the synthesis of nanomaterials and to expand the use of existing nanomaterials [81].

# 4 Parameters Influencing NP's-EOR

## *4.1 Nanoparticle Size*

The charge density and desired particle size of nanoparticles are the most significant factors that influence the strength of disjoining pressure. In case of comparable weight, small-sized nanoparticles will provide a high value of particle density and low contact angle among the fluid and rock surface. As seen in Sect. 3.3 that the higher particle density can significantly improve the structural disjoining pressure [69]. When the rock surfaces exhibit less hydrophilic, smaller nanoparticles usually propagate more readily than larger particles. McElfresh et al. [80], have summarized that small-sized particles can cause a stronger electrostatic repulsion and higher charge density.

As stated by Hendraningrat et al. [44, 46], smaller particles are significantly resulted in improving oil recovery and also increased the displacement efficiency. Many experimental studies have resolved that a smaller size of particles will result in improving oil production [30]. Kondiparty et al. [69] shown that as the nanoparticle diameter decreases from $30 \times 10^{-9}$ m to $18.5 \times 10^{-9}$ m, it would improve the structural disjoining pressure nearby 4.5 times. If the particle size is adequately small, it will not be mechanically locked in and reasonably large to prevent extra log-jamming [30]. Therefore, smaller particle sizes are usually desirable for greater ultimate oil recovery [47]. Also, it is significant to note that the high surface energy can cause significant surface adsorption and agglomeration of ultrafine nanoparticles, disturbing the stability of nanofluid. Moreover, the size of nanoparticle also can influence the surface coating process, thus, an optimal value of the nanoparticle size should be found for a specific application.

### *4.2 Nanoparticle Concentration*

One of the key elements to success and govern the EOR process is the concentration of injected nanoparticles. Chengara et al. [23], examined the impact of nanoparticles on the structural disjoining pressure when nanoparticles were dispersed on a liquid film. They noticed that the disjoining pressure increases with respect to the concentration, which also leads to an increase in the repulsion forces. Also, rising concentration would increase the displacement efficiency due to the spreading of nanoparticles on the grain surface and enhanced viscosity of nanofluid [31]. The interfacial tension between various fluids in porous media can be significantly reduced via increasing the concentration of injected nanoparticles [44, 46]. A higher level of concentration also can greatly modify the rock wettability. In conclusion, a higher concentration of nanoparticles is preferred to improve oil recovery.

However, an optimum concentration of nanoparticles should be obtained to avoid the blockage of the pore throat that will cause a reduction of the ultimate oil recovery. Generally, more concentration would enhance the wettability alteration, the hydrocarbon displacement efficiency, and reduce IFT. Moreover, during the concentration is excessively elevated, the accumulated nanoparticles will aggregate around the pore throat entry and minimize the displacement efficiency [47], Hendraningrat et al. [44, 46], stated that the formation properties of Berea sandstone sample, permeability, and porosity, were reduced by around 2% after introducing silica nanoparticles with a concentration more than 0.5 wt%. Hence, obtain an optimum injection concentration is required to achieve a higher oil recovery but this optimal concentration differs based on the porous media, kind of nanoparticle, and reservoir conditions.

## 4.3 Rock Wettability

With the increase of rock water wetness, the oil recovery can be increased [93] and vice versa [86]. MWCNT is flooded into a sand pack and Berea core samples, the improved dispersion stability is observed and propagated in high salinity and temperature conditions [59]. Some researchers have performed investigations about the effect of different nanofluids on various rock formations (sandstone, carbonate, and shale) wettability. The oil recovery can be enhanced using nanoparticles as their use fluctuates the wettability towards favorable water-wet conditions and via reducing the interfacial tension.

The change in wettability is due to the deposition of nanoparticles inside the rock pore spaces. For example, the ability of $TiO_2$ to reside inside the pore spaces has been confirmed using SEM (Scanned Electron Microscope). Therefore, it has been found that the main cause of increasing oil recovery is the wettability alteration which is made by the nanoparticles. The study conducted by Hendraningrat and Torsæter [45], concluded that the use of $TiO_2$ nanoparticles as an EOR agent can decrease the interfacial tension between oil and brine, however, the dominant mechanism is wettability alteration. It was also noted by Hendraningrat and Torsæter [45], that the oil-brine interfacial tension is high in the mixture of silica nanoparticles with 3%wt. NaCl brine as compared to its mixture of brine and stabilizer has resulted in the high oil recovery from Berea sandstone samples. Therefore, the change in wettability from water-wet to intermediate-wet in Berea sandstone is the principle mechanism to improve the oil recovery.

Polysilicon nanoparticles have been applied by Roustaei et al. [102]. These polysilicon nanoparticles are classified into three types as hydrophobic and lipophilic polysilicon nanoparticles (HLPN), neutrally wet polysilicon nanoparticles (NWPN), and lipophobic and hydrophilic polysilicon nanoparticles (LHPN). Interfacial tension and wettability measurements showed that NWP and HLP both showed an increase in oil recovery by reducing the interfacial tension and varying the wettability from strong water-wet to less water wet. NWP nanoparticles have a strong impact on rock wettability, while HLP nanoparticles affect interfacial tension. Onyekonwu and Ogolo [92] performed EOR investigations to check the effect on wettability alteration by using three different polystyrene nanoplastics (PSNPs). Their work investigated the presence of a single-layer organic compound in HLPN and NWPN resulted in a 50% increase in oil recovery in a water-wet rock.

It was showed by Ju et al. [56] that the rock samples' pore spaces can be altered to hydrophilic from hydrophobic due to the absorption of nanoparticles on the pre surfaces of the rock. This change can significantly increase the oil relative permeability ($K_{ro}$) and decrease the water relative permeability ($K_{rw}$). Ju and Fan [55], showed that the wettability can modify to water-wet from oil-wet when untreated LHPN adsorb on the surface. While some researchers [92] discovered that the adsorption of untreated LHPN is not recommended as it can change the wettability toward strong water-wet conditions, which is not favorable. The effect of carbonate rock wettability is investigated using $SiO_2$ NPs mixture [100]. The results showed that

$SiO_2$ NPs can change the wettability of carbonate formations. Al-Anssari et al. [4], results showed that that $SiO_2$ NPs change the wettability of mixed wet and oil-wet calcite formations validating the results of [100]. Also, Mohebbifar et al. [83], discovered that the nano-bio materials can alter the shale wettability to water-wet conditions from oil-wet conditions because it can utilize both nano and biomaterials simultaneously.

### 4.4 *Salinity*

By increasing the concentration of NP and the salinity of brine, the viscosity of nanofluids increases. Dispersion stability is greatly affected by the salinity of the reservoir as well as the salinity of the nanofluids. Zeta potential related to the particles can be reduced by increasing the salinity which ultimately leads to agglomeration with ease [80]. The presence of salt water causes higher ionic forces strength that can lead to less electrical repulsion between the particles and can allow van der Waals attractive forces to control. Particle–to-particle collision and attraction will happen instead of the particle to the surface because most of the surface of the rock is charged [80]. Thus to avoid the dispersion and to maintain the stability of the NPs, modification is required in a high salinity environment that can be attained by few methods like surfactant ionic control and surface modification or a combination of both [31]. Worthen et al. [119], stabilized the NPs in a high-salinity environment by attaching low molecular weight ligands to the surface of NP. They claimed the NP stabilization originated from the steric hindrance. While Hendraningrat [47] used high-stability silica NPs to alter the rock wettability toward more water-wet conditions. The adsorption of NPs can be improved at high salinity environment due to the increased physicochemical interactions [126].

Kanj et al. [62] showed that the transport of NPs cannot be hindered by the presence of high salinity fluid; however, the grain surface adsorption can be increased, which can ultimately improve the recovery of oil from the reservoir. Though a high salinity environment reduced the stability of NPs; therefore, the optimum salinity environment and surface modification aspects need utmost consideration.

### 4.5 *Reservoir Temperature*

High temperatures could negatively affect NP results; Alsaba et al. [9] showed that the thermal stability of copper oxide is better when compared to aluminum oxide and magnesium oxide in terms of rheological properties. The reservoir temperature is usually higher than the surface temperature, therefore, for the effective application of nano EOR, nanofluids should operate at relatively high temperatures [31].

According to Caldelas [20], the adsorption and desorption of NPs are less dependence on temperature, thus it has a minor effect on NP retention. However, Hendraningrat et al. [44, 46], said that oil recovery is significantly influenced by the temperature. The higher temperature is more suitable for EOR because it can change the reservoir fluid properties at the molecular level, which results in a decrease in contact angle between the fluids. Though the mechanism involves several variables, therefore, the effect of temperature is complex and difficult to clarify. For example, the zeta potential of particles can be decreased by increasing the temperature. This decrease in zeta potential decreases the stability of nanofluid, which can decrease the oil recovery due to agglomeration [80].

On the other hand, the increase in oil recovery at high temperatures can be credited to the decrease in interfacial tension because it will weaken molecular attraction, increases Brownian motion, and decrease viscosity [20]. As the temperature affects the reservoir fluids and nanofluids, therefore, the real influence on oil production cannot be comprehensive. Therefore, a comprehensive study on the impact of temperature on NP-EOR is required for better understanding.

## 5 Conclusion and the Future Challenges

This chapter presented a critical review of the most recent knowledge in the application of NPs for enhanced oil recovery. Many studies on the nanotechnology application in the EOR have shown optimistic results to improve the ultimate oil recovery. The commonly used nanomaterial for Nano-EOR is the NP, which has a number of benefits: (i) nanoscale particle size with a great surface-area-to-volume ratio and great mobility through porous rock formation; (ii) the potential to increase the fluid performance with a significant amount; (iii) improved heat and mass transfer with applicability in high-temperature conditions; and (iv) high flexibility in merging with other materials such as polymers and surfactants. To date, various categories of NPs (organic and inorganic), as represented via the most common silica, are capable to extract additional hydrocarbon up to 20%, using some well-known mechanisms such as IFT reduction, wettability alteration, disjoining pressure, and mobility ratio control. Various parameters, such as NP concentration, size, temperature, wettability, and salinity, have been confirmed to affect the performance of nanofluid flooding in nano-EOR. Despite the significantly growing interest in nano-EOR, no field-scale application has been published and existing studies are still at a small scale. Therefore, the future improvement of nanoparticles in EOR applications will have several challenges. The main challenges would be the economic aspects, limitations of technology, health and environmental concerns. Several investigations have revealed that NPs might improve the production but most of the investigations are based on a small scale (laboratory) and cannot be applied for a large scale (field) application. There are several shortcomings of implementing NPs in a field scale:

- The preparation of homogenous NP suspensions under high reservoir conditions (salinity, pressure, and temperature), is still a challenge.
- A number of mechanisms of nano-EOR have been proposed to improve oil recovery but these mechanisms and the affecting factors have not been fully understood. However, the interactions between the formation and NPs require further investigation.
- Most numerical models are used the colloidal particle model, which is not well present the performance of NPs. The proposed models have also some limitations to the NPs physical interaction and also do not involve the mineral chemical interaction.
- There is a shortage of information on the impact of different categories of NPs on the human body as the NPs size can be simply breathed by a human organ and potentially placed inside the lungs. Therefore, it is necessary to integrate NPs health and safety research to avoid the threat to surroundings and humans.

## References


1. Aaiza G, Khan I, Shafie S (2015) Energy transfer in mixed convection MHD flow of nanofluid containing different shapes of nanoparticles in a channel filled with saturated porous medium. Nanoscale Res Lett 10(1):490
2. Abdullahi MB, Rajaei K, Junin R, Bayat AE (2019) Appraising the impact of metal-oxide nanoparticles on rheological properties of HPAM in different electrolyte solutions for enhanced oil recovery. J Pet Sci Eng 172:1057–1068
3. Agista MN, Guo K, Yu Z (2018) A state-of-the-art review of nanoparticles application in petroleum with a focus on enhanced oil recovery. Appl Sci 8:871
4. Al-Anssari S, Barifcani A, Wang S et al (2016) Wettability alteration of oil-wet carbonate by silica nanofluid. J Colloid Interface Sci 461:435–442
5. Al-Maamari RSH, Qaboos S, Buckley JS (2003) Asphaltene precipitation and alteration of wetting. The potential for wettability changes during oil production. SPE Reserv Eval Eng 6(4):210–214
6. Al-Shehri AA, Ellis ES, Servin JMF, Kosynkin DV et al (2013) Illuminating the reservoir: magnetic nanomappers. In: Proceedings of the SPE Middle East oil and gas show and conference, Manama, Bahrain, 10–13 Mar 2013
7. Alomair OA, Matar KM, Alsaeed YH (2014) Nanofluids application for heavy oil recovery. In: Proceedings of the SPE Asia Pacific oil and gas conference and exhibition, APOGCE 2014, pp 1346–1363, Oct 2014
8. Alsaba MT, Al Dushaishi MF, Abbas AK (2020) A comprehensive review of nanoparticles applications in the oil and gas industry. J Pet Explor Prod Technol 10:1389–1399. https://doi.org/10.1007/s13202-019-00825-z
9. Alsaba MT, Al Fadhli A, Marafi A, Hussain A, Bander F, Al Dushaishi MF (2018) Application of nanoparticles in improving rheological properties of water based drilling fluids. In: SPE Kingdom of Saudi Arabia annual technical symposium and exhibition, Dammam, Saudi Arabia, 23–26 Apr 2018. https://doi.org/10.2118/192239-MS
10. Alvi MA, Belayneh M, Saasen A, Aadnøy BS (2018) The effect of micro-sized boron nitride BN and iron trioxide $Fe_2O_3$ Nanoparticles on the properties of laboratory bentonite drilling fluid. In: SPE Norway one day seminar, Bergen, Norway, 18 Apr 2018
11. Anderson WG (1986) Wettability literature survey part 1: rock/oil/brine interactions and the effects of core handling on wettability. J Pet Technol 38(10):1125–1144

12. Arab D, Kantzas A, Bryant SL (2018) Nanoparticle-enhanced surfactant floods to unlock heavy oil. In: SPE improved oil recovery conference, Tulsa, Oklahoma, USA, 14–18 Apr 2018
13. Barroso AL, Marcelino CP, Leal AB, Odum DM, Lucena C, Masculo M, Castro F (2018) New generation nano technology drilling fluids application associated to geomechanic best practices: field trial record in Bahia, Brazil. In: Offshore technology conference, Houston, Texas, USA, 30 Apr–3 May 2018
14. Barry MM, Jung Y, Lee JK, Phuoc TX, Chyu MK (2015) Fluid filtration and rheological properties of nanoparticle additive and intercalated clay hybrid bentonite drilling fluids. J Pet Sci Eng 127:338–346
15. Bayat AE, Junin R, Samsuri A et al (2014) Impact of metal oxide nanoparticles on enhanced oil recovery from limestone media at several temperatures. Energy Fuels 28(10):6255–6266
16. Benko H (2017) ISO technical committee 229 nanotechnologies. Metrology and standardization of nanotechnology: protocols and industrial innovations, pp 259–268
17. Bennetzen MV, Mogensen K (2014) Novel applications of nanoparticles for future enhanced oil recovery. International petroleum technology conference. Society of Petroleum Engineers, Kuala Lumpur, pp 1–14
18. Bera A, Belhaj H (2016) Application of nanotechnology by means of nanoparticles and nanodispersions in oil recovery-a comprehensive review. J Nat Gas Sci Eng 34:1284–1309
19. Cabuil V (2004) Chapter 119—Magnetic nanoparticles: preparation and properties. In: Dekker encyclopedia of nanoscience and nanotechnology. Roldan Group Publications
20. Caldelas FM, Murphy M, Huh C, Bryant SL (2011) Factors governing distance of nanoparticle propagation in porous media. In Proceedings of the SPE production and operations symposium, Oklahoma City, OK, USA, 27–29 Mar 2011. Society of Petroleum Engineers, Richardson, TX, USA
21. Chandran K (2013) Multiwall carbon nanotubes (MWNT) fluid in EOR using core flooding method under the presence of electromagnetic waves. Petronas University Technology, Malaysia
22. Charitidis CA, Georgiou P, Koklioti MA, Trompeta AF, Markakis V (2014) Manufacturing nanomaterials: from research to industry. Manuf Rev 1:11
23. Chengara A, Nikolov AD, Wasan DT, Trokhymchuk A, Henderson D (2004) Spreading of nanofluids driven by the structural disjoining pressure gradient. J Colloid Interface Sci 280(1):192–201
24. Crews JB, Huang T (2008) Performance enhancements of viscoelastic surfactant stimulation fluids with nanoparticles. In: Europec/EAGE conference and exhibition, Rome, Italy, 9–12 June 2008
25. Dahle GS (2013) The effect of nanoparticles on oil/water interfacial tension. Project thesis, NTNU
26. Ding Y, Zheng S, Meng X, Yang D (2018) Low salinity hot water injection with addition of nanoparticles for enhancing heavy oil recovery under reservoir conditions. In: SPE Western regional meeting, Garden Grove, California, USA, 22–26 Apr 2018
27. Dorcheh SK, Vahabi K (2016) Biosynthesis of nanoparticles by fungi: large-scale production. In: Fungal metabolites. Reference series in phytochemistry. Springer International Publishing
28. Dreaden EC, Alkilany AM, Huang X, Murphy CJ, El-Sayed MA (2012) The golden age: gold nanoparticles for biomedicine. Chem Soc Rev 41(7):2740–2779
29. Ehtesabi H, Ahadian MM, Taghikhani V (2014) Investigation of diffusion and deposition of $TiO_2$ nanoparticles in sandstone rocks for EOR application. In: 76th EAGE conference and exhibition. Amsterdam, Netherlands.
30. El-Diasty A, Khattab H, Tantawy M (2021) Application of Nanofluid Injection for Enhanced Oil Recovery (EOR). J Univ Shanghai Sci Technol 23(8):1007–6735
31. El-Diasty AI, Aly AM (2015) Understanding the mechanism of nanoparticles applications in enhanced oil recovery. In: Proceedings of the SPE North Africa technical conference and exhibition, Cairo, Egypt, 14–16 Sept 2015. Society of Petroleum Engineers, Richardson, TX, USA

32. Ellahi R, Hassan M, Zeeshan A, Khan AA (2016) The shape effects of nanoparticles suspended in HFE-7100 over wedge with entropy generation and mixed convection. Applied Nanosci 6(5):641–651
33. Elshawaf M (2018) Consequence of graphene oxide nanoparticles on heavy oil recovery. In: SPE Kingdom of Saudi Arabia annual technical symposium and exhibition, Dammam, Saudi Arabia, 23–26 Apr 2018
34. Ersoy H, Rybicki FJ (2007) Biochemical safety profiles of gadolinium-based extracellular contrast agents and nephrogenic systemic fibrosis. J Magn Reson Imaging 26(5):1190–1197. https://doi.org/10.1002/jmri.21135
35. Feynman RP (1960) The wonders that await a micro-microscope. Saturday Rev 43(2):45–47
36. Franco CA, Zabala R, Cortés FB (2017) Nanotechnology applied to the enhancement of oil and gas productivity and recovery of Colombian fields. J Pet Sci Eng 157:39–55
37. Franco C, Cardona L, Lopera S et al (2016) Heavy oil upgrading and enhanced recovery in a continuous steam injection process assisted by nanoparticulated catalysts. In: Proceedings of the SPE improved oil recovery conference, Tulsa, OK, USA, 11–13 Apr 2016
38. Gatoo MA, Naseem S, Arfat MY, Mahmood Dar A, Qasim K, Zubair S (2014) Physicochemical properties of nanomaterials: implication in associated toxic manifestations. Biomed Res Int 2014:498420. https://doi.org/10.1155/2014/498420
39. Hafiz A, Kamal M, Al-Harthi M, Elkatatny S, Murtaza M (2018) Synthesis and experimental investigation of novel CNT-polymer nanocomposite to enhance borehole stability at high temperature drilling applications. In: SPE Kingdom of Saudi Arabia annual technical symposium and exhibition, 23–26 April, Dammam, Saudi Arabia
40. Hamilton RL, Crosser OK (1962) Thermal conductivity of heterogeneous two-component systems. Ind Eng Chem Fundam 1(3):187–191
41. Hammond PS, Unsal E (2011) Spontaneous imbibition of surfactant solution into an oil-wet capillary: wettability restoration by surfactant contaminant complexation. Langmuir 27(8):4412–4429
42. Haroun M, Hassan SA, Ansari A et al (2012) Smart nano-EOR process for Abu Dhabi carbonate reservoirs. Presented at the Abu Dhabi international petroleum exhibition & conference, Abu Dhabi, UAE, 11–14 Nov 2012. SPE-162386
43. Hassani SS, Amrollahi A, Rashidi A, Soleymani M, Rayatdoost S (2016) The effect of nanoparticles on the heat transfer properties of drilling fluids. J Pet Sci Eng 146:183–190
44. Hendraningrat L, Li S, Torsæter O (2013) A coreflood investigation of nanofluid enhanced oil recovery. J Pet Sci Eng 111:128–138
45. Hendraningrat L, Torsæter O (2015) Metal oxide-based nanoparticles: revealing their potential to enhance oil recovery in different wettability systems. Appl Nanosci 5(2):181–199
46. Hendraningrat L, Li S, Torsæter O (2013) Effect of some parameters influencing enhanced oil recovery process using silica nanoparticles: an experimental investigation. In: Proceedings of the SPE reservoir characterization and simulation conference and exhibition, Abu Dhabi, UAE, 16–18 Sept 2013. Society of Petroleum Engineers, Richardson, TX, USA
47. Hendraningrat L (2015) Unlocking the potential of hydrophilic nanoparticles as novel enhanced oil recovery method: an experimental investigation. PhD thesis, Norwegian University of Science and Technology, Trondheim, Norway
48. Hogeweg AS, Hincapie RE, Foedisch H et al (2018) Evaluation of aluminium oxide and titanium dioxide nanoparticles for EOR applications. In: SPE Europec featured at 80th EAGE conference and exhibition, Copenhagen, Denmark, 11–14 June 2018. https://doi.org/10.2118/190872-MS
49. Hu L, Chen M, Fang X et al (2012) Oil–water interfacial self-assembly: a novel strategy for nanofilm and nanodevice fabrication. Chem Soc Rev 41(3):1350–1362
50. Hübler AW, Osuagwu O (2010) Digital quantum batteries: energy and information storage in nanovacuum tube arrays. Complex 15(5):48–55
51. IEA (2018). Whatever happened to enhanced oil recovery?—analysis, 01 Nov 2018. https://www.iea.org/commentaries/whatever-happened-to-enhanced-oil-recovery. Accessed 09 Nov 2020

52. Jagar AA, Kamal K, Abbas KhM, Amir HM (2018) Recent advances in application of nanotechnology in chemical enhanced oil recovery: effects of nanoparticles on wettability alteration, interfacial tension reduction, and flooding. Egypt J Pet 27(4):1371–1383
53. Jahagirdar SR (2008) Oil-microbe detection tool using nano optical fibers. In: Proceedings of the SPE Western regional and Pacific section AAPG joint meeting, Bakersfield, CA, USA, 29 Mar–4 Apr 2008
54. Jeevanandam J, Barhoum A, Chan YS, Dufresne A, Danquah MK (2018) Review on nanoparticles and nanostructured materials: history, sources, toxicity and regulations. Beilstein J Nanotechnol 9(1):1050–1074
55. Ju B, Fan T (2009) Experimental study and mathematical model of nanoparticle transport in porous media. Powder Technol 192:195–202
56. Ju B, Fan T, Ma M (2006) Enhanced oil recovery by flooding with hydrophilic nanoparticles. China Particuol 4:41–46
57. Ju B, Fan T (2009) Experimental study and mathematical model of nanoparticle transport in porous media. Powder Technol 192(2):195–202
58. Ju B, Dai S, Luan Z, Zhu T et al (2002) Study of wettability and permeability change caused by adsorption of nanometer structured polysilicon on the surface of porous media. In: SPE 2002
59. Kadhum MJ, Swatske DP, Chen C et al (2015) Propagation of carbon nanotube hybrids through porous media for advancing oilfield technology. In: Proceedings of the SPE international symposium on oilfield chemistry, The Woodlands, TX, USA, 13–15 Apr 2015. Society of Petroleum Engineers, Richardson, TX, USA
60. Kamal MS, Adewunmi AA, Sultan AS et al (2017) Recent advances in nanoparticles enhanced oil recovery: rheology, interfacial tension, oil recovery, and wettability alteration. J Nanomater 2017:15. https://doi.org/10.1155/2017/2473175
61. Kanj MY, Rashid MH, Giannelis E (2011) Industry first field trial of reservoir nanoagents. In: SPE Middle East oil and gas show and conference. Society of Petroleum Engineers, Manama, Bahrain
62. Kanj MY, Funk JJ, Al-Yousif Z (2009) Nanofluid coreflood experiments in the ARAB-D. In: Proceedings of the SPE Saudi Arabia section technical symposium, Al-Khobar, Saudi Arabia, 9–11 May 2009. Society of Petroleum Engineers: Richardson, TX, USA
63. Kanj MY, Rashid MH, Giannelis E (2011) Industry First field trial of reservoir nanoagents. In: SPE Middle East oil and gas show and conference, Manama, Bahrain, 25–28 Sept 2011
64. Kapusta S, Balzano L (2011) Nanotechnology applications in oil and gas exploration and production. In: IPCT 2011. https://doi.org/10.2523/15152-MS
65. Karimi A, Fakhroueian Z, Bahramian A et al (2012) Wettability alteration in carbonates using zirconium oxide nanofluids: EOR implications. Energy Fuels 26(2):1028–1036
66. Kazemzadeh Y, Eshraghi SE, Kazemi K et al (2015) Behavior of asphaltene adsorption onto the metal oxide nanoparticle surface and its effect on heavy oil recovery. Ind Eng Chem Res 54(1):233–239
67. Khan I (2017) Shape effects of $MoS_2$ nanoparticles on MHD slip flow of molybdenum disulphide nanofluid in a porous medium. J Mol Liq 233:442–451
68. Khan AI, Arasu AV (2019) A review of influence of nanoparticle synthesis and geometrical parameters on thermophysical properties and stability of nanofluids. Therm Sci Eng Prog 11:334–364
69. Kondiparty K, Nikolov A, Wu S et al (2011) Wetting and spreading of nanofluids on solid surfaces driven by the structural disjoining pressure: statics analysis and experiments. Langmuir 27:3324–3335
70. Kosynkin D, Alaskar M (2016) Oil Industry first interwell trial of reservoir nanoagent tracers. In: SPE annual technical conference and exhibition, Dubai, UAE, 26–28 Sept 2016
71. Kothari N, Raina B, Chandak KB et al (2010) Application of ferrofluids for enhanced surfactant flooding in IOR. In: SPE EUROPEC/EAGE annual conference and exhibition. Society of Petroleum Engineers, Barcelona, Spain

72. Kovscek AR, Wong H, Radke CJ (1993) A pore-level scenario for the development of mixed wettability in oil reservoirs. AIChE J 39(6):1072–1085
73. Kumar G, Kakati A, Mani E, Sangwai JS (2018) Nanoparticle stabilized solvent-based emulsion for enhanced heavy oil recovery. In: SPE Canada heavy oil technical conference. OnePetro
74. Li S, Torsæter O, Lau HC et al (2019) The impact of nanoparticle adsorption on transport and wettability alteration in water-wet Berea sandstone: an experimental study. Front Phys 7:74
75. Li S, Genys M, Wang K, Torsæter O (2015) Experimental study of wettability alteration during nanofluid enhanced oil recovery process and its effect on oil recovery. In: SPE reservoir characterisation and simulation conference and exhibition. Society of Petroleum Engineers, Sept 2015
76. Li S, Hendraningrat L, Torsaeter O (2013) Improved oil recovery by hydrophilic silica nanoparticles suspension: 2 phase flow experimental studies. In: IPTC 2013: international petroleum technology conference. European Association of Geoscientists & Engineers, pp cp–350
77. Li Y, Guo J, Wang S, Yang R, Lu Q (2019) Reducing hydroxypropyl guar gum adsorption on rock by silica nanoparticles for tight reservoir damage remediation. In: International petroleum technology conference, Beijing, China, 26–28 Mar 2019
78. Liu H, Jin X, Ding B (2016) Application of nanotechnology in petroleum exploration and development. Pet Explor Dev 43:1107–1115
79. Maghzi A, Mohebbi A, Kharrat R (2011) Pore-scale monitoring of wettability alteration by silica nanoparticles during polymer flooding to heavy oil in a five-spot glass micromodel. Transp Porous Media 3:653–664
80. Mcelfresh P, Holcomb D, Ector D (2012) Application of nanofluid technology to improve recovery in oil and gas. In: Proceedings of the SPE international oilfield nanotechnology conference and exhibition, Noordwijk, The Netherlands, 12–14 June 2012. Society of Petroleum Engineers, Richardson, TX, USA
81. Medina OE, Olmos C, Lopera SH et al (2019) Nanotechnology applied to thermal enhanced oil recovery processes: a review. Energies 12(24):4671
82. Menezes JL, Yan JN, Sharma MM (1993) Wettability alteration caused by oil-based muds and mud components. SPE Drill Complet 8(1):35–44
83. Mohebbifar M, Ghazanfari MH, Vossoughi M (2015) Experimental investigation of nano-biomaterial applications for heavy oil recovery in shaly porous models: a pore-level study. J Energy Res Technol 137(1)
84. Mokhatab S, Araujo M, Islam MR (2006) Applications of nanotechnology in oil and gas E&P. J Pet Technol 58(4):48–51
85. Morcos SK (2007) Nephrogenic systemic fibrosis following the administration of extracellular gadolinium-based contrast agents: is the stability of the contrast agent molecule an important factor
86. Morrow NR (1990) Wettability and its effect on oil recovery. J Pet Technol 42:476–484
87. Nasr-El-Din HA, de Wolf CA, Stanitzek T, Alex A, Gerdes S, Lummer NR (2013) Field treatment to stimulate a deep, sour, tight-gas well using a new, low-corrosion and environmentally friendly fluid. SPE Prod Oper 28(03):277–285
88. Nazari MR, Bahramian A, Fakhroueian Z et al (2015) Comparative study of using nanoparticles for enhanced oil recovery: wettability alteration of carbonate rocks. Energy Fuels 29(4):2111–2119
89. Negin C, Ali S, Xie Q (2016) Application of nanotechnology for enhancing oil recovery—a review. Petroleum 2:324–333
90. Ogolo NA, Olafuyi OA, Onyekonwu MO (2012) Enhanced oil recovery using nanoparticles. In: SPE. Society of Petroleum Engineers, Al-Khobar, Saudi Arabia
91. Onyekonwu MO, Ogolo NA (2010) Investigating the use of nanoparticles in enhancing oil recovery. In Nigeria Annual international conference and exhibition. OnePetro
92. Onyekonwu MO, Ogolo NA (2010) Investigating the use of nanoparticles in enhancing oil recovery. In: SPE 2010. https://doi.org/10.2118/140744-MS

93. Owens W, Archer D (1971) The effect of rock wettability on oil-water relative permeability relationships. J Pet Technol 23:873–878
94. Portela CM, Vidyasagar A, Krödel S, Weissenbach T, Yee DW, Greer JR, Kochmann DM (2020) Extreme mechanical resilience of self-assembled nanolabyrinthine materials. Proc Nat Acad Sci 17(11):5686–5693
95. Powers KW, Palazuelos M, Moudgil BM, Roberts SM (2007) Characterization of the size, shape, and state of dispersion of nanoparticles for toxicological studies. Nanotoxicology 1(1):42–51
96. Pratyush S, Sumit B (2010) Nano-robots system and methods for well logging and borehole measurements. U.S. Patent 20100242585A1, 12 Nov 2010
97. Rahmani AR, Bryant S, Huh C et al (2015) Crosswell magnetic sensing of superparamagnetic nanoparticles for subsurface applications. SPE J 20:1067–1082
98. Reincke F, Kegel WK, Zhang H et al (2006) Understanding the self-assembly of charged nanoparticles at the water/oil interface. Phys Chem Chem Phys 8(33):3828–3835
99. Richard A, Esther O (2019) Nanotechnology and global energy demand: challenges and prospects for a paradigm shift in the oil and gas industry. J Pet Explor Prod Technol 9:1423–1441
100. Roustaei A, Bagherzadeh H (2015) Experimental investigation of $SiO_2$ nanoparticles on enhanced oil recovery of carbonate reservoirs. J Pet Explor Prod Technol 5:27–33
101. Roustaei A, Saffarzadeh S, Mohammadi M (2013) An evaluation of modified silica nanoparticles' efficiency in enhancing oil recovery of light and intermediate oil reservoirs. Egypt J Pet 22(3):427–433
102. Roustaei A, Moghadasi J, Bagherzadeh H et al (2012) An experimental investigation of polysilicon nanoparticles' recovery efficiencies through changes in interfacial tension and wettability alteration. In: SPE. Society of Petroleum Engineers, Noordwijk, Netherlands
103. Salathiel RA (1973) Oil recovery by surface film drainage in mixed-wettability rocks. J Petrol Technol 25(10):1216–1224
104. Salehi M, Johnson SJ, Liang JT (2008) Mechanistic study of wettability alteration using surfactants with applications in naturally fractured reservoirs. Langmuir 24:14099–14107
105. ShamsiJazeyi H, Miller C, Wong M et al (2014) Polymer-coated nanoparticles for enhanced oil recovery. J Appl Polym Sci 131:40576 (Wiley Periodicals, Inc.)
106. Shin WK, Cho J, Kannan AG, Lee YS, Kim DW (2016) Cross-linked composite gel polymer electrolyte using mesoporous methacrylate-functionalized $SiO_2$ nanoparticles for lithium-ion polymer batteries. Sci Rep 6(1):1–10
107. Skauge T, Spildo K, Skauge A (2010) Nano-sized particles for EOR. In: SPE improved oil recovery. Society of Petroleum Engineers, Oklahoma, USA
108. Song Y, Marcus C (2007) Hyperpolarized silicon nanoparticles: reinventing oil exploration. Presented at the Schlumberger seminar, Schlumberger, College Station, TX, USA, 29–31 Jan 2007
109. Specification PA (2007) Terminology for nanomaterials. British Standards Institute, London
110. Timofeeva EV, Routbort JL, Singh D (2009) Particle shape effects on thermophysical properties of alumina nanofluids. J Appl Phys 106(1):014304
111. Tong Z, Morrow NR (2006) Variations in wetting behavior of mixed wet cores resulting from probe oil solvency and exposure to synthetic oil-based mud emulsifiers. J Pet Sci Eng 52:149–160
112. Vryzas Z, Kelessidis VC (2017) Nano-based drilling fluids: a review. Energies 10:1–34
113. Wang LY, Luo J, Maye MM et al (2005) Iron oxide-gold core-shell nanoparticles and thin film assembly. J Mater Chem 15(18):1821–1832. https://doi.org/10.1039/b501375e
114. Wang L, Wang Z, Yang H et al (1999) The study of thermal stability of the $SiO_2$ powders with high specific surface area. Mater Chem Phys 57(3):260–263
115. Wang L, Zhang G, Ge JJ, Li G, Zhang J, Ding B (2010) Preparation of microgel nanospheres and their application in EOR. In: International oil and gas conference and exhibition in China. Society of Petroleum Engineers, Jan 2010

116. Wasan DT, Nikolov A (2003) Spreading of nanofluids on solids. J Nat 423:156–159. https://doi.org/10.1038/nature01591
117. Wasan DT, Nikolov A, Kondiparty K (2011) The wetting and spreading of nanofluids on solids: role of the structural disjoining pressure. Curr Opin Colloid Interface Sci 16:344–349. https://doi.org/10.1016/j.cocis.2011.02.001
118. Wei L, Zhu JH, Qi JH (2007) Application of nano-nickel catalyst in the viscosity reduction of Liaohe extra-heavy oil by aqua-thermolysis. J Fuel Chem Technol 35(2):176–180
119. Worthen AJ, Tran V, Cornell KA et al (2016) Steric stabilization of nanoparticles with grafted low molecular weight ligands in highly concentrated brines including divalent ions. Soft Matter 12:2025–2039
120. Yaha N, Kashif M, Nasir N et al (2012) Cobalt ferrite nanoparticles: an innovative approach for enhanced oil recovery application. J Nano Res 17:115
121. Yefei W, Huaimin X, Weizhao Y et al (2011) Surfactant induced reservoir wettability alteration: recent theoretical and experimental advances in enhanced oil recovery. Pet Sci 8:463–476
122. Yeh SW, Ehrllch R, Emanuel AS (1992) Miscible-gas flood-induced wettability alteration: experimental observations and oil recovery implications. SPE Form Eval 7(2):167–172
123. Yu J, Berlin JM, Lu W et al (2010) Transport study of nanoparticles for oilfield application. In: SPE. Society of Petroleum Engineers, Aberdeen, United Kingdom
124. Zabala R, Franco CA, Cortés FB (2016) Application of nanofluids for improving oil mobility in heavy oil and extra-heavy oil: a field test. In: SPE improved oil recovery conference, Tulsa, Oklahoma, USA, 11–13 Apr 2016
125. Zakaria MF, Husein M, Harelamnd G (2012) Novel nanoparticle-based drilling fluid with improved characteristics. SPE international oilfield nanotechnology conference. Society of Petroleum Engineers, Noordwijk, pp 1–6
126. Zhang T, Murphy MJ, Yu H et al (2014) Investigation of nanoparticle adsorption during transport in porous media. SPE 20:667–677
127. Zhang J, Nguyen QP, Flaaten AK et al (2009) Mechanisms of enhanced natural imbibition with novel chemical. SPE Reserv Eval Eng 12(6):912–920
128. Zitha PLJ (2005) Smart fluids in the oilfield. Explor Prod: Oil Gas Rev 1:66–68

# Chapter 6
# Nanotechnology Application for Wireless Communication System

**Ekhlas Kadum Hamza and Shahad Nafea Jaafar**

## 1 Introduction

Nanotechnology has given improved and outstanding results in different kinds of applications, such as biomedical, industrial, agriculture, and military. This science has brought about the development of nanomachines that are little parts made up of an organized set of molecules carrying out preset functions [28]. The gadgets that use wireless networks vary from television receivers, radio frequency identification tags, cellular phones to satellites [17]. Access to network communication for cellular phones and tablets is rising more and more rapidly. This has resulted in higher requirements on the efficiency of mobile devices and the internet. In this era of envisioned unprecedented nanotechnology role in multidisciplinary domains such as environmental, industrial, biomedical, and military, one of the emerging social and scientific impacts of such technology would be in healthcare and bioengineering applications. As a promising alternative to current medical technologies like catheters and endoscopes, the nano-enabled devices could reach delicate body sites such as the spinal cord, gastrointestinal, or inside the human eye, non-invasively, which have not been possible yet with current technologies [35]. With the characteristics of iniquitousness and variety of nanodevices, different kinds of information can be sensed and gathered together to complete complicated tasks. The connectivity and links between nanodevices lead to the idea of nanonetworks followed by the nano-communication proposal, which will expand the capabilities of these devices in terms of enhancement in features and range of operations [1]. The use of nanotechnology in communication systems enables manufacturers to produce integrated circuits and sensors for computers that are substantially small in size, run speedily, saves power, and are also cost-effective as compared to modules of today. The other issue in a communication system based on nanotechnology is discovering

E. K. Hamza (✉) · S. N. Jaafar
Control & System Engineering Department, University of Technology, Baghdad, Iraq
e-mail: ekhlas.k.hamza@uotechnology.edu.iq



M. Mujawar et al. (eds.), *Nanotechnology for Electronic Applications*, Materials Horizons: From Nature to Nanomaterials, https://doi.org/10.1007/978-981-16-6022-1_6

new materials on the nanometer length scale expected to play an important role in future challenges in the field of communication systems such as in ultra-high-speed devices for long and short-range communications links, power-efficient computing devices, high-density memory and logics, and ultra-fast interconnects. Also, the use of molecules, instead of electromagnetic or acoustic waves, to encode and transmit the information represents a new communication paradigm that demands novel solutions such as molecular transceivers, channel models, or protocols for nanonetworks [16]. Molecular transceivers will be easy to integrate into nanodevices due to their size and domain of operation. These transceivers are able to react with specific molecules and release others as a response after performing some type of processing. Recent advancements in molecular and carbon electronics have applied a new generation of electronic nanocomponents such as nanobatteries, nanomemories, logical circuitry in the nanoscale, and even nanoantennas [2]. An outline of numerous problems relating to nanotechnology in the frameworks of communication is considered in this chapter. Further, this chapter also provides a synopsis of the different possible approaches for the advancements in nanotechnology in the transmission set-ups as well as the likely prospects for research studies in the future, which can result in enhancement of communication systems and applications and future technologies in the field of telecommunication based on nanotechnology. This chapter's objective is to learn about applications the use of nanotechnology in wireless communication, illustrate how nano plays a role in mobile devices, explain the most important applications that use nanotechnology in the Internet of Things (IoT) and body area network, discuss how to establish a bidirectional wireless nano-communication, and learn the benefits of using nanotechnology in 5G and know 5G nano core.

## 2 The Concept of Nanotechnology

Nanotechnology is primarily "an engineering function of processes on the scale of atoms and molecules". The actual meaning of this technology is to build components upwards starting from the base, using current methods and devices to develop products that are complete and highly efficient [28]. Nanotechnology [1, 2] has given different high-performing results to numerous applications in the real world in sectors like biotechnology, biomedical, industrial, agriculture, and the military by making tools on a scale of one to a few hundred nanometers. Nanotechnology seems to be arriving rather faster than the technologies from the past, such as steam engines and digital computers [24]. Nanotechnology has the ability and capability to radically change the medical field and its techniques so as to be more customized, mobile, inexpensive, secured as well as easily manageable. Below are some instances of significant developments in these sectors [5, 13, 34, 36, 37]. The embracing of new developments in nanotechnology and specifically its performance on a broader scale may only be realized via contact and discussions among the scientists, industry, government, and the public. This has frequently been sidelined and resulted in incorrect information and misinterpretation of the threats and advantages of these up-to-date developments. The requirement for this dialog has been acknowledged not only by governments but by research and industry sponsor organizations as well. Currently, these agencies have the advantage of actively pursuing discussions with scientists and the public who are interested [10]. In communication systems, mobile devices with a huge volume of calculations and communication while connecting with people of the world like the home, workplace, and open areas need intelligent technologies of sensing, computing, and communication specifically, when these tools become implanted in networks. The chief demands for intelligent mobile devices are that they must be independent, strong, and easy to install and last without explicit management or care. Further, the portability of these gadgets also implies restricted size and limitations on power usage. Other demands for these intelligent mobile networks are intelligent communication and interconnecting with other devices and the surroundings, sensing, context, awareness, and high information rates that need higher storage and processing capacity. Owing to these higher demands, the present technologies are unable to cope or find a solution. On the other hand, nanotechnology can give results for sensing, actuation, radio, implanting intelligence into the surroundings as well as high-performing processing power and storage.

## 3 How Can Nano Help?

The core reasons for using nanotechnology in wireless devices are performance improvement of wireless communication devices parts in contrast to digital electronics. The key issues that should be considered are the transition from semiconductor-based wireless products to nanotechnology-based wireless communication products and the need for inexpensive, tiny, and lower energy-consuming devices [17].

## 4 Advantages of Using Nanotechnology in Wireless Communications

1. The primary reasons for using nanotechnology in wireless systems are due to better implementation, lower energy usage, tiny size, and the latest characteristics like adaptability or transparency. Although many of these have been verified to be technologically viable, however, several technological problems remain unresolved.
2. Lower expense is a significant part: the latest technologies will have difficulties in acquiring a market niche unless they can compete in pricing with current technologies. The obstacles to market the latest technologies for radio frequency electronics are lower compared to digital electronics. Thus, it is plain to see that there is a requirement for new technologies.

3. Referring to professional opinion, the impact on economic limitation is from medium to high, on technological and policy limitations is medium, and on the surroundings, health, security, and community limitations are low.
4. The impact of nanomaterials in wireless network tools on the surroundings, health, and security is anticipated to be close to that of other integrated circuits. The candidate nanoparticles for wireless systems are normally included in the host matrix and totally packaged into the circuits. Thus, users will be protected from nanomaterials when using their devices. Nonetheless, the advancement of independent wireless sensors needs cautious analysis to study the effects on people and their surroundings, specifically if sensors are built in a way that they are in touch with the human physique or embedded in the body or disposable.
5. The general characteristics of the ever-present data access and surrounding intelligent networks and systems have not been adequately studied. Specifically, privacy and safety questions require cautious evaluation [17].

## 5 Nanotechnology Applications

As shown in Fig. 1, there are numerous applications of nanotechnology today. However, in reference to the electrical and electronics sectors, the viable uses of technology are in communications, bio-engineering, medical electronics, and robotics [1]. Nanotechnology plays a significant part in the area of telecommunication engineering. It creates a huge impact in several ways, such as the handling of communication technologies and characteristics. Nanotechnology has a broad scope of uses and has made a footprint in the wireless network industry in many aspects.

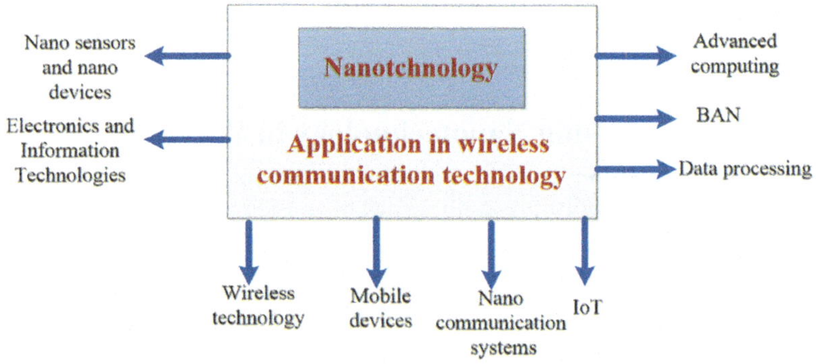


**Fig. 1** Applications of nanotechnology

## 5.1 Wireless Technology

Wireless technology will be completely reshaped into the latest technology known as nanotechnology, which will influence the performance of the mobile as well as the core network. Additionally, due to its excellent safety features and improved sensor effects, nanotechnology has become the biggest application compared to the earlier conventional technologies [30].

During the application of intelligent operations, wireless technology industries have ensured that computation and communication are to be carried out according to the needs. The arrival of intelligent and nanotechnology ideas for mobile devices can help in implanting these tools into the human world that will make a brand new way to permit ubiquitous detecting and computing [30]. The nanodevices can be uploaded with some capabilities such as self-powering, sensing of the surroundings, or smart interconnection with other networks [23]. In mobile phones, the improved carbon nanotubes will be included in the near future, which is under nanotechnology [23]. In the 5G mobile network, the cells known as nanodevices use nanotechnology. One of the most fitting innovations of the wireless industry is to create nano-intelligent technologies that enable a person to be serviced in an intelligent way. This simply means that the mobile and network devices in combination with the intelligence are implanted in human surroundings like the home, office, or open areas to build the latest platform that permits sensing and computing in nano-transmission networks [15].

## 5.2 Mobile Devices

Mobile devices for computation and detection have become an important necessity for distant enterprises to obtain information, which should always be available for the benefit of and service to customers [26]. These gadgets, for instance, the cellular phone can be connected to the human environment such as the home, workplace, and open areas [26]. One of the main prerequisites for inserting tools into physical objects is that they need to be flexible in their environment and become a part of the network of devices encircling them. For instance, the organic frameworks were made in such a way that they are adjusted to nature on their own. Nanotechnology assists in the development of the latest kinds of nanodevices and nanosensors that have the capability to interface with these organic frameworks. Nanotechnology gives options that can be complex like assisting to advance our natural surroundings, or as fundamental, like being aware if an organic product is of good condition or spoilt [21].

An alternative use of nanotechnology is in wireless devices like distant sensors which is a large area of study, especially in the enhancement of army nanotechnologies. Nanosensors in observation applications similarly play an important role in improving the accuracy of arms by raising the fatality of attacks, for example, by providing information to alter control levels. Nano wireless devices can be spread over a conflict by picking up data, such as temperature, weight, vibration, rate of speed, light, magnetic, or acoustics and send out this data continuously [24].

## 5.3 Nano-Communication Systems

Nanomachines are defined as mechanical gadgets that rely on nanometer-scale components. A nuclear machine is a mechanical tool that displays an acceptable limit using pieces of nanometer-scale and outlines the sub-nuclear framework. It has the ability to carry out conveyance, processing, generate data as well as able to recognize or has the capability to activate other networks.

The most fundamental plan of action to interrelate microelectronic devices is to connect through electromagnetic waves, which can be generated in wires or wirelessly with a little loss. To set up bidirectional wireless nano-transmissions, an RF system must fit into the nanomachine, which is needed to build nanoscale antennas for higher frequencies [22]. The information exchange between nanoscale machines is called molecular communication that constitutes the sending and receiving of data encrypted in molecules. Molecular transmission can be used to interrelate several nanomachines, which results in nanosystems that use messages encrypted in molecules. The coding method that represents the data in nanosystems is known as molecular encoding. It uses parameters of the molecules that are situated inside to encrypt the data like molecular geometrics and the placing of molecular elements or polarization. The receiver should have the ability to identify these particular molecules to decrypt the data. This method is the same as that used for encoded packets in transmission systems, whereby the intended receiver has the capability of reading the data. As depicted in Fig. 2, molecular encryption is used in phenomenal communication whereby only members of the transmitter side can decrypt the message sent [11]. Typed messages, audio, and video are normally sent over conventional transmission systems. On the other hand, in nanosystems as the text is molecular, the data that has been sent is more connected to phenomena, chemical states, and processes. The exchange of information between nanomachines is similar to conventional systems. This means networks that send messages or data to the receiver via a carrier are first encrypted on the transmitter side and subsequently decrypted on the receiver side. However, messages are communicated in molecular form. These molecular messages have a predetermined outer framework that can be easily recognized at the receiver end. It is non-functioning, meaning that molecular messages are not inclined to respond with other molecules in the medium. Further, these molecular messages can simply be removed without any undesirable consequences if they are decrypted by the nanomachine at the receiver end [3].

Carriers are specific molecules that have the ability to carry chemo-signals or molecular structures that are having the data. Using molecules as data carriers in molecular communication has been noticed in biological networks [14]. The carriers that are used can be molecular motors or calcium ions. Molecular motors such as

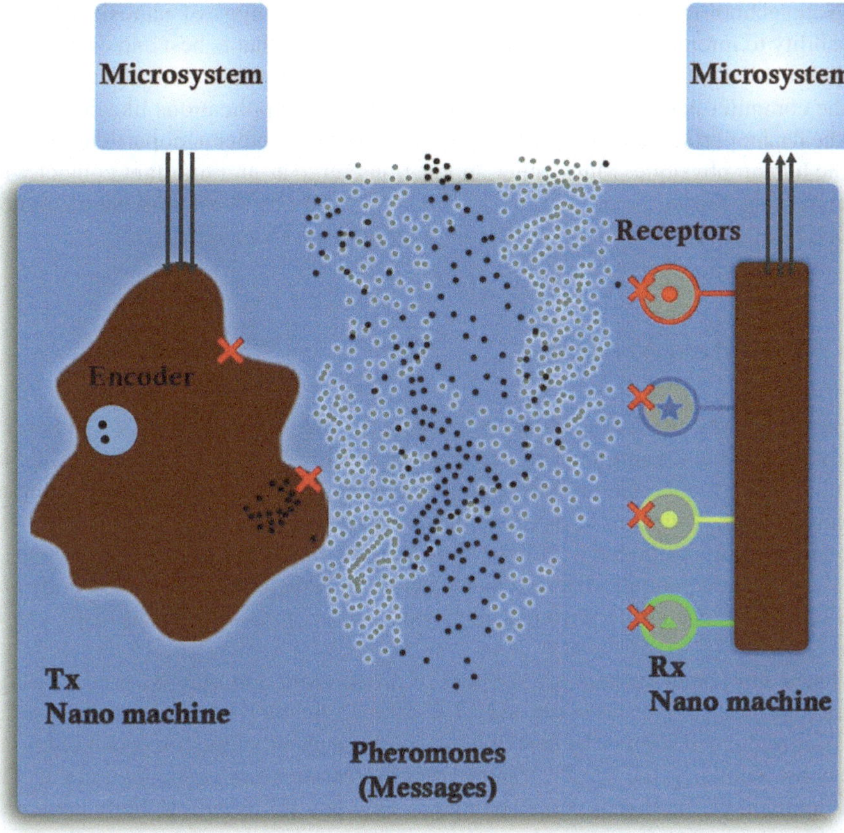


**Fig. 2** Molecular communication based on pheromones encoding [11]

kinesin, dynein, and myosin are proteins, which give rise to movements and use chemical power that carries an information molecule packet from transmitter to receiver [32].

## *5.4 Internet of Things (IoT)*

IoT is the organization of physical objects found in machines, programming, sensors, and networks so that it is able to attain the necessary arrangement by an exchange of information between the executive and other associated things [19]. Nano-intelligent things are the next likely feat, which may strike soon in the forthcoming future [12]. Nano biochips are made in such a way that data can move among themselves or to the equipment or the population in general. They can self-learn, and each time the nano biochips play out their tasks, they improve themselves [7, 8].

Nanotechnology together with IoT, which supplies things of nanosize, has the capability to interact with people or equipment in a way that is best and effective. In addition, things of nano-intelligent are clever as they dynamically link all nano things to the computer network or to other nanonetworks and applications like connecting medical nanoparticles devices to the computer network or other nanoparticles devices to the human physique. Many other applications serve as an important contact with humans [29]. Further, there are several smart applications like nanoscale machines, which have the ability to link to the computer network from afar.

## 5.5 *Body Area Network*

Presently, BAN devices can be connected to the garment or body. Several study groups have completed research on enhancing intelligent nanomaterials by combining with microelectronics to produce into a dress or inserted into the human physique [18]. Therapeutic devices like pacemakers, prostheses, and stents have become a reality in the medical field. For example, the use of sensors is for congestive heart failure patients. These sensors can be embedded into the human body and can liaise with each other through inter-body nano-communication as displayed in Fig. 3 [9].

These implanted sensors are the size of a rice grain and may be used to gauge numerous medical metrics in the body, for instance, the blood flow rate in the arteries and a complicated surgery by carrying out an inside check of the important organs of the body. Similarly, the sensors can be used to treat nerve or tissue incitement. The development of small-scale implants unlocks the possibility of intrabody networks, whereby the body area network sections are reduced and vanishes into the body. Sensors and actuators are recognized by implanting devices from a distance via the tissue. In the near future, it is expected that body area networks and intrabody networks will use the different bio nanotechnologies. Biotechnology is defined as any application of technology that uses biological systems, living organisms, or their derivatives to create or change products or operations for a certain application. Bio nanotechnology is identified as a part of nanotechnology due to the fact of using natural structures, such as proteins, deoxyribonucleic acid, and so on, to create components of nanoscale devices like nano engines.

## 5.6 *Advanced Computing (Quantum Computing)*

Substituting conventional computers with advance and faster quantum computers brings about extraordinary changes, which means adding the latest traits for processing and computing in an intelligent manner [3]. In quantum computers, the binary codes are recurring by quantum bits also known as qubits, which are usually in a state of 0 and 1. They can also achieve a mixed state called superposition where

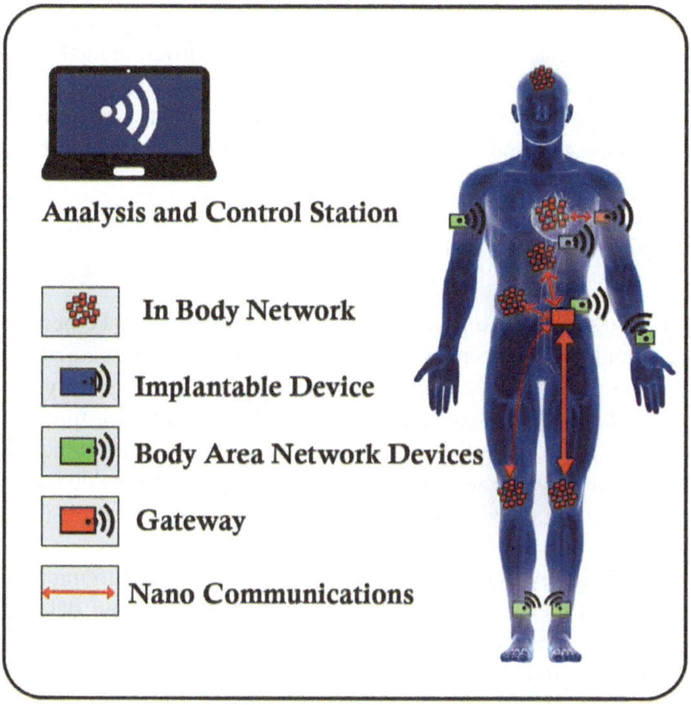


**Fig. 3** Nanosensors for interbody communications [9]

the qubits are 0 and 1 at the same time. As quantum computers can be in numerous states concurrently, it can be presumed that these computers have the capability to carry out millions of calculations at one time. As a result, this enables the computer to perform quicker than before. However, the advancement of quantum computers is currently under study.

To build a quantum computer, five key standards known as the DiVincenzo checklist [26] has to be followed:

1. An expandable physical system with identifiable quantum bits.
2. Setting the quantum bits to a fiducial state. Primarily this simply means initializing the quantum bits close to the quantum states prior to carrying out any computations. For instance, initializing all the spins in a system to be similar.
3. The decoherence times for quantum bits must be longer compared to the gate operations. This is because it commands the duration of time and the quantum bits can be entangled without losing any data, which means that any calculation must be completed prior to the quantum bits losing any data.
4. A universal set of quantum gates. In a quantum computer, two kinds of gate operations are needed to make a universal computer, which can be created and programmed in such a way that it is able to finish any computing function. A computer can be produced for operations like factorization, database searching,

or quantum system simulation from these single and two-quantum bit gate operations. Both the gate operations can be executed by different methods depending on the particular two-state system used for every quantum bit.
5. There must be a quantum bit-specific measurement capability. When the computation is finished, the data has to exit from the quantum bits [24].

## 5.7 *Data Storage and Processing*

With regards to data processing and communication, the advancement in the electronic, optical, and optoelectronic parts could result in the production of transmission devices, which are speedy and accurate. The photonic crystals are likely to be used for constructing exclusively optical circuits based on light mainly for future data operations. The idea of nanotechnology in a nanoscale storage device, which is based on complementary metal oxide semiconductor technology that uses quantum dabs and carbon nanotubes is expected to be a huge potential for keeping a substantial volume of information.

The latest technology like the flexible three-dimensional nano microchip, which securely inter-links memory and has a great effect in reducing data information blockages, assures speedy and accurate information processing. Continued research and advancement in this area could pave the road for unusually important steps in the operation, capability, and the potentiality of quickly managing a very huge volume of information called big data compared to the normal microchips. Further, the three-dimensional integration memory and growing nanotechnologies such as the carbon nanotube transistors are encouraging strides for creating a future generation of extreme proficiency and highly advanced electronic networks, which have the ability to work on a large volume of information [27].

## 5.8 *Nanotechnology in Electronics and Information Technologies*

The goal of nanotechnology is to generate highly efficient and economic materials and gadgets as well as continue to make significant progress in developing the electronics and information technologies sectors. As a result of using nanotechnology, systems have become quicker, small-sized, and increasingly mobile with the ability to a huge volume of data. A foremost instance is a basic switch or transistor, which operates all the latest computers and plays a significant part in the advancement of computer technology. Transistors are electrical circuit parts, which control voltage or current source and another voltage or current source. The electrical properties of carbon nanotubes (CNTs) that have a width of a millionth of a millimeter and are much favored can be distinct and beneficial in contrast to semiconductors like silicon. International Business Machines, which is the biggest data technology corporation

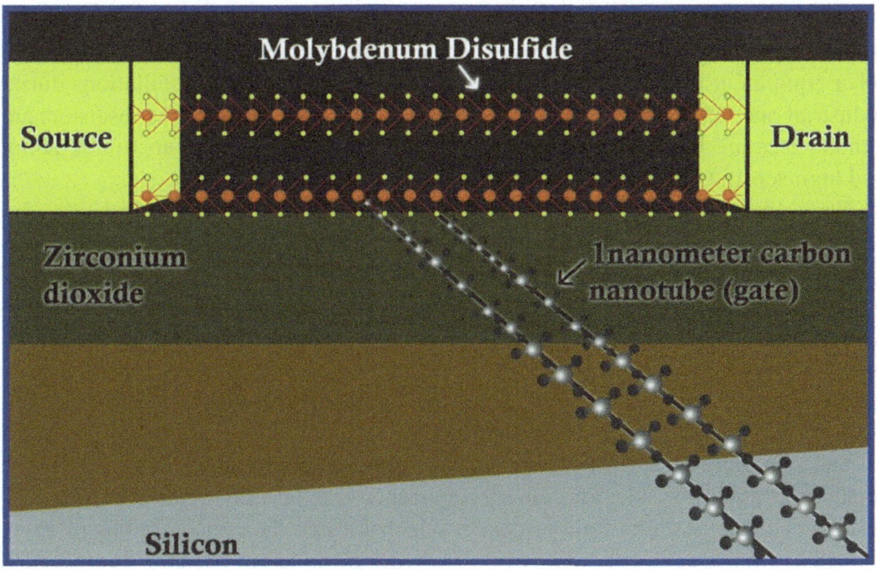


**Fig. 4** The tiniest transistor in the globe [10]

in the globe, is mindful of the capability of the carbon nanotube and describes it as the "foundation of the future beyond silicon". Additionally, as CNT is tiny, billions of transistors can be inserted in a one-centimeter square section. As a result, the speed of computers increases and, in turn, they perform more efficiently. Briefly, the tiny, quicker, and efficient transistors indicate that all of the computer memory can be kept in one small microchip. With the manufacture of nano-scaled electrical circuit parts, computers produced using nanotechnology are anticipated to be small-sized, speedier, and have bigger volumes with reduced power usage compared to those computers manufactured using the present technical know-how. Figure 4 illustrates the tiniest transistor in the globe, which is made by using CNTs and molybdenum disulfide that are substitute silicon [10].

## *5.9 Nanosensors and Nanodevices*

Nanosensors and nanodevices offer up-to-date results in numerous environments like detecting the surroundings as well as biological sensing. These sensors and devices also provide a high level of detection and are accessible in circumstances that are either stationary or dynamic in several applications, for example, in the medical and security fields as well as in surveillance. There is a pressing requirement to make new kinds of sensors and devices, which have the ability to find and ascertain quickly the origin of pollutants and other menacing agents at any time due to the growing number of industrial applications and worldwide supply [31]. On the other hand, there is a need to produce sensors and devices, which have the ability to interconnect with other equipment in production as well as locate several kinds of oscillations during industrial operation. The other significant areas that need the latest nanosensors and nanodevices are healthcare, especially inside the human body. Thus, quick action and high sensing are required in nanoscale sectors [6].

Nanosensors are sensory points in the biological, chemical, or surgical areas that are used to send data with regards to nanoparticles to the macroscopic world. Nanodevices are used mostly for different medicinal causes and used as doorways to create other kinds of nanoproducts like microchips for the nanoscale and nanorobots. In the human physique, nanosensors interconnect in real time and send data on the antibodies to the harmful substances called antigens, on the cellular receptors to their glands, and on the deoxyribonucleic acid and ribonucleic acid to the nucleic acid with a complimentary sequence [20].

The technique to convert can be optical, mass, or electrochemical. In the optical method, several types of extraordinary occurrences can be used to find different types of chemical metrics like luminescence, which locates the concentration of $H_2O_2$ by using luminescent optical sensors, absorption, polarization, and fluorescence. Through this mass technique, the connection occurs due to acoustic wave, microbalance, and resonant. In the nearby future, nanosensors will bring about several latest applications, for instance, allowing customized pictures for viruses and pathogens or fabricating screen geomatics deoxyribonucleic acid for a big set of single nucleotide polymorphisms. Further new applications that can be created in the future will assure an impressive interconnectivity between physics, materials, computing, and communication by combining the nanoscale sensors with the nanoprocessors. Optical transmission and nano-micro-electromechanical networks will together create a structure of a new group of nanosatellites, which will have the capability to behave as a bio explorer with a very high level of detection or the ability to keep track of rough surroundings [4]. Table 1 shows components of a wireless device that deal with nanotechnology [17].

## 6 Nanotechnology in 5G

The nanomachine found in the 5G nanocore is the cellular phone itself, which is well prepared for nanotechnology. The wireless industry primarily targets the application of intelligence that makes sure the computation and transferring of data are at hand as needed. By introducing intelligence in cellular phones, it can assist in implanting the devices in human surroundings, which can build an up-to-date platform to carry out the ever-present sensing, computing, and communication. The nanomachines are generally uploaded with some of the key characteristics, such as the ability to clean itself, self-powered, and sensitive to the surroundings with which it has been associating, adaptable as well as transparent. Developing the graphene's transistor was a significant step as well as an important achievement [33]. This transistor was made by

**Table 1** Components of a wireless device [17]

| Components of a wireless device | | Examples of nanotechnology enablers |
|---|---|---|
| Radio transmitters and receivers | Antenna | Nanomaterials for antenna cavities to reduce loss, use of nanoparticle inks for printable antennae, (CNT) antennae for in chip links |
| | Analog signal processing | Graphene-based high-frequency radio frequency circuits, printable electronics, spintronics, plasmonics |
| | Radio frequency front-end circuits | Graphene nano-electromechanical devices resonators |
| | Digital baseband operation | Photonics, spintronics, nanowire transistors, nanoparticle inks for printed integrated circuits |
| Power | Power scavenging | Nano-electromechanical device networks, thermoelectric materials |
| | Storage | Storage of nanomaterials for super capacitors of cells |
| Sensors | Optical organic | Photovoltaic, quantum dots, nanophotonics, plasmonics |
| | Chemical | Nanowires/nanotubes, nano arrays |
| | Mechanical | Nano-electromechanical devices |
| | Electrical/electrochemical | Nanowire/nanotube transistors, ion selective field effect transistors |
| Memory | | New materials for ferroelectric random access memory, CTS memory, phase change memory, and molecular memories |

using a new kind of material known as graphene. It is chiefly composed of graphite, which is made up of one layer of carbon atoms laid out in a honeycomb lattice structure. This specific formation enables the electrons to move through very rapidly and provides better performance than the current transmitter and receiver microchip material. The newest frequency attained by graphene's transistor is 26 GHz that is so much ahead compared to the present technology levels. Frequencies more than 1 THz are being used by the military for observing hidden arms as well as for medical purposes for imaging, thus, avoiding harmful X-rays. For the normal frequencies, the transmitters and receivers using graphene have the ability to enable the mobile phone and base stations to be more sensitive in detecting feeble signals. The primary challenge is to differentiate radio waves from other signals nearby. A highly sensitive cellular phone with an improved SNR has a greater benefit of obtaining the signal from the closest cellular base station. Mobile phones improved with carbon nanotubes are nanotechnology that will soon be available in the market. In the 5G nanocore, these cellular phones are known as nanoequipment as they have been produced using nanotechnology. A main innovation of the wireless industry is to target environmental intelligence. This means that computation and communication are readily accessible and prepared to provide service to the consumer in an intelligent manner. Thus, it needs the devices to be mobile. Mobile devices work hand in hand with intelligence, which can be implanted in human surroundings such as the home, workplace, and open areas will build an up-to-date platform that will have the capability of ever-present sensing, computing, and communication. A specification of nanoequipment is listed below:

1. Self-cleaning: the ability of the mobile phone to clean by itself.
2. Self-powered: the ability of the mobile phone to obtain energy from the sun, water or air.
3. Detect the surroundings: the ability of the mobile phone to predict the weather, the volume of air pollution, etc.
4. Adaptable: ability to bend but not break [25].

## 7 Conclusion

The most significant technological advancement of the twenty-first century is nanotechnology whose applications commence from the disciplines of science like chemistry, physics, and biology and expands to other kinds of disciplines such as health, engineering, food, and electronics. Today, wireless communication usage is growing faster and advancing speedily. Nanotechnology is on the verge of making a strong impact on communication systems, which will result in a lesser need to adhere to corresponding developments or huge data volumes or restriction in the use of devices or increase in the user devices or increase in execution registering. Another area of communication systems is molecular communication which attempts to duplicate the natural communication standards and uses these for different intriguing applications. Molecular communication has high prospects and is at a significant level whereby some practical applications of this type of communication are needed. This can only be viable if scientists from various disciplines of science try to cooperate, particularly those from the field of telecommunication engineering. Nanonetworking is a unique concept that is able to expand the characteristics of nanomachines by linking them with a single nanocore system. Further, through accessing the internet, the nanonetwork is not only a promise of the future but this idea has the capability of using the Internet of Things in a smarter way. Day-to-day, the application of the latest nanotechnology materializes and creates a better life and the number of such applications will keep on growing in the future.

## References


1. Akyildiz I, Brunetti F, Blázquez C (2008) Nanonetworks: a new communication paradigm. Comput Netw 52(12):2260–2279. https://doi.org/10.1016/j.comnet.2008.04.001
2. Akyildiz I, Jornet JM (2010) Electromagnetic wireless nanosensor networks. Nano Commun Netw (Elsevier Ltd.) 1(1):3–19. https://doi.org/10.1016/j.nancom.2010.04.001
3. Alfano G, Miorandi D (2006) On information transmission among nanomachines. Paper presented at the 1st international conference on nano-networks and workshops, nano-net, Lausanne, Switzerland, 14–16 Sept 2006. https://doi.org/10.1109/NANONET.2006.346231
4. Balasubramaniam S, Kangasharju J (2012) Realizing the internet of nano things: challenges, solutions, and applications. Comput Innov Technol Comput Prof 46:62–68
5. Boisseau P, Loubaton B (2011) Nanomedicine, nanotechnology in medicine. C R Phys 12(7):620–636
6. Burmistrova N, Kolontaeva O, Dürkop A (2015) New nanomaterials and luminescent optical sensors for detection of hydrogen peroxide. Chemosensors 3:253–273. https://doi.org/10.3390/chemosensors3040253
7. Carrara DS (2020) Nano-bio-chip technologies. http://si.epfl.ch/page-34869-en.html. Accessed 6 Oct 2020
8. Carrara S (2010) Nano-bio-technology and sensing chips: new systems for detection in personalized therapies and cell biology. Sensors 10:526–543
9. Dressler F, Fischer S (2015) Connecting in-body nano communication with body area networks: challenges and opportunities of the internet of nano things. Elsevier
10. Ersöz PDM, Işitan DA, Balaban M (2014) Nanotechnology—1. Universal nanotechnology skills creation and motivation development, 1st edn
11. Firdous R (2012) Future of wireless mobile communication with nanotechnology and application of CNT in MOSFETs (nano transistors). Int Res J Eng Technol (IRJET) 5(12):503–507. www.irjet.net
12. Fonseca M (2016) Nanotechnology. The internet of nano-things and its promises for jobs creation. http://www.intelligenthq.com/technology/what-can-nanotechnology-do-for-entrepreneurs-and-job-creation/. Accessed 12 Oct 2016
13. George S (2015) Nanomaterial properties: implications for safe medical applications of nanotechnology. Springer, pp 45–69
14. Govil J, Govil J, Gupta A, Gupta V (2012) Nano technology/networks in molecular communication: an advance step of electrical communications. Appl Mech Mater 110–116:3770–3776. https://doi.org/10.4028/www.scientific.net/AMM.110-116.3770
15. Hema Latha D et al (2014) A study on 5th generation mobile technology—future network service. (IJCSIT) Int J Comput Sci Inf Technol 5(6):8309–8313
16. Hossain S (2013) 5G wireless communication systems. Am J Eng Res (AJER) 2(10):344–353
17. ICT (2011) ICT: nanotechnology for wireless communications. ObservatoryNANO, Briefing no 25, pp 1–4. https://www.google.com/search?sxsrf=ALeKk02HYQ-pjIVhiaEb1rqjxEbmWjeW8w%3A1603813327038&ei=zz-YX9PgAdLWkwXWmLXwDw&q=ICT%3A+N
18. Jones V (2006) From BAN to AmI-BAN: micro and nano technologies in future Body Area Networks. ACM Trans Model Comput Simul (TOMACS)
19. Kalyani P (2015) IoT—Internet of things artificial intelligence and nano technology a perfect future blend. J Manag Eng Inf Technol (JMEIT) 2(2)
20. Ku T-H, Zhang T, Luo H, Yen T, Chen P-W, Han Y, Lo Y-H (2015) Nucleic acid aptamers: an emerging tool for biotechnology and biomedical sensing. Sensors (Basel, Switzerland) 15(7):16281–16313. www.mdpi.com/journal/sensors. https://doi.org/10.3390/s150716281
21. Lokhande S, Pate R (2014) Role of nanotechnology in shaping the future of mobile and wireless devices. Int J Sci Res (IJSR) 3(1):212–215
22. Massimiliano Pierobon JM, Akkari N, Almasri S, Akyildiz IF (2014) A routing framework for energy harvesting wireless nanosensor networks in the Terahertz Band. Wirel Netw (Springer) 20(5). https://doi.org/10.1007/s11276-013-0665-y

23. Mattson MP, Haddon RC, Rao AM (2000) Molecular functionalization of carbon nanotubes and use as substrates for neuronal growth. J Mol Neurosci 14(3):175–182. https://doi.org/10.1385/jmn:14:3:175
24. Minoli D (2006) Nanotechnology applications to telecommunications and networking. Wiley-Interscience, Hoboken
25. Mohamed AF, Mustafa DABAN (2012) Nanotechnology for 5G. Int J Sci Res (IJSR) 5(2):1044–1047. www.ijsr.net
26. Mousa DAM (2012) Challenges of future R&D in mobile communications. Int J Adv Comput Sci Appl (IJACSA) 3(10). https://doi.org/10.14569/IJACSA.2012.031001
27. nanowerk (2020) A quantum computer based on five atoms. https://www.nanowerk.com/nanotechnology-news/newsid=42777.php. Accessed 10 Aug 2020
28. Nayyar A, Puri V, Le D-N (2017) Internet of nano things (IoNT): next evolutionary step in nanotechnology. Nanosci Nanotechnol 7(1):4–8. https://doi.org/10.5923/j.nn.20170701.02
29. Omanović-Mikličanin E, Maksimović M, Vujović V (2015) The future of healthcare: nanomedicine and internet of nano things. Folia Med Fac Med Univ Saraev 50(1)
30. Padmavathi G, Shanmugapriya D, Valliammal N, Geetha G, Kabila Kandhasamy CJ (2016) UGC sponsored two day national conference on internet of things. World Sci News 41:1–315
31. Pirbadian S, Barchinger SE, Leung KM, Byun HS, Jangir Y, Bouhenni RA, Reed SB, Romine MF, Saffarini DA, Shi L, Gorby YA, Golbeck JH, El-Naggar MY (2014) Shewanella oneidensis MR-1 nanowires are outer membrane and periplasmic extensions of the extracellular electron transport components. Proc Natl Acad Sci U S A 111(35):12883–12888. https://doi.org/10.1073/pnas.1410551111
32. Rauta P, Sarwade N (2013) Establishing a molecular communication channel for nano networks. Int J VLSI Des Commun Syst (VLSICS) 4(2)
33. Ryhänen T, Uusitalo M, Ikkala O, Kärkkäinen A (2010) Nanotechnologies for future mobile devices, 1st edn. Cambridge University Press
34. Schulte P, Geraci C, Murashov V, Kuempel E, Zumwalde R, Castranova V, Martinez K (2014) Occupational safety and health criteria for responsible development of nanotechnology. J Nanoparticle Res 16(1):1–17
35. Sitti M, Ceylan H, Hu W, Giltinan J, Turan M, Yim S, Diller E (2015) Biomedical applications of untethered mobile milli/microrobots. Proc IEEE Inst Electr Electron Eng 103(2):205–224. https://doi.org/10.1109/jproc.2014.2385105
36. Weiss PS (2015) Where are the products of nanotechnology. ACS Nano 9(4):3397–3398
37. Yashveer S, Vikram S, Kaswan V, Kaushik A, Tokas J (2014) Green biotechnology, nanotechnology and bio-fortification: perspectives on novel environment-friendly crop improvement strategies. Biotechnol Genet Eng Rev 30(2):113–126

# Chapter 7
# Nanomaterials-Based Chemical Sensing

**Neethu Joseph and B. Manoj**

## 1 Introduction

Industrialization and exponential growth in populace developing rapid urbanization cause spurt in contamination at an alarming rate [1–3]. Betterment in the quality of water, air and soil is considered to be a gigantic task in the present age. Effective treatment of contaminations is critical in the security of the environment. Material science assumes an indispensable part in understanding and making the environment clean and benign. The role of nanomaterials in containing environmental pollutants is indispensable [4, 5]. Environmental and water contamination can be due to the presence of organics, microorganisms, heavy metal ions such as zinc, arsenic, lead, nickel, cadmium, chromium, copper and mercury, which are non-biodegradable in nature causing extraordinary danger to human wellbeing [6, 7]. The effect of contaminants like Cr, Cd, Hg, Zn, As, Cu, Ni and Pb which cause harm to human wellbeing and climate are currently viewed as major ecological issues all through the world [8, 9]. Thinking about this peril, numerous procedures have been set up to control levels of poisonous heavy metal particles in drinking water (different frameworks about logical poisonousness principles and presentation rules). A similar study was conducted by World Health Organization (WHO) and Environmental Protection Agency (EPA) to check on the toxicity levels of each heavy metal ion is portrayed in Table 1 [10, 11].

Heavy metal particles can prompt numerous intense possible dangers to human life by causing infections and diseases, like kidney harm, hepatitis, unnatural birth cycles, anaemia, encephalopathy and nephritic disorder [6, 7, 12, 13]. The source of metal contamination of water bodies is illustrated in Fig. 1. Lead particles are delivered in a large amount from mining centres, enterprises manufacturing corrosive batteries,

N. Joseph · B. Manoj (✉)
Department of Physics & Electronics, CHRIST (Deemed to be University), Bangalore, India
e-mail: manoj.b@christuniversity.in

N. Joseph
e-mail: neethu.joseph@res.christuniversity.in

 
M. Mujawar et al. (eds.), *Nanotechnology for Electronic Applications*, Materials Horizons: From Nature to Nanomaterials, https://doi.org/10.1007/978-981-16-6022-1_7

**Table 1** Toxicity level in mg/L for heavy metals in drinking water resources

| Toxic heavy metal ions | Toxicity levels (mg/L) | |
|---|---|---|
| | World Health Organization | Environmental Protection Agency |
| As | 0.010 | 0.010 |
| Cd | 0.003 | 0.005 |
| Cu | 2 | 1.3 |
| Hg | 0.001 | 0.002 |
| Ni | 0.07 | 0.04 |
| Zn | 3 | 5 |
| Pb | 0.010 | 0.015 |

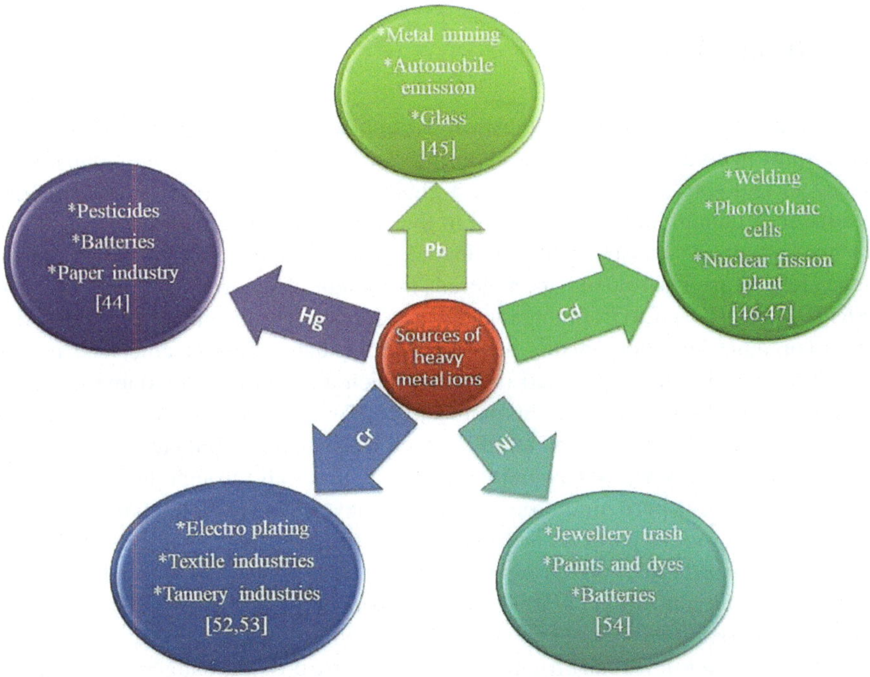


**Fig. 1** Heavy metal ion and its sources in environment contamination

glass and paper and cleaning ventures to the water sources. They have the potential of causing skin illnesses in contact with coins, zips, watches and so forth. Cadmium is significantly detected in water released from textile plants, during the electroplating process of batteries, photovoltaic cells and metallurgy cycle [14].

Chromium metal particles (VI) lead to illnesses such as liver harm, stomach upsets, nephritis etc. and Cr (VI) particles create an ulcer in the nasal area [15]. These heavy metal particles if spill into our surroundings can reach living beings in numerous manners, legitimately or in a roundabout way. The riskiest part lies with the way that, because of the biological amplification process, the poisonousness levels of these toxins can increase a few times and enter living creatures. It likewise has a negative effect where the poisonous substance can even be moved to people in the future causing inherited trade of wellbeing impacts as appeared in Table 2.

Until this point, strategies to identify heavy metal particles at the very minute levels such as in the ppt and ppq range have been set up utilizing advanced analytical instruments like inductively coupled plasma-mass spectrometry (ICPMS), X-ray fluorescence spectroscopy (XPS), atomic absorption spectroscopy (AAS) and mass spectroscopy (MS) [20–24]. These procedures have disadvantages such as high cost, difficulty in portability, low yield, sophisticated pre-treatment steps and very importantly the requirement of well-trained manpower [11, 16–24].

The reported strategies to eliminate toxic heavy metal ions are sorption [25], ion exchange [26], membrane filtration [27], chemical precipitation [28], photocatalytic degradation [29], coagulation [30], oxidation/reduction [31] and solvent extraction [32]. Among these techniques, sorption is broadly investigated and evaluated as a successful procedure because of its exceptional highlights, like minimal cost, easy procedure, environmental-friendly, biocompatibility and ease in absorbent recovery [33].

Nanomaterials having magnificent sorption properties, mild steadiness, ecological-friendly execution have empowered gigantic advancements in detecting heavy metal particles [34]. Novel nanomaterials having better performances are continually applied to enhance their applications considering the ecological contamination remediation process.

## 2 Nanocarbon Materials-Based Metal Ion Sensing

With all three dimensions under 100 nm, nanomaterials are classified as zero-dimensional nanomaterial (0-D), which includes quantum dots (QDs) and fullerene [36]. Nanomaterials with one of its dimensions larger than 100 nm and the other two dimensions lesser than 100 nm are named as one-dimensional nanomaterials (1-D), which includes titanium and carbon nanotubes (CNTs) [37, 38]. Nano-sized materials with two of its dimensions more prominent than 100 nm are named two-dimensional nanomaterials, a well-known model is graphene. All its three dimensions measuring more than 100 nm in size are named as three-dimensional nanomaterials (3-D), with the most attractive models like graphite falling under this category along with different composites of nanomaterials [38]. Figure 2 exhibits the structure of various carbon materials depending on the above-explained dimensions.

**Table 2** Effects of heavy metal ions to human health

| $Hg^{1+}$ | $Pb^{2+}$ | $Cd^{3+}$ | $As^{3+}$ | $Cd^{4+}$ | $Zn^{2+}$ |
|---|---|---|---|---|---|
| • Liver damage | • Neural damage | • Kidney failure | • Cancer, liver and kidney failure | • Cancer of respiratory tract | • Stomach cramps |
| • Neural damage | • Kidney damage | • Bone disorders | • Respiratory problems | • Stomach cancer | • Fever |
| • Intestinal toxicity | • Intestinal diseases | • Chronic-anaemia | • Diabetics | • Liver damage | • Vomiting |
| • Neuro-toxicity | • Decrease in sperm count | • Prostate cancer | • Neural damage | | • Nausea |
| • Nephro-toxicity | • Abortions in women | • Hematopoietic cancer | • Cardio vascular problems | | • Diarrhoea |
| | | • Liver cancer | | | |
| [39–41] | [42, 43] | [44, 45] | [46–48] | [49, 50] | [51] |

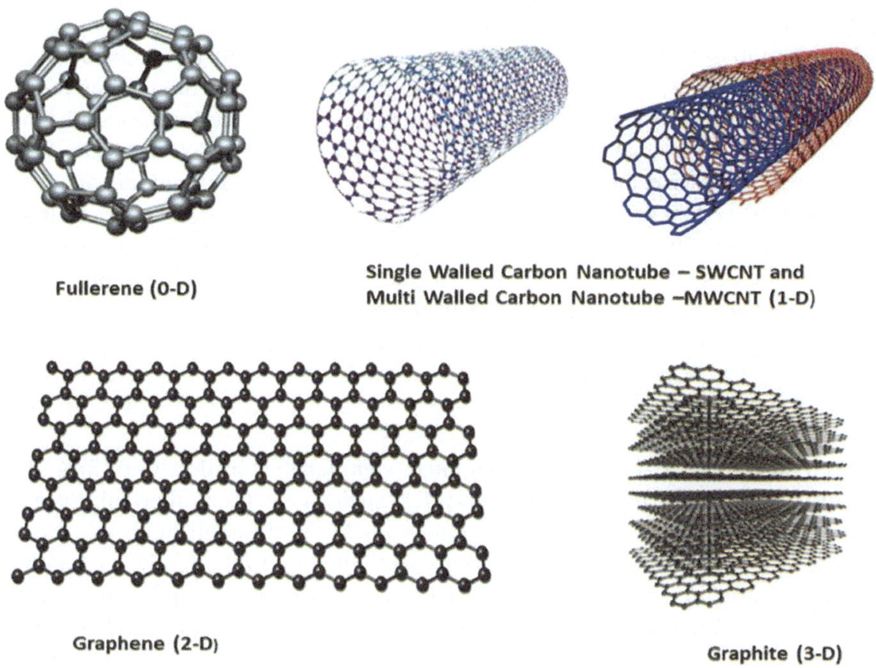


**Fig. 2** Classifications of carbon nanomaterials based on their dimensions

## *2.1 Adsorption of Heavy Metals Based on Carbon Nanomaterials*

### 2.1.1 Fullerene

With a confined ring structure, fullerene which was first discovered in 1985 possesses pentagonal and hexagonal shape within its structure [52]. It also possesses high electron proclivity, high surface to volume ratio, along with its hydrophobic nature. These unique properties possessed by fullerene can be utilized for various applications like semiconductors, gadgets, solar cells, biomedical sciences, beauty care products, sensors, artificial photosynthesis etc. [53–55]. Fullerenes are clean green materials, which are efficient in hydrogen storage as fullerene particles can, without much of an effort, be changed over from its C–C bonds to an altered scheme, i.e. C–H bond, in view of lower bond energies, both hydrogen and carbon [56]. The fullerenes accounted for its higher storage of 6.1% hydrogen, which is also due to the above-mentioned unique atomic structure properties. The structure of fullerene, without much of an effort, is turned back to the initial structure in light of the higher C–C bond energies [57, 58]. Studies have reported that fullerenes adsorbed particles by the infiltration of adsorbents in the carbon nanoclusters, which was present in between its spaces/defects between the clusters. Notwithstanding the lower aggregation, deformities, and large surface area make it ideal to be applied for remediation of water from heavy metal ions [59, 60]. On the grounds of studies directed on relative investigations of nanocomposite-polystyrene film and fullerene for the expulsion of $Cu^{2+}$ particles, Saito et al. reported that better efficiency for the same was indicated by fullerene. Likewise, they have also discovered that the Langmuir model of adsorption was followed by fullerene for $Cu^{2+}$ adsorption [61]. From these reported studies we can conclude that fullerene can be the best choice of material for adsorption of heavy metal ions, along with its added advantage of high hydrogen storage, which can be exploited for an expanded study using the material.

### 2.1.2 Carbon Nanotubes (CNTs)

Carbon nanotubes (CNTs) have exhibited significant consideration since their discovery because of their unique properties, such as electrical, thermal, mechanical as well as current capacity, which are valuable in different fields of material science and innovation [62, 63]. Nanotubes also belong to the family of fullerene due to their structural similarities and are partitioned into two classifications as single-walled carbon nanotubes (SWCNT) and multi-walled carbon nanotubes (MWCNTs) [64]. SWCNTs and MWCNTs provide an appealing possibility for detecting surface contaminants. The presence of metallic layers in MWCNTs prompts metallic highlights, which makes SWCNTs to be semiconductive or metallic; the latter is more helpful in electrode modification [65]. With the diameter ranging from 1 nm to a few nanometres with an enormous explicit surface region (150–1500 $m^2/g$), they have mesoporous structures making them an ideal contender for the elimination of heavy metal particles through the method of adsorption [66, 67]. Likewise, CNTs can be functionalized without much of a stretch with different organic molecules, making them explicit for the choice of adsorbents, and thus making its adsorption capacity improved [68]. The adsorption component of heavy metals using CNTs depends on its electrochemical potential, external features as well as ion exchange process [66–69].

Alijani et al. planned SWCNT centred composite with the infusion of magnetite cobalt sulphide as the composite material. These nanocomposites were used for the evacuation of $Hg^{2+}$, which resulted in a 99.56% of high adsorption rate within a time span of 7 min [70]. In contrast, 45.39% was the adsorption rate of mercury using SWCNTs alone [70]. Anitha et al. carried out another study where molecular dynamic simulation of SWCNTs with various functionalization, e.g. SWCNT-$NH_2$, SWCNT-OH and SWCNT-COOH, enhances adsorption limits of heavy metal particles such as $Hg^{2+}$, $Cu^{2+}$, $Pb^{2+}$, $Cd^{2+}$ etc. from water media. The outcomes uncovered that the SWCNTs-COOH has a lot of adsorption limits of around 150–230% higher, in contrast to the exposed SWCNTs. The SWCNTs-NH and SWCNTs-OH composites demonstrated a higher adsorption rate of 10–47% when compared with the performance of SWCNTs alone [71]. Gupta et al. fabricated SWCNTs-polysulfone composite-based layer for metal ion detection. The designed composite showed high elimination rate for metal particles which rejected 94.2% $Pb^{+2}$, 96.8% $Cr^{+6}$, 87.6% $As^{+3}$ particles. The layer with no SWCNTs indicated just 28.3% $Pb^{+2}$, 28.5% $As^{+3}$ and 30.3% $Cr^{+6}$ particles separately. This outcome shows better results in the productivity of the film because of the infusion of SWCNTs [72].

The MWCNTs display novel properties, like high electrical-thermal conductivity, better elasticity and larger surface area [73]. In view of these properties, the same is applied broadly in gadgets, sensors, solar cells, biomedical applications etc. [74–76]. MWCNTs after undergoing oxidation have been accounted for possessing a high adsorption limit as well as proficiency for the $Cd^{2+}$, $Cr^{6+}$, $Pb^{2+}$particles from the water [77, 78]. Hence the adsorption of metal ions relies on the pH scale to an extent, and there is a possibility of desorption and also re-usage of MWCNTs by altering its pH scale [79]. The MWCNTs-$Al_2O_3$, MWCNTs$MnO_2$-$Fe_2O_3$, MWCNTs-$ZrO_2$, MWCNTs-$Fe_3O_4$ and MWCNTs-$Fe_2O_3$ nano-structured composites have been effectively useful for the expulsion of substantial particles like $Cu^{2+}$, $Pb^{2+}$, $Ni^{2+}$, $As^{3+}$ and $Cr^{6+}$particles from water [80–83]. The adsorption proficiency of functionalized MWCNTs expanded similar to different constituents of organics oxides, which is likewise anticipated as functionalized MWCNTs are multiple times further viable in adsorption of metal ions than un-oxidized MWCNTs [84]. Oxidized MWCNTs have indicated an incredibly larger sorption limit and proficiency for $Cd^{2+}$, $Cr^{6+}$, and $Pb^{2+}$. Adsorption viability of MWCNTs assisting corrosive treatment builds the possibility to eliminate chromium, cadmium and lead particles with oxygen functional groups forming precipitation of the particles, else elimination of salts on surfaces [85, 86].

### 2.1.3 Graphene and Graphene-Based Materials

Graphene-based nano-adsorbents remain astounding progressed constituents for expulsion pollutant substances from aqueous media on account of their unique size peculiarities, $\pi$–$\pi$ stacking interactions, high surface area and hydrogen bonding [87]. Nanomaterials have been discovered to be promising materials when contrasted with customary materials, both regarding cost and effectiveness [88]. Graphene-centred materials are reported by means of an ideal material for the detection and removal of vaporous pollutants [89]. Graphene oxide has been additionally used for the evacuation of dyes, specifically crystal violet (CV), rhodamine B (RhB) and methylene blue (MB) from water sources. According to the reported studies, the adsorption rate is proved to be higher with a higher concentration of dye content, on a scale of adsorption with adsorption limits of 195.4, 199.2 and 154.8 mg $g^{-1}$ for CV, MB and RhB, respectively [90]. In addition, GO has been effectively useful for the expulsion of anionic dyes, like Direct Red 23, and also Corrosive Orange 8 from watery media [91].

#### 2.1.4 Graphene-Based Photocatalytic Materials Used for the Treatment of Contaminated Water

Despite the fact that adsorption can eliminate the foreign substance from water, the adsorption procedure cannot devastate/corrupt the toxins and a removal step is required [96]. Photocatalysis is a helpful methodology for wastewater management for the total removal of natural pollutants [97]. Graphene-based photocatalysts accounted for its amended action due to its larger surface area, nanosize when evaluated with the generally used materials [98, 99]. In different investigations, graphene oxide created with $TiO_2$ and ZnO displayed a lot of photodegradation of methylene blue when compared with $TiO_2$/ZnO alone [100]. The impact of different nanocarbon materials on heavy metal ion adsorption and remediation is recorded in Table 3.

Adding to these findings, Manoj et al. (2017) synthesized organic dots from lignite, which is reported to hold a remarkably better $Cu^{2+}$ ion sensitivity at a scale of 0.0089 nM, proving that the material is ideal for metal ion sensing. The highlighting part of this study lies in where this fluorescent probe or these semiconducting dots (OSDs) are synthesized from low-grade lignite, which is much cheaper compared to the other carbonaceous precursors like coal or graphite. The main fascinating part of this study is that for the first time a lignite-derived carbon dots exhibit ratiometric fluorescence sensing up to the picomolar range [27].

From these reported studies, it is evident that nanocarbon materials-based heavy metal ion sensing is grabbing research interest considering its unique and vast potentials in various sensing strategies. The above-mentioned nanomaterials along with their hybrids as well as by the formation of composites with various elements can be used as an impending probe for sensing and removal of heavy metal ions from water sources.

## 3 Nanocarbon Materials for Gas Sensing

### *3.1 CNTs-Based Gas Sensing*

Gas sensors have pulled in intense research ideas because of the need for sensitive, quick reaction, and stable sensors for industry, environment check, biomedicine etc. The improvement of nanotechnology made tremendous impending to construct profoundly sensitive, ease, convenient sensors with lesser energy utilization. The incredibly high surface to volume proportion and empty structure makes the nanomaterials an ideal probe for gas molecules adsorption. The invention of carbon nanotubes is a stepping stone in the development of gas sensors, due to their unique structural properties, material characteristics, and morphology. Upon presentation to specific gases, the variation in CNTs' properties can be identified by different strategies. Accordingly, CNTs-based gas sensors and their systems have gained interest of late. Various types of CNT-based materials for gas sensing is described as follows.

**Table 3** Effect of different nanocarbon-based materials for metal ion adsorption

| Nanocarbon material used as absorbent | Metal ion | Ideal pH for sorption | Adsorption capacity in (mg/g)/ Efficiency (%) | References |
|---|---|---|---|---|
| SWCNT | $Hg^{2+}$ | 7.94 | 41.66 | Alijani et al. 2018 [70] |
| SWCNT-Fe3O4-CoS | $Hg^{2+}$ | 5.26 | 1666 | Alijani et al. 2018 [70] |
| SWCNT | $Cr^{+6}$ | 2.5 | 2.35 | Dehghani et al. 2015 [99] |
| MWCNT | $Cr^{+6}$ | 2.5 | 1.26 | Dehghani et al. 2015 [99] |
| Functionalized MWCNTs | $Pb^{2+}$ | 9 | 93% | Farghali et al. 2017 [100] |
| | $Ni^{2+}$ | | 83% | |
| | $Cu^{2+}$ | | 78% | |
| | $Cd^{2+}$ | | 15% | |
| | $Cr^{3+}$ | | 99.83% | |
| $Al_2O_3$-MWCNTs | $Pb^{2+}$ | 7.0 | 90% | Gupta et al. 2011 [66] |
| Fullerene (C6) | $Cu^{2+}$ | No remarks | 14.6 nmol/g | Alekseeva et al. 2016 [61] |
| Porous graphene | $As^{3+}$ | 7.0 | 90% | Tabish et al. 2018 [93] |
| rGO-$Fe_3O_4$ | $Pb^{2+}$ | 6.0 | 373.14 | Guo T et al. (2018) [94] |
| GO-$Fe_3O_4$ | $Pb^{2+}$ | 6.1 | 126.6 | Mousavi et al. 2018 [101] |
| GO-iminodiacetic acid | $Hg^{2+}$ | 5.0 | 230 | Zhang et al. 2018 [102] |
| Reduced GO-sulfophenylazo (rGOS) | $Pb^{2+}$ | 5.0 | 689 | Zheng et al. 2018 [103] |
| | $Cu^{2+}$ | | 59 | |
| | $Ni^{2+}$ | | 66 | |
| | $Cd^{2+}$ | | 267 | |
| | $Cr^{3+}$ | | 191 | |
| Activated carbon | $Zn^{2+}$, $Cd^{2+}$, $Pb^{2+}$ | No remarks | 78.43% | Bali et al. 2019 [104] |
| | $Cu^{2+}$ | | 64.75% | |

### 3.1.1 SWCNTs-Based Gas Sensing

Kong et al. (2000) created small size gas sensors incorporated with SWCNT, which displayed quick reaction at room temperature. This discovery was in contradiction with the semiconducting, oxide incorporated gas sensors which ideally works at higher temperatures [106, 107]. The investigation was supported on the basis when $NO_2$ atoms were absorbed by semiconducting SWCNTs, charge flows from SWCNT to $NO_2$ due to the propensity of $NO_2$ to catch electrons. In the case of $NH_3$, since it has a lone pair of electrons, it will be attracted by the semiconducting SWCNT. Therefore, the conductivity of the SWCNTs is immediately enhanced and lowered as they are presented to $NO_2$ and $NH_3$, individually. From that point forward, gas sensors based on SWCNTs developed a lot of attention among researchers. For instance, Goldani et al. introduced work wherein they reported that SWCNTs could be used for gas-sensing applications [108]. Additionally, An et al. stated a comparative method to deal with functionalized SWCNTs by incorporation of conductive polymers, such as polypyrrole (PPy), to create gas sensors [109]. Studies revealed that the PPy was consistently joined the outside of the SWCNTs, which brought about an expansion in an explicit surface region of the crossovers around multiple times. With the composite at a ratio of 50:50 SWCNTs-PPy, it was noticed to be exceptionally sensitive to $NO_2$ gas sensing.

### 3.1.2 MWCNTs-Based Gas Sensing

Considering the great breakthrough of SWCNTs in gas-sensing applications, MWCNTs have also acquired much research interest for gas-sensing strategies. Initial studies by Chung et al. on gas sensors based on MWCNT discovered that the sensitivity of gas sensors increases with the increasing width of MWCNTs. Studies affirm that gas sensors based on MWCNT were delicate to oxygen because of the existence of oxidized junctions in the carbon nanotubes [110]. Cho et al. testified another study wherein gas sensors based on MWCNTcomprising diaphragm, heater and contact anode were found to be sensitive to $NO_2$. To develop the retrieval efficiency of the gas sensors, the MWCNTs used for the experiment were fabricated using either the CVD technique or thermally protected dielectric diaphragm. The sensor with a heating process kept at 130 °C is shown to have a reversible reaction for a couple of minutes. The sensitivity was diminished from 1.8 to 0.8 MΩ in around 2 min when it was presented to 100 ppm $NO_2$, relating to a sensitivity of 2.5. No critical sensitivity reduction was noticed when the setup was cycled in the middle of 10–100 ppm $NO_2$ [111]. Xie et al. reported fabricated gas sensors through using self-aligned and self-welded MWCNTs, developed between two leading posts, possessing high mechanical quality and promising electrical conductivity. Additionally, the sensor creation cycle could be effortlessly industrialized, considering the fact that no external applied powers were required. Furthermore, in light of the fact that the MWCNTs were self-aligned and self-welded throughout the deposition cycle, the detecting gadgets prepared using this method had great dependability and cost-viability [112].

Apart from this, many other composites and hybrid materials on nanocarbon materials like graphene, graphene-hybrids etc. are also used for gas-sensing applications which with their performance parameters is [108–115] listed in Table 4.

**Table 4** Nanocarbon-based materials and their performance parameters on gas sensing

| Nanocarbon materials | Detected gas molecules | Sensitivity | Response time | Recover time | References |
|---|---|---|---|---|---|
| Graphene (mechanical exfoliation) | $CO_2$ | 25% (100 ppm) | 10 s | 10 s | Yoon et al. (2011) [109] |
| Graphene (mechanical exfoliation) | $NH_3$ | 6% (200 ml/min) | 13 s | 20 s | Ahmadi et al. (2015) [110] |
| Graphene (rGO) | $NO_2$ | 1.5% (100 ppm) | 5 min | 20 min | Ocola et al. (2009) [108] |
| Graphene (rGO) | $H_2$ | 4.5% (160 ppm) | 20 s | 10 s | |
| Graphene oxide (GO) film | $SO_2$ | 47% (50 ppm) | 50 s | 50 s | Kwak et al. (2012) [111] |
| Graphene film (CVD) | $NO_2$ | 25% (200 ppm) | 20 min | Recovered at 200 °C | Lee et al. (2012) [112] |
| Graphene film (CVD) | $NH_3$ | 90% (1000 ppm) | 180 min | Recovered at 200 °C | Dutta et al. (2015) [113] |
| Single SWCNT | $NO_2$ | 1000 fold (200 ppm) | 2–10 s | 12 h (RT) | Mao et al.et al. (2014) [113] |
| MWCNT | $NO_2$ | 2.5 fold (100 ppm | 30 min (RT) | 10 min (RT) | Cho et al. (2005) [114] |
| MWCNT (CVD) | $NO_2$ | 5400 fold (100 ppm) | 4 min (80 °C) | 8 min (80 °C) | Sharma et al. (2012) [115] |

Nanocarbon materials, including graphene and CNTs, are extensively considered as gas-sensing materials owing to their unique electrochemical and structural peculiarities. Gas sensors based on nanocarbon materials uphold incredible advantages such as selectivity, fast response/recovery and high sensitivity, reduced operation temperature, lesser power intake, and ease procedure using user-friendly and less-sophisticated instruments with comparatively lesser expenditure, and potentially perfect candidates for subsequent gas or chemical-sensing strategies. Nevertheless, nanocarbon-based commercial gas sensors are in need of inventions in sensor research for refining the sensor enactment or reducing the sensor cost or for inspecting a less sophisticated production method [113].

## 4 Conclusion

In summary, nanomaterials have been widely used in eliminating heavy metals as well as in the detection of toxic gas and for their removal strategies. These nanomaterials which include fullerene, graphene, CNT-based materials and so on are acquiring a great interest in chemical-sensing strategies due to their unique properties like low aggregation, high absorption rate, toxin-free and less sophisticated synthesis methods. In addition to that, these materials are reported to have higher sensitivity towards the toxic elements present in the environment which make them the best candidate for various types of chemical-sensing applications. Along with that, due to their simpler and cost-effective preparation routes, these materials can be effortlessly industrialized without compromising on the expected outcome, compared to non-carbonaceous materials. These reported unique properties of the materials can be exploited in various fields of research in [116–118] the future.

## References


1. Poorva Mehndiratta AJ, Srivastava S, Gupta N (2013) Environmental pollution and nanotechnology. Environ Pollut 2(2):103
2. Ayawei Nimibofa EAN, Cyprain AY, Donbebe W (2018) Fullerenes: synthesis and applications. J Mater Sci Res 7(3):164
3. Zhang X et al (2017) Mussel-inspired fabrication of functional materials and their environmental applications: progress and prospects. Appl Mater Today 7:222–238
4. Wu Y et al (2019) Environmental remediation of heavy metal ions by novelnanomaterials: a review. Environ Pollut 246:608–620
5. Mauter MS, Elimelech M (2008) Environmental applications of carbon-based nanomaterials. Environ SciTechnol 42(16):5843–5859
6. Sardans J, Montes F, Peñuelas J (2011) Electrothermal atomic absorption spectrometry to determine As, Cd, Cr, Cu, Hg, and Pb in soils and sediments: a review and perspectives. Soil Sediment ContamInt J 20(4):447–491
7. Baby Shaikh R, Saifullah B, Rehman FU (2018) Greener method for the removal of toxic metal ions from the wastewater by application of agricultural waste as an adsorbent. Water 10(10):1316
8. Vallee BL, Ulmer DD (1972) Biochemical effects of mercury, cadmium, and lead. Ann Rev Biochem 41(1):91–128
9. Hamilton JW, Kaltreider RC, Bajenova OV, Ihnat MA, McCaffrey J, Turpie BW, Rowell EE, Oh J, Nemeth MJ, Pesce CA, Lariviere JP (1998) Environ Health 106:1005
10. U.S. Environmental Protection Agency (1989) Risk assessment, management and 600 communication of drinking water contamination. US EPA 625/4-89/024, 601 EPA: Washington, DC
11. Aragay G, Pons J, and Merkoçi A (2011) Recent trends in macro-, micro-, and nanomaterial-based tools and strategies for heavy-metal detection. Chem Rev 111(5):3433–3458
12. Yang X et al (2018) Removal of Mn (II) by sodium alginate/graphene oxide composite double-network hydrogel beads from aqueous solutions. Sci Rep 8(1):10717–10717
13. Bali M, Tlili H (2019) Removal of heavy metals from wastewater using infiltration-percolation process and adsorption on activated carbon. Int J Environ Sci Technol 16(1):249–258
14. Xiangtao Wang YG, Yang L, Han M, Zhao J (2012) Cheng X (2012) Nanomaterials as sorbents to remove heavy metal ions in wastewater treatment. J Environ Anal Toxicol 2:154

15. Moreno-Castilla C et al (2004) Cadmium ion adsorption on different carbon adsorbents from aqueous solutions. Effect of surface chemistry, pore texture, ionic strength, and dissolved natural organic matter. Langmuir 20(19):8142–8148
16. McNeill FE, O'Meara JM (1999) The in vivo measurement of trace heavy metals by K x-ray fluorescence Adv. X-Ray Anal 41:910–921
17. Flamini R, Panighel A (2006) Mass spectrometry in grape and wine chemistry. Part II: the consumer protection. Mass Spectrom Rev 25(5):741–774
18. Merkoc A, Alegret S (2007) Comprehensive analytical chemistry. Elsevier B.V., 603, Amsterdam, pp 143
19. Wang J, Tian B (1992) Screen-printed stripping voltammetric/potentiometric electrodes for decentralized testing of trace lead. Anal Chem 64(15):1706–1709
20. Oehme I, Otto S (2009) Wolfbeis. Optical sensors for determination of heavy metal ions. Microchimica Acta 126(3–4):177–192. Wallace KJ (2009) Supramol Chem 21:89
21. Knecht MR, Sethi M (2009) Bio-inspired colorimetric detection of Hg 2+ and Pb 2+ heavy metal ions using Au nanoparticles. Anal Bioanal Chem 394(1):33–46
22. Zhang L, Fang M (2010) Nanomaterials in pollution trace detection and environmental improvement. Nano Today 5(2):128–142
23. Pierce DT, Zhao JX (eds) (2044) Trace analysis with nanomaterials. Wiley
24. Mohan AN, Manoj B (2019) Surface modified graphene/SnO2 nanocomposite from carbon black as an efficient disinfectant against Pseudomonas aeruginosa. Mater Chem Phys 232:137–144. https://doi.org/10.1016/j.matchemphys.2019.04.074Tjrdj
25. Mohan AN, Manoj, Panicker S (2019) Facile synthesis of graphene-tin oxide nanocomposite derived from agricultural waste for enhanced antibacterial activity against Pseudomonas aeruginosa. Sci Rep 9:4170. https://doi.org/10.1038/s41598-019-40916-9 .
26. Manoj B, Raj AM, Chirayil GT (2017) Tunable direct band gap photoluminescent organic semiconducting nanoparticles from lignite. Sci Rep 7(1):1–9. https://doi.org/10.1038/s41598-017-18338-2
27. Krishnan R, Balachandran M (2018) Transformation of hydrocarbon soot to graphenic carbon nanostructures. Biointerface Res Appl Chem 8(3):3187–3192
28. Thomas R, Unnikrishnan J, Nair AV et al (2020) Antibacterial performance of GO–Ag nanocomposite prepared via ecologically safe protocols. Appl Nanosci. https://doi.org/10.1007/s13204-020-01539-z
29. Joy J, Gurumurthy MS, Thomas R Balachandran M (2020) Biosynthesized Ag Nanoparticles: a Promising Pathway for Bandgap Tailoring. Biointerface Res Appl Chem 11(2):8875–8883. https://doi.org/10.33263/BRIAC112.88758883
30. Mohan A, Manoj B (2020) Extraction of graphene nanostructures from Colocasia esculenta and Nelumbo nucifera leaves and surface functionalization with tin oxide: Evaluation of their antibacterial properties. Chem A Eur J. https://doi.org/10.1002/chem.202000590
31. Raj AM, Balachandran M (2020) Coal-based fluorescent zero-dimensional carbon nanomaterials: a short review. Energy Fuels; Wu Y, Pang H, Liu Y, Wang X, Yu S, Fu D, Chen J, Wang X (2019) Environmental remediation of heavy metal ions by novel-nanomaterials: a review. Environ Pollut 246:608–620. https://doi.org/10.1016/j.envpol.2018.12.076
32. Li J, Wang X, Zhao G, Chen C, Chai Z, Alsaedi A, Hayat T, Wang X (2018) Metal–organic framework-based materials: superior adsorbents for the capture of toxic and radioactive metal ions. Chem Soc Rev 47(7):2322–2356
33. Han X, Li S, Peng Z, Al-Yuobi A-R O, Bashammakh AO, Leblanc RM (2016) Interactions between carbon nanomaterials and biomolecules. J Oleo Sci: ess15248
34. Lee K, Mazare A, Schmuki P (2014) One-dimensional titanium dioxide nanomaterials: nanotubes. Chem Rev 114(19):9385–9454
35. Xiao Z, Kong LB, Ruan S, Li X, Yu S, Li X, Jiang Y et al (2018) Recent development in nanocarbon materials for gas sensor applications. Sens Actuators B Chem 274:235–267
36. Joint FAO, WHO Expert Committee on Food Additives and World Health Organization (2004) Evaluation of certain food additives and contaminants: sixty-first report of the Joint FAO/WHO Expert Committee on Food Additives. World Health Organization

37. Jaishankar M, Tseten T, Anbalagan N, Mathew BB, Beeregowda KN (2014) Toxicity, mechanism and health effects of some heavy metals. InterdiscipToxicol 7(2):12
38. Sharma P, Dubey RS (2005) Lead toxicity in plants. Braz J Plant Physiol 17:18
39. Sullivan JB, Krieger GR (eds) (2001) Clinical environmental health and toxic exposures. Lippincott Williams & Wilkins
40. Jaishankar M et al (2014) Toxicity, mechanism and health effects of some heavy metals. InterdiscipToxicol 7(2):60–72
41. Wilbur S, Abadin H, Fay M, Yu D, Tencza B, Ingerman L, Klotzbach J, James S (2012) Toxicological profile for chromium
42. ATSDR US (2012) Toxicological profile for chromium. In: US Department of Health and Human Services, Public Health Service
43. Brita TADS, Karel M, Janssen AC, Colin R (2006) Mechanisms of chronic waterborne Zn toxicity in Daphnia magna. Aquat Toxicol 77(4):9
44. Fosmire GJ (1990) Zinc toxicity. Am J Clin Nutr 51(2):225–227
45. Han X et al (2016) Interactions between carbon nanomaterials and biomolecules. J Oleo Sci 65(1):1–7
46. Jacobs JT, Testa SM (2005) Overview of chromium (VI) in the environment: background and history. In: Chromium (VI) handbook. CRC Press, Boca Raton, p 22
47. Velma V, Vutukuru SS, Tchounwou PB (2009) Ecotoxicology of hexavalent chromium in freshwater fish: a critical review. Rev Environ Health 4(2):16
48. Kroto HW et al (1985) C60: Buckminsterfullerene. Nature 318:162
49. Sariciftci NS, Smilowitz L, Heeger AJ, Wudl F (1992) Photoinduced electron transfer from a conducting polymer to buckminsterfullerene. Science 258(5087):1474–1476
50. Bakry R, Vallant RM, Najam-ul-Haq M, Rainer M, Szabo Z, Huck CW, Bonn GK (2007) Medicinal applications of fullerenes. Int J Nanomed 2(4):639
51. Mohan Gokhale M, Ravindra Somani R (2015) Fullerenes: chemistry and its applications. Mini-Rev Org Chem 12(4):355–366
52. Mohajeri A, Omidvar A (2015) Fullerene-based materials for solar cell applications: design of novel acceptors for efficient polymer solar cells—a DFT study. Phys Chem Chem Phys 17(34):22367–22376
53. Brunet L et al (2009) Comparative photoactivity and antibacterial properties of C60 fullerenes and titanium dioxide nanoparticles. Environ Sci Technol 43(12):4355–4360
54. Pickering KD (2005) Photochemistry and environmental applications of water soluble fullerene compounds. In: Civil and environmental engineering 2005. Houstan Texas, Rice University, p 113
55. Vidyaev DG, Savostikov DV, Boretsky EA, Verkhorubov DL (2016) Hydrogen sorption by carbon nanostructured materials. Jr Indust Pollut Control 32:4
56. Kaneko K, Ishii C, Arai T, Suematsu H (1993) Defect-associated microporous nature of fullerene C60 crystals. J Phys Chem 97(26):6764–6766
57. Lucena R, Simonet BM, Cárdenas S, Valcárcel M (2011) Potential of nanoparticles in sample preparation. J Chromatogr A 1218(4):620–637
58. Alekseeva OV, Bagrovskaya NA, Noskov AV (2016) Sorption of heavy metal ions by fullerene and polystyrene/fullerene film compositions. Protect Metals Phys Chem Surfaces 52(3):443–447
59. Wang X, Li Q, Xie J, Jin Z, Wang J, Li Y, Jiang K, Fan S (2009) Fabrication of ultralong and electrically uniform single-walled carbon nanotubes on clean substrates. Nano Lett 9(9):3137–3141
60. 中西毅, 1999. Saito R, Dresselhaus G, Dresselhaus MS (1999) Physical properties of carbon nanotubes. Imperial College Press, London, 1998, xii+ 259p, 22× 15.5 cm,\10,560 [学部・大学院向, 専門書]. 日本物理学会誌, 54(10), pp 832–833
61. Beitollahi H, Movahedifar F, Tajik S, Jahani S (2019) A review on the effects of introducing CNTs in the modification process of electrochemical sensors. Electroanalysis 31(7):1195–1203
62. Dresselhaus MS, Dresselhaus G, Jorio A (2004) Annu Rev Mater Res 34:247

63. Gupta VK, Agarwal S, Saleh TA (2011) Chromium removal by combining the magnetic properties of iron oxide with adsorption properties of carbon nanotubes. Water Res 45(6):2207–2212 72.
64. Burakov AE et al (2018) Adsorption of heavy metals on conventional and nanostructured materials for wastewater treatment purposes: a review. Ecotoxicol Environ Saf 148:702–712
65. Vinod Kumar Gupta IT, Sadegh H, Shahryari-Ghoshekandi R, Makhlouf ASH, Maazinejad B (2015) Nanoparticles as adsorbent; a positive approach for removal of noxious metal ions: a review. Sci Technol Dev 34:20
66. United Nations Children's Fund (UNICEF) Progress on household drinking water, sanitation and hygiene 2000–2017: and hygiene 2000–2017: Special focus on inequalities. UNICEF, New York, pp 1–138
67. Alijani H, Shariatinia Z (2018) Synthesis of high growth rate SWCNTs and their magnetite cobalt sulfidenanohybrid as super-adsorbent for mercury removal. Chem Eng Res Des 129:132–149
68. Anitha K, Namsani S, Singh JK (2015) Removal of heavy metal ions using a functionalized single-walled carbon nanotube: a molecular dynamics study. J Phys Chem A 119(30):8349–8358
69. Gupta S, Bhatiya D, Murthy CN (2015) Metal removal studies by composite membrane of polysulfone and functionalized single-walled carbon nanotubes. Sep Sci Technol 50(3):421–429
70. Poulsen SS et al (2017) Multi-walled carbon nanotube-physicochemical properties predict the systemic acute phase response following pulmonary exposure in mice. PLoS One 12(4):e0174167
71. Tan CM, Baudot C, Han Y, Jing H (2012) Applications of multi-walled carbon nanotube in electronic packaging. Nanoscale Res Lett 7(1):1–7
72. Andrews R, Jacques D, Qian D, Rantell T (2002) Multiwall carbon nanotubes: synthesis and application. Acc Chem Res 35(12):1008–1017
73. Hung Thang B, Le Quang D, Manh Hong N, Khoi PH, Minh PN (2014) Application of multiwalled carbon nanotube nanofluid for 450 W LED floodlight. J Nanomater
74. Robati D (2013) Pseudo-second-order kinetic equations for modeling adsorption systems for removal of lead ions using multi-walled carbon nanotube. J Nanostruct Chem 3(1):55
75. Farghali AA, Tawab HA, Moaty SA, Khaled R (2017) Functionalization of acidified multi-walled carbon nanotubes for removal of heavy metals in aqueous solutions. J Nanostruct Chem 7(2):101–111
76. Yu XY, Luo T, Zhang YX, Jia Y, Zhu BJ, Fu XC, Liu JH, Huang XJ (2011) Adsorption of lead (II) on O2-plasma-oxidized multiwalled carbon nanotubes: thermodynamics, kinetics, and desorption. ACS Appl Mater Interfaces 3(7):2585–2593
77. Yang S, Li J, Shao D, Hu J, Wang X (2009) Adsorption of Ni (II) on oxidized multi-walled carbon nanotubes: effect of contact time, pH, foreign ions and PAA. J Hazard Mater 166(1):109–116
78. Ntim SA, Mitra S (2012) Adsorption of arsenic on multiwall carbon nanotube–zirconia nanohybrid for potential drinking water purification. J Colloid Interface Sci 375(1):154–159
79. Tang W-W, Zeng G-M, Gong J-L, Liu Y, Wang X-Y, Liu Y-Y, Liu Z-F, Chen L, Zhang X-R, Tu D-Z (2012) Simultaneous adsorption of atrazine and Cu (II) from wastewater by magnetic multi-walled carbon nanotubes. Chem Eng J 211:9
80. Luo C, Tian Z, Yang B, Zhang L (2013) Yan S Manganese dioxide/iron oxide/ acid oxidized multi-walled carbon nanotube magnetic nanocomposite for enhanced hexavalent chromium removal. Chem Eng J 234:11
81. Tran HN, Chao HP (2018) Adsorption and desorption of potentially toxic metals on modified biosorbents through new green grafting process. Environ Sci Pollut Res 25(13):12808–12820
82. Moosa AA, Ridha AM, Abdullha IN (2015) Chromium ions removal from wastewater using carbon nanotubes. Int J Innov Res Sci Eng Technol 4(2):8
83. Li H, Ha C-S, Kim I (2009) Fabrication of carbon nanotube/SiO (2) and carbon nanotube/SiO (2)/ag nanoparticles hybrids by using plasma treatment. Nanoscale Res Lett 4(11):1384–1388

84. Bradder P, Ling SK, Wang S, Liu S (2011) Dye adsorption on layered graphite oxide. J Chem Eng Data 56(1):138–141
85. Lee ZH, Lee KT, Bhatia S, Mohamed AR (2012) Post-combustion carbon dioxide capture: Evolution towards utilization of nanomaterials. Renew Sustain Energy Rev 16(5):2599–2609
86. Ghosh A, Subrahmanyam KS, Krishna KS, Datta S, Govindaraj A, Pati SK, Rao CNR (2008) Uptake of H2 and CO2 by graphene. J Phys Chem C 112(40):15704–15707
87. Jin QQ, Zhu XH, Xing XY, Ren TZ (2012) Adsorptive removal of cationic dyes from aqueous solutions using graphite oxide. Adsorpt Sci Technol 30(5):437–447
88. Konicki W et al (2017) Adsorption of anionic azo-dyes from aqueous solutions onto graphene oxide: equilibrium, kinetic and thermodynamic studies. J Colloid Interface Sci 496:188–200
89. Mohan VB et al (2018) Graphene-based materials and their composites: a review on production, applications and product limitations. Compos Part B 142:200–220
90. Tabish TA et al (2018) A facile synthesis of porous graphene for efficient water and wastewater treatment. Sci Rep 8(1):1817
91. Guo T et al (2018) Efficient removal of aqueous Pb(II) using partially reduced graphene oxide-Fe3O4. Adsorpt Sci Technol 36(3–4):1031–1048
92. Pan G et al (2018) Preparation of modified graphene oxide nanomaterials for water and wastewater treatment. IOP Conf Series Earth Environ Sci 170:032074
93. Li B et al (2012) ZnO/graphene-oxide nanocomposite with remarkably enhanced visible-light-driven photocatalytic performance. J Colloid Interface Sci 377(1):114–121
94. Mohammad HadiDehghani MMT, Bajpai AK, Heibati B, Tyagi I, Asif M, Agarwal S, Gupta VK (2015) Removal of noxious Cr (VI) ions using single-walled carbon nanotubes and multi-walled carbon nanotubes. Chem Eng J 279:8
95. Farghali AA et al (2017) Functionalization of acidified multi-walled carbon nanotubes for removal of heavy metals in aqueous solutions. J Nanostructure Chem 7(2):101–111
96. Mousavi SM et al (2018) Pb(II) removal from synthetic wastewater using KombuchaScoby and graphene oxide/Fe3O4. Phys Chem Res 6(4):759–771
97. Zhang C-Z, Chen B, Bai Y, Xie J (2018) A new functionalized reduced graphene oxide adsorbent for removing heavy metal ions in water via coordination and ion exchange. Sep Sci Technol 53(18):2896–2905
98. Zheng S et al (2018) Tea polyphenols functionalized and reduced graphene oxide-ZnO composites for selective Pb2+ removal and enhanced antibacterial activity. J Biomed Nanotechnol 14(7):1263–1276
99. Bali MT, Tili H (2019) Removal of heavy metals from wastewater using infiltration-percolation process and adsorption on activated carbon. Int J Environ Sci Technol 16(1):249
100. Kong J, Franklin NR, Zhou CW, Chapline MG, Peng S, Cho KJ et al (2000) Nanotube molecular wires as chemical sensors. Science 287:622–625
101. Tans SJ, Verschueren ARM, Dekker C (1998) Room-temperature transistor based on a single carbon nanotube. Nature 393:49–52
102. Martel R, Schmidt T, Shea HR, Hertel T, Avouris P (1998) Single- and multi-wall carbon nanotube field-effect transistors. Appl Phys Lett 73:2447–2449
103. Goldoni A, Larciprete R, Petaccia L, Lizzit S (2003) Single-wall carbon nanotube interaction with gases: sample contaminants and environmental monitoring. J Am Chem Soc 125:11329–11333
104. An KH, Jeong SY, Hwang HR, Lee YH (2004) Enhanced sensitivity of a gas sensor incorporating single-walled carbon nanotube-polypyrrole nanocomposites. Adv Mater 16:1005–1009
105. Cho WS, Moon SI, Lee YD, Lee YH, Park JH, Ju BK (2005) Multiwall carbon nanotube gas sensor fabricated using thermomechanical structure. IEEE Electron Device Lett 26:498–500
106. Tabib-Azar M, Xie Y (2007) Sensitive NH4OH and HCl gas sensors using self-aligned and selfweldedmultiwalled carbon nanotubes. IEEE Sens J 7:1435–1439
107. Hwang S, Lim J, Park HG, Kim WK, Kim DH, Song IS et al (2012) Chemical vapor sensing properties of graphene based on geometrical evaluation. Curr Appl Phys 12:1017–1022
108. Lu GH, Ocola LE, Chen JH (2009) Reduced graphene oxide for room-temperature gas sensors. Nanotechnology 20:445502

109. Yoon HJ, Jun DH, Yang JH, Zhou ZX, Yang SS, Cheng MMC (2011) Carbon dioxide gas sensor using a graphene sheet. Sens Actuators B Chem 157:310–313
110. Ahmadi SSS, Raissi B, Riahifar R, Yaghmaee MS, Saadati HR, Ghashghaie S et al (2015) Fabrication of counter electrode of electrochemical CO gas sensor by electrophoretic deposition of MWCNT. J Electrochem Soc 162:D3101–D3108
111. Wang J, Kwak Y, Lee IY, Maeng S, Kim GH (2012) Highly responsive hydrogen gas sensing by partially reduced graphite oxide thin films at room temperature. Carbon 50:4061–4067
112. Lee C, Ahn J, Lee KB, Kim D, Kim J (2012) Graphene-based flexible NO2 chemical sensors. Thin Solid Films 520:5459–5462
113. Mao S, Lu G, Chen J (2014) Nanocarbon-based gas sensors: progress and challenges. J Mater Chem A 2(16):5573–5579
114. Cho WS et al (2005) Multiwall carbon nanotube gas sensor fabricatedusing thermomechanical structure. IEEE Electron Device Lett 26(7):498–500
115. Sharma A, Tomar M, Gupta M (2012) Room temperature trace level detectionof $NO_2$ gas using $SnO_2$ modified carbon nanotubes based sensor. J Mater Chem 22(44):23608–23616
116. Raj AM, Manoj B (2021) Cost-effective route to nanodiamonds fromlow-rank coal and their fluorescent & dielectric characteristics. Ceram Int
117. Ramya AV, Joseph N, Balachandran M (2021) Facile synthesis of few-layergraphene oxide from Cinnamomum camphora. Nanobiotechnology Rep 16(2):183–187
118. Ramya AV, Balachandran M (2021) Valorization of agro-industrial fruitpeel waste to fluorescent nanocarbon sensor: ultrasensitive detection ofpotentially hazardous tropane alkaloid. Front Environl Sci Eng 16(3):1–11

# Chapter 8
# Application of Nanomaterials in Fuel Cell and Photovoltaic System

**Riya Thomas and B. Manoj**

## 1 Introduction

Rising alarms of environmental concomitant issues and escalation of oil prices urged the demand for an alternative supply of clean energies. Material innovations and advancements in the field of nanotechnology offered the liberty to develop fuel cell and photovoltaic technologies to meet the requirements in this energy-driven era. The paradigm of nanotechnology is conceivable because of the physical property of the high surface-to-volume ratio of nano-architecture materials. Nanomaterials can take different forms of configurations such as nanoplatelets, nanowires, nanofibers, nanopagodas, nanocones, nanorods, nanopillars, nanocombs, quantum dots, and so on [1, 2]. It is the manipulation of atomic arrangement to curtail the electrostatic energy from the charges on the polar surface leading to the formation of various nanostructures [3]. Nanotechnology is now realized in fuel cell and solar cell fabrications to optimize the materials, performance, and robustness.

Fuel cell has been the favorite contender in the fuel crisis since hydrogen and oxygen being used as fuels and are also zero emissive in nature. It has got important courtesies in transportation and portable power systems due to its high power density, enhance efficiency, and quiet operation [4]. The fuel cell is an electrochemical cell that consists of two electrodes and an electrolyte through which hydrogen ions are allowed to pass through (Fig. 1). Depending upon the electrolytes used, fuel mixture, operating conditions, and fields of applications, fuel cells can be categorized as given in Table 1.

---

R. Thomas · B. Manoj (✉)
Department of Physics and Electronics, CHRIST (Deemed to be University), Bangalore 560029, India
e-mail: manoj.b@christuniversity.in

R. Thomas
e-mail: riya.thomas@res.christuniversity.in

 
M. Mujawar et al. (eds.), *Nanotechnology for Electronic Applications*, Materials Horizons: From Nature to Nanomaterials, https://doi.org/10.1007/978-981-16-6022-1_8

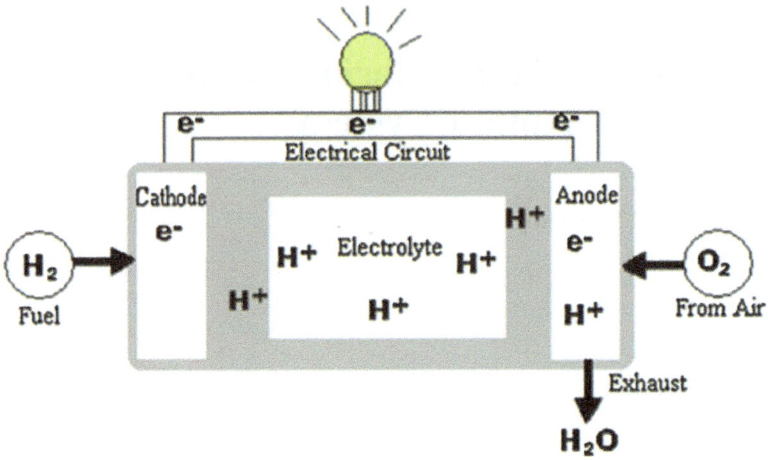


**Fig. 1** Schematic diagram of a fuel cell [5] (*Reprinted with permission*)

**Table 1** Various types of solar cells

| Fuel cell | Electrolyte | Operating temperature (°C) | Fuel mixture | Applications |
|---|---|---|---|---|
| Alkaline fuel cell | KOH solution | Room temperature up to 90 | $H_2 + O_2$ | Military, mobile, space |
| Proton exchange membrane fuel cell | Proton exchange membrane | Room temperature up to 80 | $H_2 + O_2$ or air | Portable transportation |
| Direct methanol fuel cell | Proton exchange membrane | Room temperature up to 130 | $CH_2OH + O_2$ or air | Stationary, portable devices |
| Phosphoric acid fuel cell | Phosphoric acid | 160–220 | Natural gas, biogas, $H_2 + O_2$ or air | Intermediate scale power stations |
| Molten carbonate fuel cell | Molten mixture of alkali metal carbonates | 620–660 | Natural gas, biogas, $H_2 + O_2$ or air | Large scale Power stations |
| Solid oxide fuel cell | Oxide ion-conducting ceramic | 800–100 | Natural gas, biogas, $H_2 + O_2$ or air | Intermediate scale power stations |

Hydrogen infrastructure, flexible electrodes, durability, the substitution of costly noble metals like Pt, Ru, Au, etc., as electrocatalysts, and elimination of electrode poisoning are the major deterrents in the integration of fuel cells in advanced electronics. To address these issues, nanostructured fuel cell electrodes, nanosized catalysts, novel materials of low-cost, high efficiency are developed considering fuel cells as a sustainable energy alternative. The size of fuel cells is huge and cannot be utilized as internal combustion engines in vehicles. Nanotechnology can also aid in reducing the dimension of fuel cells for easier handling and thus the power crisis can be resolved without causing pollution to the atmosphere.

Harnessing the solar radiation with an efficient solar cell is a promising technology for a limitless supply of clean and sustainable energy. The solar energy incident on the earth is technically speaking much higher than that of the current world energy demand. A solar cell or photovoltaic cell is an electronic device that can convert sunlight into electricity via the photoelectric effect (Fig. 2). It is important to understand that all materials are not apt for solar cell applications. The capacity to absorb photons and quick generation of electron–hole pairs remain the major criteria for the implementation of photovoltaic technology [3]. Based on the evolution, solar cells are classified into three generations namely first, second, and third (Fig. 3). Among them, only first-generation solar cells are fully commercialized. The second generation is employing its position in the market; whereas the third generation is still an emerging technology for the near future.

Modern photovoltaic cells have seen rapid growth in the utilization of nanomaterials in reaping green energy from solar radiations. Nanomaterials play a vital role in enhancing the effectiveness of solar cells through higher photon trapping and photocarrier collection without adding an extra cost [6, 7]. The in-built photon pathways of nanostructures augment the probability of electron–hole pair creation that benefits light trapping. Henceforth, the applications of nanotechnology in a solar cell are in terms of processing low-cost semiconductors and efficient photocatalysts, developing membranes for separations, improving energy and power proficiencies.

In this review, we discuss how nanostructured materials are exploited in refining the proficiency of fuel cells and solar cells in different facets. We provide a summary

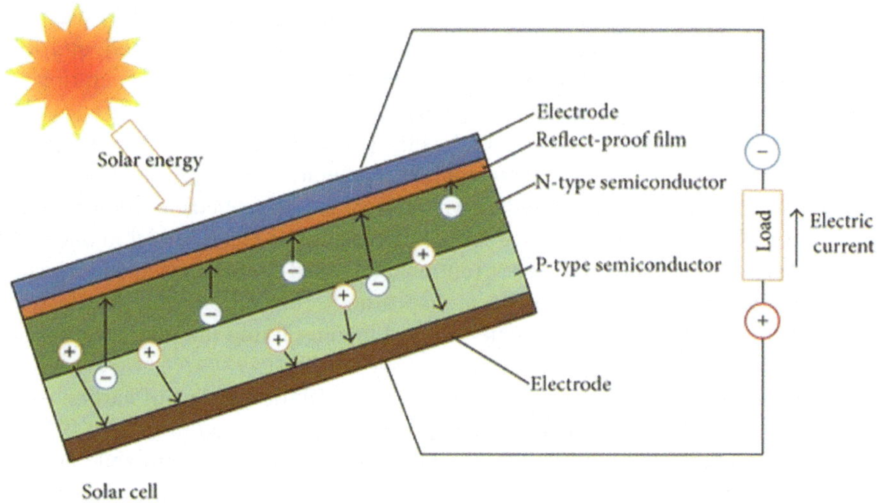


**Fig. 2** Schematic diagram of a solar cell (*Reprinted with permission*)

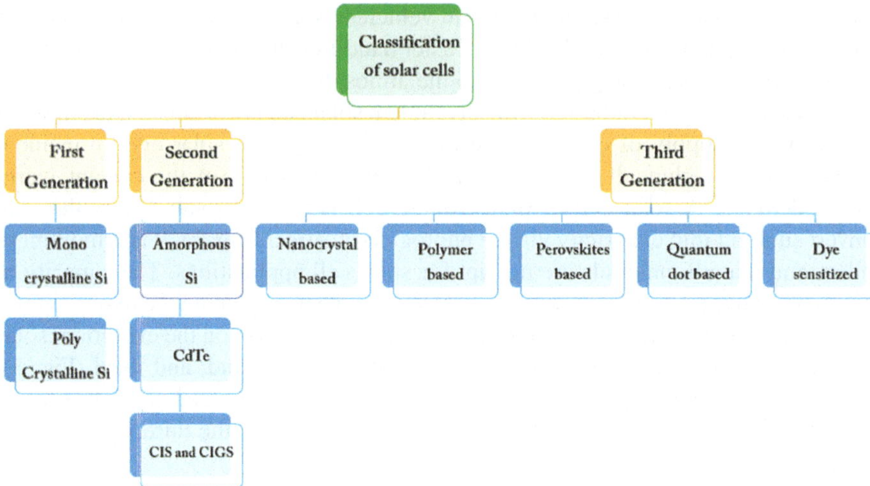


**Fig. 3** Classifications of solar cells

of synthesis techniques, processing, and applications of nanomaterials in energy harvesting technologies.

## 2 Nanomaterials-Based Fuel Cells

The progression of fuel cells in a growing energy market goes hand in hand with the development of nanotechnology. Among the various types of widely used fuel cells, polymer electrolyte membrane fuel cell, molten carbonate fuel cell, and solid oxide fuel cell are presently receiving significant attention due to easy operation and maintenance and hence are under active research.

Polymer electrolyte membrane fuel cells (PEMFCs), also known as proton exchange membrane fuel cells are typical low-temperature fuel cells mainly used for power generation owing to their high power density and controlled emission of greenhouse gases [8]. However, large-scale deployment of fuel cells is hindered due to the high cost and low durability of the conventional Pt-based catalyst layers. Novel nanoarchitectures consisting of various sizes, shapes, and compositions play a crucial role in developing high-performance materials and thereby fixing the above-mentioned bottlenecks.

There are numerous reports on the promising catalytic support of carbon nanomaterials owing to their good electron conductivity, stability, and low cost [9]. The endeavors of scientists found that the higher surface-to-volume ratio and chemical stability of carbon nanotubes (CNT) can be exploited toward superior support for metal catalysts [10]. Yang et al. proposed the utilization of SWNT and MWNT-based support for Pt in proton exchange membrane fuel cells [11]. Pt nanoparticles (3–5 nm) were deposited on the nanotube surface by a self-assembly process that occurred due to the hydrogen bonding between CNT and Pt surface. The electrochemical characterization revealed higher electrocatalytic activity and greater power density (340 $mW/cm^2$ at 0.6 V) of SWNT/Pt in comparison to MWNT/Pt composite (Fig. 4). Recently, Xue and his co-workers prepared a zigzag carbon as an efficient and stable metal-free electrocatalyst [12]. They prepared zigzag-edged graphene nanoribbon by unzipping an MWNT but retaining the nanotube backbone structure. Both density functional theory (DFT) calculation and experimental verification disclose the availability of more electrochemically active sites at zigzag pattern contrast to other carbon defects. This carbon-based electrocatalyst for the $H_2/O_2$ PEMFCs obtained a gravimetric power density of 520 W $g^{-1}$ (Fig. 5).

The management of water that is evolved as a by-product in the electrochemical reactions is a serious concern in fuel-cell technology. Graphene-based microporous layer (MPL) has shown great promise in managing humidification of proton membrane and removal of excess water [13]. Ozden et al. prepared a graphene MPL with a peak power density of 0.98 W $cm^{-2}$ at 100% RH (hydrogen relative humidity) that is 7% higher than commercial Vulcan-based MPL [14]. Laser-induced graphene (LIG) based MPL presents a 20% increment in power density at 80% RH, compared to conventional carbon black MPL [15].

Multi-metallic nanosheets of few atomic thicknesses are also gaining momentum in PEMFCs technology due to their high aspect ratio, elevated electron mobility, unique physicochemical characteristics, and unsaturated surface coordination [16]. Mori et al. have developed a non-Pt cathode catalyst for simultaneous reduction

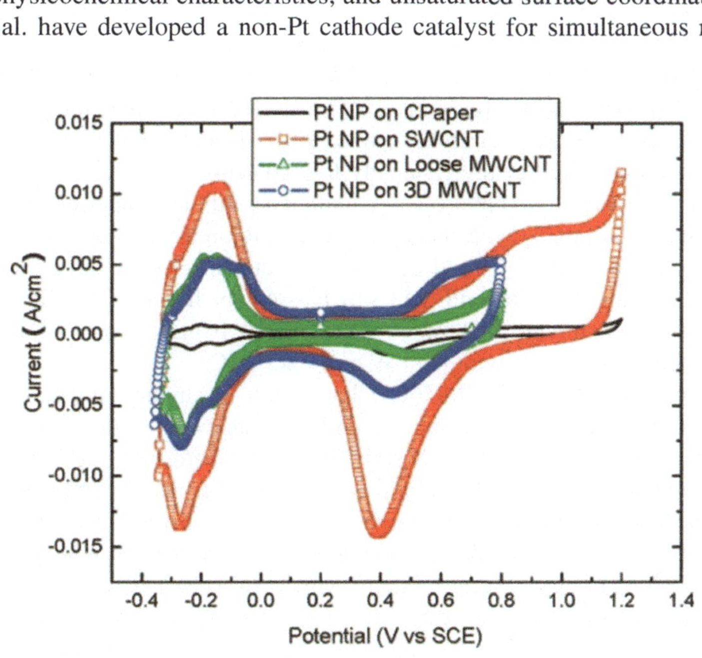


**Fig. 4** Electrochemical performance of self-organized Pt-nanoparticles on various carbon substrates [11] (*Reprinted with permission*)

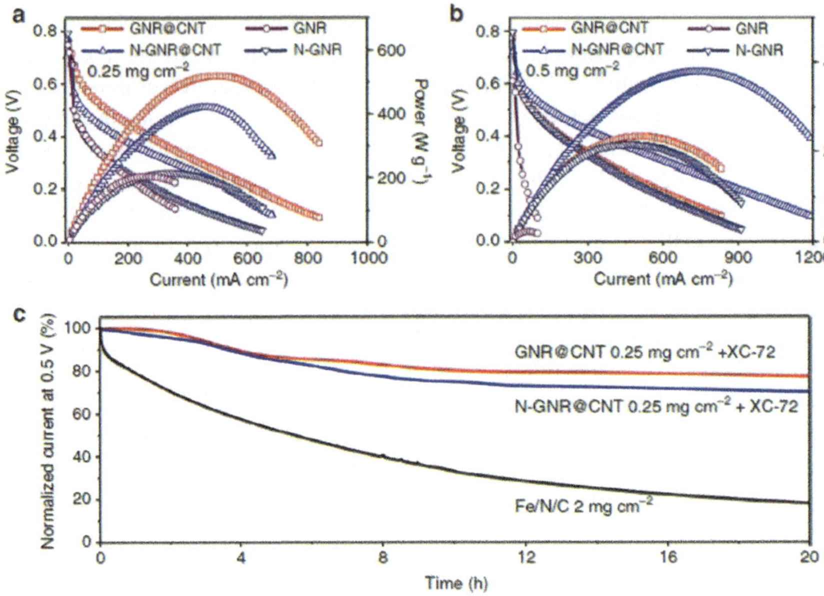


**Fig. 5** Polarization and power density curves of a graphene nanoribbon with a catalyst loading of **a** 0.25 mg cm$^{-2}$ and **b** 0.50 mg cm$^{-2}$ **c** stability of the indicated catalysts measured at 0.5 V [12] (*Reprinted with permission*)

of $O_2$ and $H_2O_2$ [17]. The fuel cell was fabricated with $Ni^{II}Ru^{II}$ and $Ni^{III}r^{III}$ as cathode and anode catalysts, respectively. The work demonstrated the possibility of utilizing $O_2$ and $H_2O_2$ as electron acceptors, while $H_2$ and CO as fuels at the anode. Non-platinum core–shell catalyst, namely IrNi@PdIr/C prepared by Qin and team demonstrated an augmented hydrogen oxidation activity and superior stability than that of commercial Pt/C [18]. Nanocomposite membranes based on sulfonated poly (ether ether ketone) (SPEEK) and perovskites nanoparticles are suggested to be a potential alternative material in PEMFCs by Hosseinabadi et al. [19]. The membrane with an optimized amount of $BaZr_{0.9}Y_{0.1}O_{3-\delta}$ (BZY10) as the perovskite material attained a peak power density of 0.44 W cm$^{-2}$ at 80 °C.

Molten carbonate fuel cells (MCFCs) that operate at a higher temperature greater than 600 °C are functioning on the principle of transport of carbonate ions from the cathode to anode. Since MCFCs work at high temperatures, they can convert a variety of hydrocarbon fuels like natural gas, coal gas, biogas, etc., into hydrogen by themselves through the process of internal reforming and hence eliminating the need for an external reformer [20]. It has an energy efficiency of 60%, which can further increase to 85% at cogeneration mode in which waste heat is recovered and used [21]. Besides, MCFCs are more resilient to impurities and are also not susceptible to poisoning by $CO_2$ or CO. But to be commercially successful, MCFCs still need to fix the difficulties of long cell life, high operating temperature, corrosion of electrolytes, and heavyweight instrumental setup.

The MCFCs anode made out of a porous sintered Ni-base suffer from higher creep strain, increased contact resistance, decreased active surface area, and lower cell performance [22]. Nguyen and the group have coated a layer of nano Ni and nano Ni-$Al_2O_3$ layer at anode surface to produce a highly porous surface suitable for enhanced electrochemical reactions [23]. In the study, both Ni and Ni-$Al_2O_3$ nanoparticle-coated MCFCs anodes showcase decreased charge transfer resistance and amended cell stability. Ni-coated anode was able to enhance the power density by 7% from the standard cell performance (Fig. 6).

To obtain a balance between the mechanical and electrochemical properties of electrode material in MCFCs, Accardo et al. prepared nano-ceria and nano-zirconia reinforced anodes and cell performances are compared [24]. These hard metal oxide nanoparticles help to resolve the difficulties of corrosion, sulfur poisoning, and large creep formation thereby eliminating the need for third alloy metal in anode strengthening. Though nano-ceria has superior mechanical properties, it is the nano-zirconia that exhibited higher polarization and power density curves (Fig. 7). Similarly, Frattini et al. investigated the effect of $ZrO_2$ nanoparticles in the mechanical strength and the cell performance of a NiAl alloy anode matrix [25]. The results showed that zirconia nanoparticles strongly adhere to the metal particles and hence exhibited improved mechanical properties. In comparison with the standard nickel-aluminium anode, cells assembled with 3% of $ZrO_2$ nanoparticles exhibited higher bending strength and lower creep resistance. Thus, the nano-zirconia modified morphology

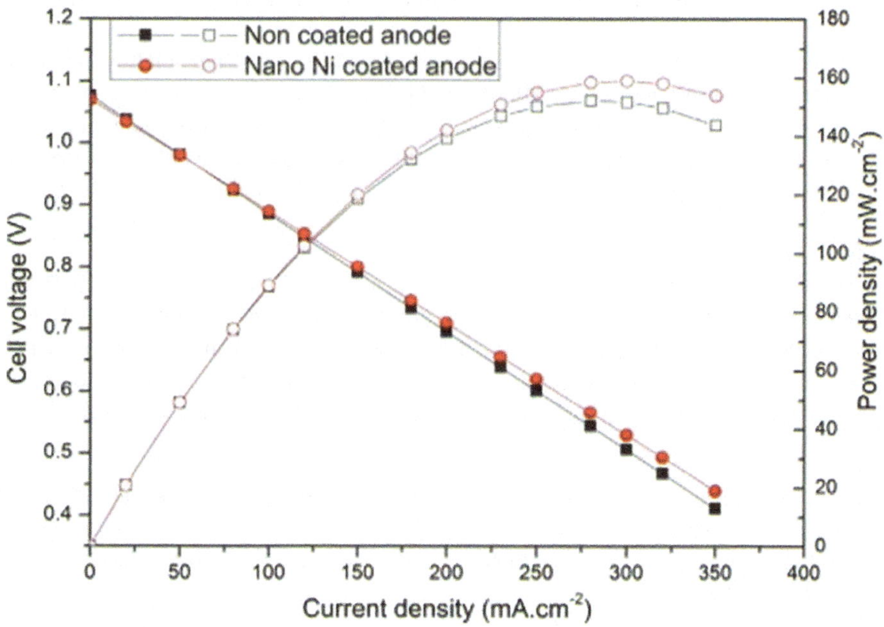


**Fig. 6** Current density–voltage curves of anodes at the operating temperature of 600 °C [23] (*Reprinted with permission*)

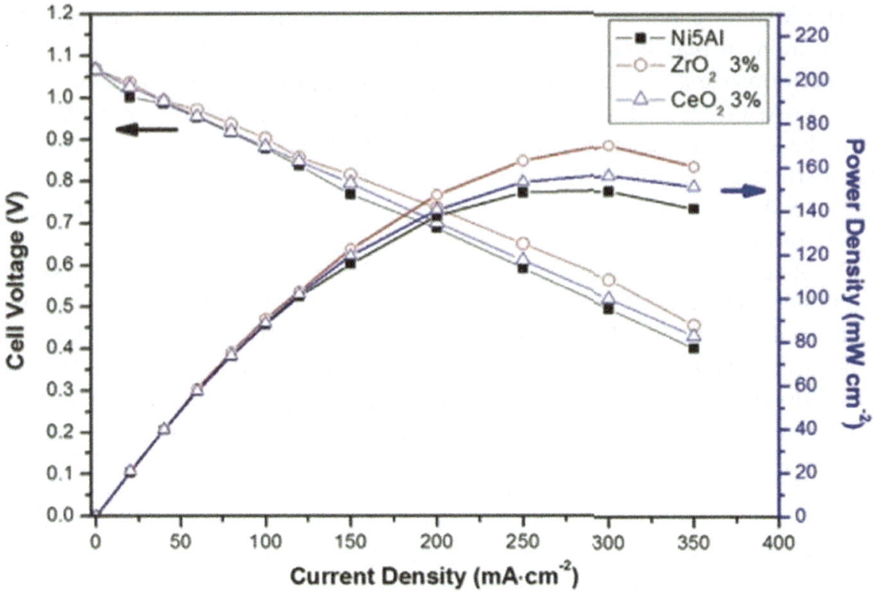


**Fig. 7** Polarization and power curves after 500 h [24] (*Reprinted with permission*)

and microstructure at the grain level facilitating better cell performance at a minimal cost.

Liu et al. have proposed the potential of rare metal ceria nanoparticles as a catalyst for graphite electrooxidation in MCFCs [26]. Though ceria is known for its redox properties ($Ce^{4+}/Ce^{3+}$), it is quite unstable. Hence, the ceria nanoparticles prepared via the sol–gel method are mixed with graphite to improve the physical contact between reactant and catalyst. Results indicated that the addition of $CeO_2$ significantly decreased the activation energy required for graphite electrooxidation via indirect electrooxidation pathways (Table 2.).

Solid oxide fuel cells (SOFCs) operating at higher temperatures (600–1000 °C) doesn't require expensive catalyst like platinum and are also not susceptible to CO catalyst poisoning. The output power of SOFCs is ranging from 100 W to 2 MW, which can be utilized for power units in vehicles to generators [27]. However, conventional SOFCs still severely suffer from sulfur poisoning, solid electrolyte cracking,

**Table 2** The apparent activation energy of graphite mixed with 0, 10, 30, and 50% ceria at −0.8 V and −0.4 V [26] (*Reprinted with permission*)

| Polarization potential/V | $E_a$/kJ mol$^{-1}$ (0 wt.%) | $E_a$/kJ mol$^{-1}$ (10 wt.%) | $E_a$/kJ mol$^{-1}$ (30 wt.%) | $E_a$/kJ mol$^{-1}$ (50 wt.%) |
|---|---|---|---|---|
| −0.8 | 76.3 | 65.0 | 44.6 | 42.9 |
| −0.4 | 42.8 | 32.5 | 16.6 | 14.9 |

declined ionic conductivity, and expensive sealing for operating in high temperatures. So, the art of nanotechnology presented new designs for overcoming critical challenges and also developed novel nanomaterials for functioning at lower temperatures.

Lim et al. fabricated a nickel-samaria-doped ceria nanocomposite anode via a co-sputtering technique for low-temperature SOFCs (LT-SOFCs) [28]. This nanocomposite provided expanded reaction sites for enhanced ion electronic conductivity yielding a maximum power density of 178 mW/cm$^2$ and polarization resistance of 0.55 Ωcm$^2$ at 450 °C (Fig. 8). Mixed metal oxide nanocomposites were used as cathode and anode for symmetrical triple-layered LT-SOFCs by Shaheen and his co-workers [29]. $Cu_{0.5}Sr_{0.5}$ (CS) and $La_{0.2}$ doped $Cu_{0.4}Sr_{0.4}$ (LCS) nanocomposites were prepared by a cost-effective pechini method. The LCS nanocomposite tested over the temperature range of 500–600 °C with $H_2$ fuel exhibited a power density of 782 mW/cm$^2$ and electrical conductivity of 4.70 S/cm. A nanocomposite of samarium (Sm) and germanium (Ge) co-doped ceria $Ce_{0.7}Sm_{0.15}Ge_{0.15}O_{2-\delta}$ (SGeDC) prepared via co-precipitation method was also applied as an electrolyte for LT-SOFCs [30]. The conductivity and cell performance of the composite electrolyte were enhanced on the addition of Ge. At 600 °C, peak power density and maximum open-circuit voltage of the cell were measured to be 600 mW/cm$^2$ and 0.95 V, respectively (Fig. 9).

Heat balance management is an added advantage for SOFCs in improving fuel cell competence. Shao has researched on developing Mg-based nanomaterials as stationary hydrogen storage materials to provide electricity and heat [31]. The expelled heat was supplied to hydrogen desorption for $MgH_2$ materials and a net efficiency of 82% (electrical power efficiency ~68.6%) could be achieved. The Mg nanoparticle of smaller dimensions prepared through plasma synthesis was found to be effective in hydrogen storage kinetics and overall performance.

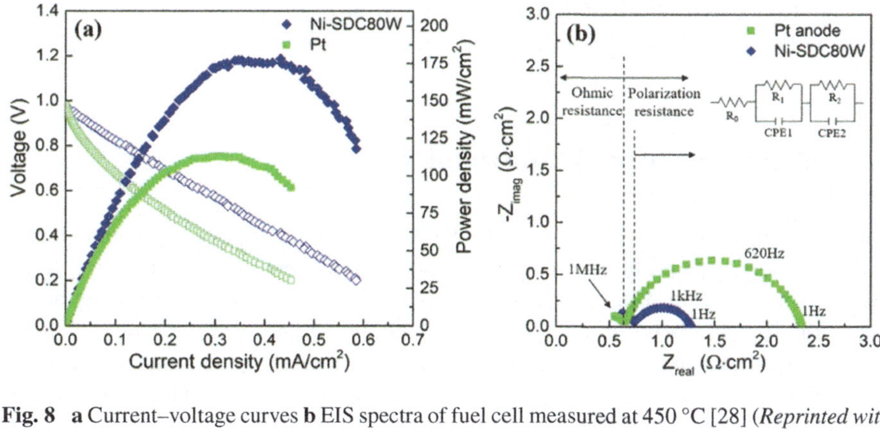


**Fig. 8** **a** Current–voltage curves **b** EIS spectra of fuel cell measured at 450 °C [28] (*Reprinted with permission*)

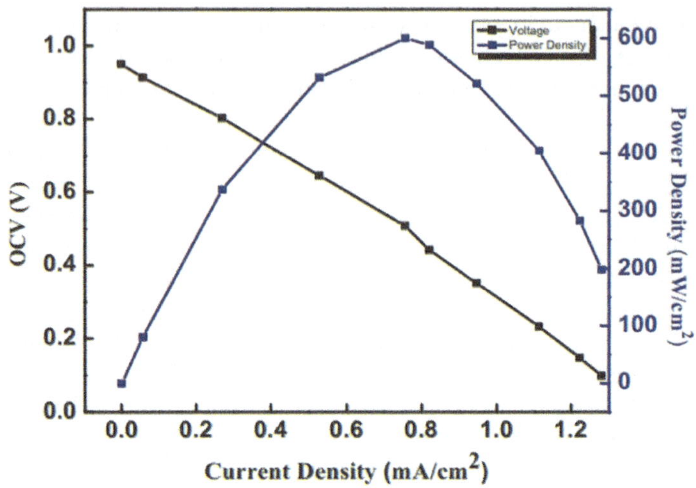


**Fig. 9** I-V/I-P fuel cell measured at 600 °C [30] (*Reprinted with permission*)

Baldi and his co-workers have proposed a hybrid cogeneration system based on SOFCs and PEMFCs suitable for off-grid applications that can supply both electricity and heat (Fig. 10) [6]. A purification unit was introduced to SOFCs for the production of hydrogen and the stored fuel was given to PEMFCs during high demand. The exploitation of PEMFCs in combination with SOFCs for energy conversion technology was far economically viable and proficient (electrical efficiency > 60%) compared to the stand-alone mode of SOFCs.

Several efforts have been incorporated in utilizing the astounding properties of nanomaterials for engineering a proficient fuel cell. The quality of nanostructures preserves the desired surface chemistry for overcoming the intrinsic limitations of a fuel cell. Nanochemistry plays a vital role in boosting performance, stability, viability along with cost-effective synthesis strategies for realizing industrial relevance.

## 3 Nanoarchitectured Photovoltaics

The voluntary change to cleaner energies fetched an immense advancement in photovoltaic technology and thereby replacing the pollution-causing conventional fossil fuels. An increase in efficiency and reduction in cost are the major exemplars in the chronological evolution of solar cells. The engineered nanomaterials of metal and metal oxides, perovskites, polymers, carbon derivates, and nanocomposites are evolving research topics in photovoltaics that can compete with conventional Si-based PV cells.

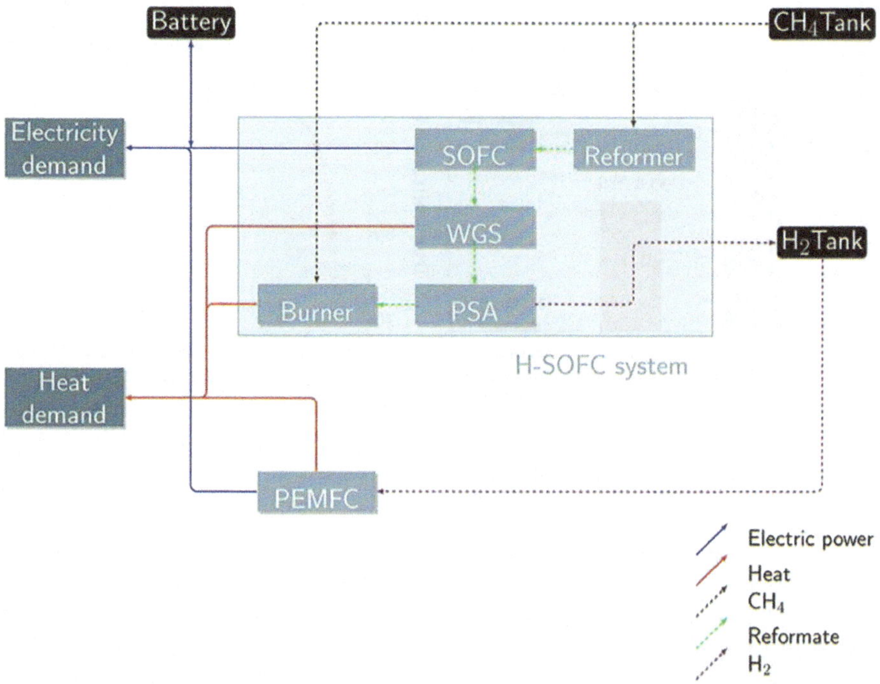


**Fig. 10** Schematic representation of the proposed energy system [6] (*Reprinted with permission*)

Owing to the exciting and outstanding characteristics, two-dimensional nanomaterials have gained momentum for future-generation photovoltaic technology. Atomically thin structure, large surface area, higher flexibility, the opportunity for artificial nanostructuring, and novel functionalization contributed to the scaling trends of 2D-based solar photovoltaics. 2D metal architectures like nanomesh and nanogrid offer consistent transportation of charge carriers deprived of losing their optical transmittance [32]. Gong et al. fabricated a transparent nanomesh electrode using an ultrathin gold nanowire using a self-assembly approach [33]. The metal nanostructure of mesh was synthesized in an ambient condition eliminating the need for sophisticated techniques. The nanomesh with a pore size of 8–52 μm was decidedly flexible, easily transferrable, and washable. A platinum nanogrid electrode prepared from a new molecular complex $Pt(NO_3)_2(ACN)_2$ via nanoimprint lithography has been reported [34]. On thermal decomposition, this molecular complex gives rise to grid patterns of platinum linewidth down to 40 nm. Wherein, the electrode exhibits optical transmittance up to 90% and resistance below 100 Ω.

2D transition metal dichalcogenides (TMDCs) have also attracted interest in the growing solar community due to their suitable energy bandgap and admirable light absorption properties [35]. Kang et al. have fabricated efficient heterojunction solar cells using $MoS_2$ thin films and Au nanomesh electrodes [36]. The highly uniform $MoS_2$ thin films were prepared on large scale through thermal decomposition of

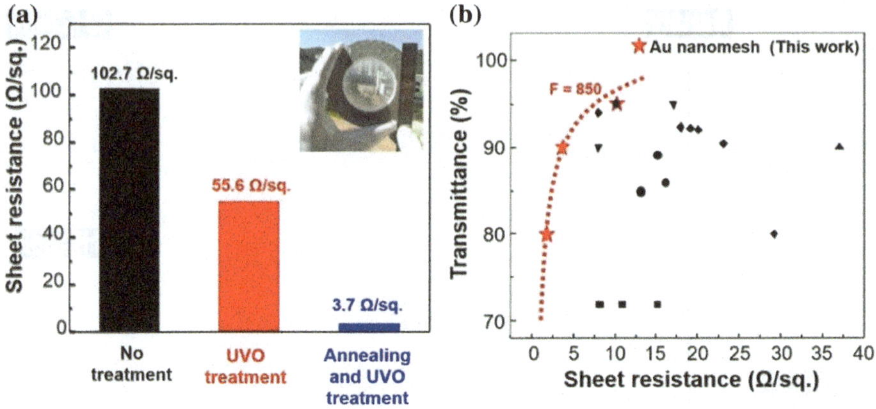


**Fig. 11** **a** Sheet resistance reduction of Au nanomesh on pre-treatment **b** sheet resistance versus transmittance from 400 to 800 nm [36] (*Reprinted with permission*)

solution precursors. The Au nanomesh obtained by vacuum deposition of Au over polymer nanomesh was of high transparency and lower sheet resistance (3Ω/sq) (Fig. 11). P-n heterojunction was formed by transferring $MoS_2$ wafers on a p-Si substrate and then Au nanomesh was transferred over $MoS_2$ as transparent top electrodes. The implemented heterojunction solar cells exhibited a maximum efficiency of 5.96% for a circular diameter of 0.3 in.

Carbon nanomaterials are known for their good electrical conductivity, chemical, and thermal stability, high catalytic property, and low cost compared to that of Pt. Casaluci and his group illustrated the application of spray-coated graphene ink as a counter electrode in dye-sensitized solar cells [37]. The deposited graphene ink had a large area (43.2 $cm^2$ active area) and a transparency of 44%. Considering the deposition of graphene ink over rigid and flexible substrates, transparent, all printed, flexible graphene-based solar cells are a viable alternative to current photovoltaic technology. Several groups have investigated on edge functionalization of graphene nanoplatelets using nonmetals, halogens, semimetals, etc., to augment the power conversion efficiency and stability of solar cells [38]. Owing to the variable oxygen functional groups, graphene oxide (GO) has good dispersal stability and enhanced surface energy that be exploited in the modification of interface at perovskites solar cells [39]. Sun et al. constructed heterojunction perovskites solar cells using GO sheets as nucleation centers [40]. GO introduced between indium tin oxide (ITO) and $CH_3NH_3PbI_3$ optimize the perovskites crystallization process and also improved the energy alignment amid interfacial layers. The increased charge carrier transportation with a reduced rate of recombination significantly improved the efficiency rate to 6.62%.

In the context of comparative effectiveness and lower cost to that of silicon solar cells, perovskites solar cells (PSCs) are recognized as a promising candidate avenue in the photovoltaic research scene. In contrast to the bulk counterpart, 2D perovskites have got improved stability and defect passivation effect [41]. Zuo et al. have also

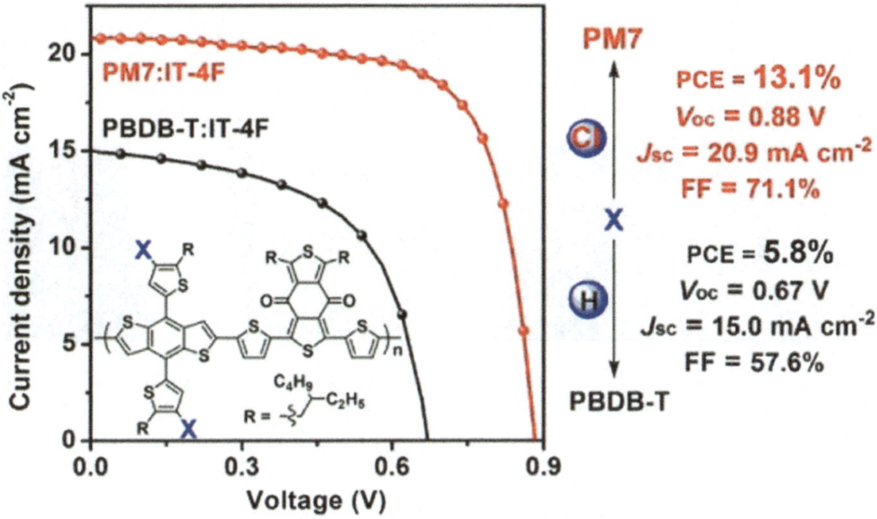


**Fig. 12** Current density versus voltage curve of polymer solar cell before and after toluene processing [43] (*Reprinted with permission*)

achieved an efficiency of 14.9% by constructing self-assembled and uniform 2D perovskites film [42]. More interestingly, the highly oriented 2D perovskites layers were constructed via a direct drop cast method in the air. Furthermore, the efficiency of PSCs was not subjected to any degradation for more than 5 months after placing them in the glove box.

The promise of lightweight, large-area fabrication, and less expense are drawn a great deal of attention for polymer solar cells. Fan and his coworkers studied the performance of chlorine substituted 2D conjugated polymer solar cells [43]. By chlorine substitution, the absorption, crystallinity, and carrier mobility were drastically enhanced. Besides, the authors also reported that via toluene processing polymer photovoltaic cells achieved an efficiency of 13.1% with a fill factor of 71.1% (Fig. 12).

One-dimensional nanostructure is suggested to be a foremost building block in producing dynamic improvement in the performance of solar cells. Largely 1D nanomaterials are used in the fabrication of transparent electrodes since this nanoscale configuration can maintain a noteworthy balance between transparency and conductivity [44]. The anisotropic architecture is most favorable for direct charge transport and thereby reducing the electrical resistance in comparison to other nanoparticles [45].

Silver metal is known for its high conductivity and an appropriate work function is obscured in solar energy conversion due to its limitation of having both transmittance and conductivity. When silver is ascribed to 1D configuration, nanowire largely enhanced the optical transparency and electrical conductivity [44]. The optical and electrical properties are lifted on decreasing the diameter of silver nanowires, but then

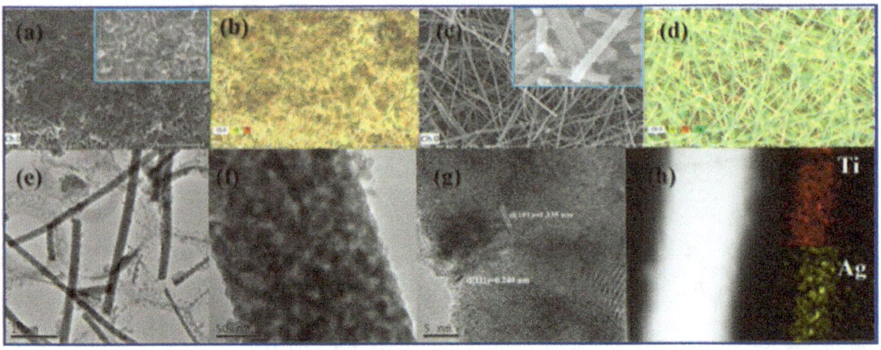


**Fig. 13** **a, c** TEM images of nanofibers with different reduction time **b, d** corresponding mapping **e–h** HR-TEM images of nanofiber on reduction with $AgNO_3$ solution for 45 s [48] (*Reprinted with permission*)

the stability issue arises. Lee et al. have resolved these limitations through electrodeposition of silver onto silver nanowire transparent conducting electrodes [46]. The electrodeposition technique facilitated the precise manipulation of nanowire diameter and also reduced junction resistance. The performance of electrodeposited nanowire networks was found to be equivalent to that of a standard sputtered transparent conducting oxide. Due to their unique properties, metal oxides are attracted intensive investigations for applications in solar cells. Among all metal oxide nanoparticles, $TiO_2$ is widely recognized as a semiconductor for solar cells. But $TiO_2$ thin films are having the critical issue of hindered electron transport due to several surface defects and grain boundaries. 1D $TiO_2$ nanofiber could overcome these constraints of charge transport and light-harvesting via straight pathways and photon scattering, respectively [47]. Sun et al. studied the importance of noble metals in improving the performance of a dye-sensitized solar cell (Fig. 13) [48]. It is found that the decoration of Ag over $TiO_2$ nanofiber boosted the light absorption qualities and diminished the recombination rate. Upon modification of photoanode using Ag metal, the solar energy conversion efficiency is increased by 18%.

Nanostructured carbon materials like carbon nanotubes and nanofibers are extensively utilized in the fabrication of solar energy harvesting devices. Wang et al. adopted a carbon nanotube (CNT) bridging method to form a superior carbon/perovskite interface in carbon-based PSCs [49]. These CNT bridges penetrate both $CH_3NH_3PbI_3$ and carbon layers to promote higher charge extraction, transport, and conductivity. The carbon paste for the counter electrode was prepared from three different graphite powders (CE A, CE B, CE C) and the performances were compared. The fabricated C-PSCs with CE C+CNT composite accomplished an efficiency of 15.75% with a fill factor of 0.72 and admirable stability for 90 days under high humid and temperature conditions (Table 3).

Out of numerous polymer nanostructures, polymer nanowires (PNWs) have got a long-range $\pi$ connectivity, charge transport, and high interfacial area between acceptor and donor [50]. Lee et al. have reported a PNWs-based organic solar cell with

**Table 3** Photovoltaic parameters of C-PSCs with different counter electrodes [49] (*Reprinted with permission*)

| Sample | $J_{sc}$ (mA/cm$^2$) | $V_{oc}$ (mV) | FF | PCE (%) |
|---|---|---|---|---|
| CE A | 19.44 ± 0.39 | 935.82 ± 18.72 | 0.6444 ± 0.0129 | 11.72 ± 0.23 |
| CE B | 19.76 ± 0.40 | 941.08 ± 18.382 | 0.6790 ± 0.0136 | 12.63 ± 0.25 |
| CE C | 20.92 ± 0.42 | 950.41 ± 19.01 | 0.6971 ± 0.0140 | 13.86 ± 0.28 |
| With CNT | 22.36 ± 0.45 | 974.37 ± 19.49 | 0.7222 ± 0.0144 | 15.73 ± 0.31 |

an efficiency of 10.62% and a high short-circuit current density of 19.45 mA/cm$^2$. The nanowire synthesized through the film processing method at a mild blending temperature was highly crystalline in nature and few nanometres in width (6 nm). PNWs possess a HOMO level of −5.46 eV and a lower bandgap of 1.59 eV.

The utilization of zero-dimensional quantum dots in solar cells can boost the efficiency owing to its intrinsic tunable bandgap, multiple exciton generation possibility, and large dipole moment [51]. But there is an imperative need for developing less toxic QDs for expanding the commercial applications by minimizing the environmental and health concerns. Peng et al. synthesized "green QDs" from the I-III-$VI_2$ group, Cu-In-Ga-Se (CIGSe) QDs through simultaneous nucleation and growth method [52]. He studied the phenomenon of optimizing the alloying strategy to augment the photoinduced electron extraction in quantum dot sensitized solar cell (QDSSC) and thereby the efficiency of 11.49% with a fill factor of 0.62 was obtained (Fig. 14).

Selopal et al. have explored the light-harvesting proficiency of specially designed "giant" core/shell CdSe/$(CdS)_x$ QDs in QDSSCs [53]. Keeping the core size (1.65 nm) fixed, the shell thickness of CDS was varied and optimized the maximum photoconversion efficiency (PCE ~ 6.86%) at a thickness of 1.96 nm. To further

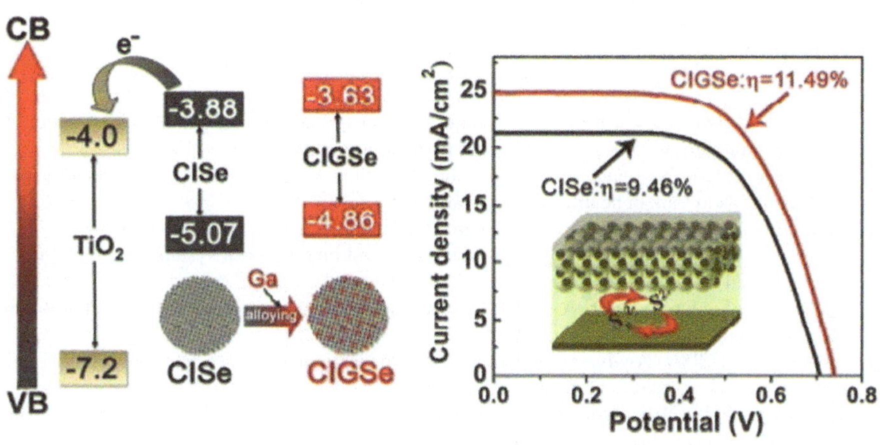


**Fig. 14** Diagram of the energy level of J-V curve [52] (*Reprinted with permission*)

enhance the PCE and enduring stability, a $CdSe_xS_{1-x}$ interfacial was incorporated between the CdSe core and CdS shell. The synergetic effect of graphene quantum dots (GQDs) and perovskite strontium ruthenate ($SrRuO_3$) film in the counter electrode for dye-sensitized solar cells. was analyzed by Liu and his group [54]. The decoration of GQDs over $SrRuO_3$ film enabled faster ion diffusion providing more active catalytic sites. The synergistic effect of hybrid facilitates achieving an efficiency of 8.05% and longstanding electrochemical stability which can be utilized as a Pt-free counter electrode.

The implementation of nanomaterials in photovoltaics holds promise in augmenting the capabilities of light harvesting, charge transfer, and collection efficiencies. In summary, the appropriate nanochemistry can tailor the physicochemical properties of materials to enhance the electrogenic capabilities suitable for energy scavenging applications.

## 4 Conclusion

The wide-ranging choices of resources enable nanotechnology to resolve major energy-related problems. Its substantial contribution lies in developing new industries for commercializing energy harvesting and storage devices with enhanced competence. The modern fuel cell and photovoltaic technology incorporated nanomaterials effectively and successfully for ease of manufacturability, performance improvement, and low-cost device fabrication. Despite all the developments, many challenges are yet to be resolved for real-life applications as per the predicted characteristics. Nevertheless, to pursue the unseen opportunities of nanomaterials in solar cell and photovoltaic systems comprehensive research is still entailed.

## References

1. Murr LE (2015) Classifications and structures of nanomaterials. In: Handbook of materials structures, properties, processing and performance. Springer, Cham
2. Khan FA (2020) Nanomaterials: types, classifications, and sources. In: Applications of nanomaterials in human health. Springer, Singapore
3. Tala-Ighil R (2015) Nanomaterials in solar cells. Handb Nanoelectrochem 2015:1–8
4. Wee JH (2007) Applications of proton exchange membrane fuel cell systems. Renew Sustain Energy Rev 11:1720e1738
5. Mekhilef S, Saidur R, Safari A (2012) Comparative study of different fuel cell technologies. Renew Sustain Energ Rev 16(1):981–989
6. Hu L, Chen G (2007) Analysis of optical absorption in silicon nanowire arrays for photovoltaic applications. Nano Lett 7(11):3249–3252
7. Han SE, Chen G (2010) Optical absorption enhancement in silicon nanohole arrays for solar photovoltaics. Nano Lett 10:1012–1016
8. Jawaid M, Thariq M, Saba N (2018) Durability and life prediction in biocomposites, fibre-reinforced composites and hybrid composites. Woodhead Publishing

9. Zhang M, Dai L (2012) Carbon nanomaterials as metal-free catalysts in next generation fuel cells. Nano Energ 1(4):514–517
10. Dicks AL (2006) The role of carbon in fuel cells. J Power Sources 156(2):128–141
11. Yang DQ, Sun S, Dodelet JP, Sacher E (2008) A facile route for the self-organized high-density decoration of Pt nanoparticles on carbon nanotubes. J Phys Chem C 112(31):11717–11721
12. Xue L, Li Y, Liu X, Liu Q, Shang J, Duan H, Dai L, Shui J (2018) Zigzag carbon as efficient and stable oxygen reduction electrocatalyst for proton exchange membrane fuel cells. Nat Commun 9(1):1–8
13. Leeuwner MJ, Wilkinson DP, Gyenge EL (2015) Novel graphene foam microporous layers 20 for PEM fuel cells: interfacial characteristics and comparative performance. Fuel Cells 15:790–801
14. Ozden A, Shahgaldi S, Li X, Hamdullahpur F (2018) A graphene-based microporous layer for proton exchange membrane fuel cells: characterization and performance comparison. Renew Energ 126:485–494
15. Tiliakos A, Trefilov AM, Tanasă E, Balan A, Stamatin I (2020) Laser-induced graphene as the microporous layer in proton exchange membrane fuel cells. Appl Surf Sci 28:504:144096
16. Sial MA, Din MA, Wang X (2018) Multimetallic nanosheets: synthesis and applications in fuel cells. Chem Soc Rev 47(16):6175–6200
17. Mori Y, Ando T, Matsumoto T, Yatabe T, Kikkawa M, Yoon KS, Ogo S (2018) Multifunctional catalysts for H2O2-resistant hydrogen fuel cell. Angew Chem 130(48):16018–16022
18. Qin B, Yu H, Jia J, Jun C, Gao X, Yao D, Sun X, Song W, Yi B, Shao Z (2018) A novel IrNi@ PdIr/C core–shell electrocatalyst with enhanced activity and durability for the hydrogen oxidation reaction in alkaline anion exchange membrane fuel cells. Nanoscale 10(10):4872–4881
19. Hosseinabadi P, Hooshyari K, Javanbakht M, Enhessari M (2019) Synthesis and optimization of nanocomposite membranes based on SPEEK and perovskite nanoparticles for polymer electrolyte membrane fuel cells. New J Chem 43(41):16232–16245
20. Dincer I, Rosen MA (2012) Exergy: energy, environment and sustainable development. Newnes 2012
21. Dinçer I, Zamfirescu C (2011) Hydrogen and fuel cell systems. In: Sustainable energy systems and applications. Springer, Boston, pp 519–632
22. Dyer CK, Moseley PT, Ogumi Z, Rand DAJ, Scrosati B, Garche J (2009) Encyclopedia of electrochemical power sources. Elsevier, Amsterdam, 3,4538
23. Nguyen HV, Othman MR, Seo D, Yoon SP, Ham HC, Nam SW, Han J, Kim J (2014) Nano Ni layered anode for enhanced MCFC performance at reduced operating temperature. Int J Hydrog Energy 39(23):12285–12290
24. Accardo G, Frattini D, Yoon SP, Ham HC, Nam SW (2017) Performance and properties of anodes reinforced with metal oxide nanoparticles for molten carbonate fuel cells. J Power Sources 370:52–60
25. Frattini D, Accardo G, Moreno A, Yoon SP, Han JH, Nam SW (2017) Strengthening mechanism and electrochemical characterization of ZrO2 nanoparticles in Nickel–Aluminum alloy for molten carbonate fuel cells. J Ind Chem Eng 56:285–291
26. Liu J, Ye K, Cheng K, Wang G, Yin J, Cao D (2014) The catalytic effect of CeO2 for electrochemical oxidation of graphite in molten carbonate. Electrochim Acta 135:270–275
27. Garche J (2009) Encyclopedia of electrochemical power sources
28. Lim Y, Lee H, Hong S, Kim YB (2019) Co-sputtered nanocomposite nickel cermet anode for high-performance low-temperature solid oxide fuel cells. J Power Sources 412:160–169
29. Shaheen K, Shah Z, Gulab H, Hanif MB, Faisal S, Suo H (2020) Metal oxide nanocomposites as anode and cathode for low temperature solid oxide fuel cell. Solid State Sci 102:106162
30. Arshad MS, Raza R, Ahmad MA, Abbas G, Ali A, Rafique A, Ullah MK, Rauf S, Asghar MI, Mushtaq N, Atiq S (2018) An efficient Sm and Ge co-doped ceria nanocomposite electrolyte for low temperature solid oxide fuel cells. Ceram Int 44(1):170–174
31. Baldi F, Wang L, Pérez-Fortes M, Maréchal F (2018) A cogeneration system based on solid oxide and proton exchange membrane fuel cells with hybrid storage for off-grid applications. Front Energy Res 6:139

32. Huang PS, Gao T (2018) Current development of 1D and 2D metallic nanomaterials for the application of transparent conductors in solar cells: fabrication and modelling. Nano-Struct Nano-Objects 15:119–139
33. Gong S, Zhao Y, Yap LW, Shi Q, Wang Y, Bay JA, Lai DT, Uddin H, Cheng W (2016) Fabrication of highly transparent and flexible nanomesh electrode via self-assembly of ultrathin gold nanowires. Adv Electron Mat 2(7):1600121
34. Wisser FM, Eckhardt K, Nickel W, Böhlmann W, Kaskel S, Grothe J (2018) Highly transparent metal electrodes via direct printing processes. Mat Res Bull 8:231–234
35. Yu X, Sivula K (2016) Toward large-area solar energy conversion with semiconducting 2D transition metal dichalcogenides. ACS Energy Lett 1(1):315–322
36. Kang SB, Kwon KC, Choi KS, Lee R, Hong K, Suh JM, Im MJ, Sanger A, Choi IY, Kim SY, Shin JC (2018) Transfer of ultrathin molybdenum disulfide and transparent nanomesh electrode onto silicon for efficient heterojunction solar cells. Nano Energy 50:649–658
37. Casaluci S, Gemmi M, Pellegrini V, Di Carlo A, Bonaccorso F (2016) Graphene-based large area dye-sensitized solar cell modules. Nanoscale 8(9):5368–5378
38. Kweon DH, Baek JB (2019) Edge-functionalized graphene nanoplatelets as metal-free electrocatalysts for dye-sensitized solar cells. Adv Mat 31(13):1804440
39. Loh KP, Bao Q, Eda G, Chhowalla M (2010) Graphene oxide as a chemically tunable platform for optical applications. Nat Chem 2(12):1015
40. Sun X, Lin T, Song Q, Fu Y, Wang Y, Jin F, Zhao H, Li W, Su Z, Chu B (2017) Improved performance of hole-transporting layer-free perovskite solar cells by using graphene oxide sheets as the nucleation centers. RSC Adv 7(72):45320–45326
41. Lin Y, Bai Y, Fang Y, Chen Z, Yang S, Zheng X, Tang S, Liu Y, Zhao J, Huang J (2018) Enhanced thermal stability in perovskite solar cells by assembling 2D/3D stacking structures. J Phys Chem Lett 9(3):654–658
42. Zuo C, Scully AD, Vak D, Tan W, Jiao X, McNeill CR, Angmo D, Ding L, Gao M (2019) Self-assembled 2D perovskite layers for efficient printable solar cells. Adv Energ Mat 9(4):1803258
43. Fan Q, Zhu Q, Xu Z, Su W, Chen J, Wu J, Guo X, Ma W, Zhang M, Li Y (2018) Chlorine substituted 2D-conjugated polymer for high-performance polymer solar cells with 13.1% efficiency via toluene processing. Nano Energy 48:413–420
44. Song M, You DS, Lim K, Park S, Jung S, Kim CS, Kim DH, Kim DG, Kim JK, Park J, Kang YC (2013) Highly efficient and bendable organic solar cells with solution-processed silver nanowire electrodes. Adv Funct Mater 23(34):4177–4184
45. Sun X, Chen T, Yang Z, Peng H (2013) The alignment of carbon nanotubes: an effective route to extend their excellent properties to macroscopic scale. Acc Chem Res 46(2):539–549
46. Lee S, Jang J, Park T, Park YM, Park JS, Kim YK, Lee HK, Jeon EC, Lee DK, Ahn B, Chung CH (2020) Electrodeposited silver nanowire transparent conducting electrodes for thin-film solar cells. ACS App Mater Interfaces 12(5):6169–6175
47. Sun H, Deng J, Qiu L, Fang X, Peng H (2015) Recent progress in solar cells based on one-dimensional nanomaterials. Energy Environ Sci 8(4):1139–1159
48. Sun J, Yang X, Zhao L, Dong B, Wang S (2020) Ag-decorated TiO2 nanofibers for highly efficient dye sensitized solar cell. Mater Lett 260:126882
49. Wang Y, Zhao H, Mei Y, Liu H, Wang S, Li X (2018) Carbon nanotube bridging method for hole transport layer-free paintable carbon-based perovskite solar cells. ACS App Mater Interfaces 11(1):916–923
50. Jo SB, Lee WH, Qiu L, Cho K (2012) Polymer blends with semiconducting nanowires for organic electronics. J Mater Chem 22(10):4244–4260
51. Semonin OE, Luther JM, Choi S, Chen HY, Gao J, Nozik AJ, Beard MC (2011) Peak external photocurrent quantum efficiency exceeding 100% via MEG in a quantum dot solar cell. Science 334:1530–1533
52. Peng W, Du J, Pan Z, Nakazawa N, Sun J, Du Z, Shen G, Yu J, Hu JS, Shen Q, Zhong X (2017) Alloying strategy in Cu–In–Ga–Se quantum dots for high efficiency quantum dot sensitized solar cells. ACS App Mater Interfaces 9(6):5328–5336

53. Selopal GS, Zhao H, Wang ZM, Rosei F (2020) Core/shell quantum dots solar cells. Adv Funct Mater 30(13):1908762
54. Liu T, Yu K, Gao L, Chen H, Wang N, Hao L, Li T, He H, Guo Z (2017) A graphene quantum dot decorated SrRuO 3 mesoporous film as an efficient counter electrode for high-performance dye-sensitized solar cells. J Mater Chem A 5(34):17848–17855

# Chapter 9
# Nanocellulose for Gas Sensor Applications

**Vijaykiran N. Narwade, Hanuma Reddy Tiyyagura, Yasir Beeran Pottathara, Madhuri A. Lakhane, Indrani Banerjee, Vipul V. Kusumkar, Eva Viglašová, Michal Galamboš, Ravindra U. Mene, and Kashinath A. Bogle**

## 1 Introduction

Sensors gain the worldwide interest of R&D due to factors such as monetary investment, the declared literature and the figure of relevant scientists acting in environmental issues in the last few decades [27, 28, 32, 45, 97]. Mostly, sensor devices take the help of a wide range of transducer and signal transformation units with subsequent modifications in technological problems. Sensor technology includes a wide spectrum of sensor devices from simple temperature measurement thermocouple to erudite optical systems for bacteria or virus detection and identifications [9, 42, 80, 82]. The sensing techniques like mechanical, semiconductor, electrochemical, biosensing and optical make their important impression in the prominent

---

V. N. Narwade (✉) · K. A. Bogle (✉)
School of Physical Sciences, Swami Ramanand Teerth Marathwada University, Nanded, Maharashtra 431606, India

H. R. Tiyyagura · Y. B. Pottathara
Faculty of Mechanical Engineering, University of Maribor, Smetanova ul. 17, 2000 Maribor, Slovenia

M. A. Lakhane
St. John College of Engineering and Management, Palghar, Maharashtra, India

I. Banerjee
School of Nano Sciences, Central University of Gujarat, Gandhinagar 382030, India

V. V. Kusumkar · E. Viglašová · M. Galamboš
Faculty of Natural Sciences, Department of Nuclear Chemistry, Comenius University, Mlynska dolina, Ilkovičova 6, Bratislava 84215, Slovakia

R. U. Mene
PDEA'S Annasaheb Magar Mahavidyalaya Hadapsar, Pune 411028, India



M. Mujawar et al. (eds.), *Nanotechnology for Electronic Applications*, Materials Horizons: From Nature to Nanomaterials, https://doi.org/10.1007/978-981-16-6022-1_9

fields that include health and environment [13, 22, 50] such that most of the government bodies have mandatory approval for the sensors required in the field of eco-friendly observing and other related issues. Sensors play a dynamic role in tracking the hazardous gases or vapours in the workplaces and industries as well as it helps us in monitoring impurities in normal waters by industrial wastes and pesticides overflow from agriculture arenas as the function of a sensor is to provide up-to-date information on our surroundings [7, 68]. Thus, a near-future sensor industry revolution may swiftly bring innovations in sensor devices for medical, environmental and many more fields. A chemical sensor device comprises a transducer unit with chemical bodies and a chemical layer provides data about its surroundings. Thus, nowadays bio polymer-based sensors have been investigated more by many of the researchers from sensing areas.

Biopolymers are sustainable polymers synthesized from renewable supplies as an alternative to conventional fossils [60]. The carbon-neutral nature and carbon offset are the two key features of biopolymers that have zero impact on the environmental $CO_2$ level after their combustion. Due to the wide availability of cellulose biopolymer in nature, the utilization of cellulose attracted many areas of research. Cellulose is the major constituent of plant physiology present in the plant cell walls [24, 40]. It is nearly more than 70% of the main constituent of cotton, jute and also have content in agricultural by-products [15, 75]. Cellulose polymer is primarily a polysaccharide, having molecular repeat unit couple of d-anhydroglucose ring units linked by β-1 → 4 glycoside oxygen linkages which are flexible bonds that can twist and bend the polymeric chain [5]. Trees, bacteria, plants and algae are commonly known wide sources of cellulose. The extraction and processing methods change the form of cellulose II to cellulose III. Considering technological pertinence, cellulose II has been the most stable [10, 100]. This material shows extra perks such as its readily biodegradable nature; it has low carbon waste and it does not cause any nuisance to the ecosystem [19, 78]. Currently, cellulose has exhibited its capabilities in various engineering applications.

Cellulose which is mass-produced from the plant is chosen to be an ideal source due to less cost, physical–chemical properties as well as sustainable characteristics [94]. The network structure of cellulose supports the external force applied to it. In addition, it possesses physical properties like good toughness, high strength nature and elevated thermal stability [23]. Due to the noteworthy characteristics of cellulose biopolymer, it is broadly widening its area to develop from laboratory research to industrial applications. The present cellulose research findings mostly stressed on functionalization by means of different ways such that it can efficiently produce superior economic usages and having a wide array of application sectors including sensor fields.

Cellulose at its nano level is termed nanocellulose (NC) which has a diameter of up to a hundred nanometres, whereas the length may vary from nanometres to some micrometre scale [52]. The network structure of cellulose as mentioned earlier is the key factor for the use of cellulose in the sensor industry.

Cellulose (Fig. 1—chemical structure) having the chemical formula (C6H10O5)n consists of repeated units of the monomeric glucose [83]. Nano form of cellulose

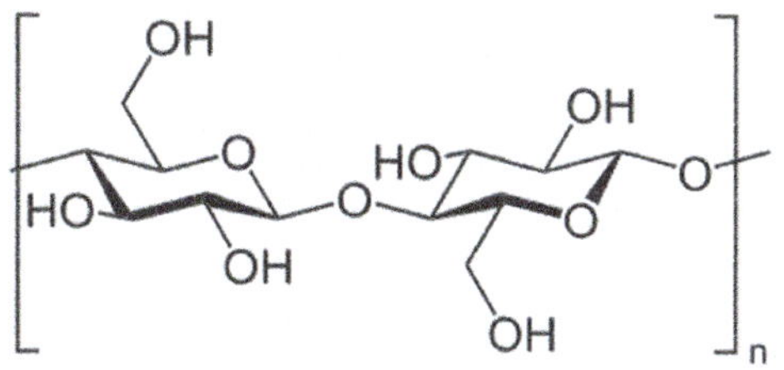

**Fig. 1** Cellulose chemical structure

causes to possess great properties over bulk cellulose form in terms of extraordinary strength and large surface to volume ratio, larger surface area, which makes NC potential and efficient matrix material for sensor devices over the metal oxides and carbon materials. Mechanical and chemical treatments procedures [69] are well-known methods used for the conversion of bulk cellulose materials into the different forms of NC (Fig. 2—classification of nanocellulose), viz., (i) cellulose nanocrystals (CNC), (ii) cellulose nanowhiskers (CNWs), (iii) cellulose nanofibers (CNF) and (iv) bacterial cellulose (BNC).

In this chapter, the application of cellulose, its derivatives and composites for the detection and sensing of gases such as volatile organic compounds (VOC), ammonia ($NH_3$), nitrogen dioxide ($NO_2$) etc. are studied, as well as the chapter extensively focuses on the sensing mechanism of the cellulose-based materials.

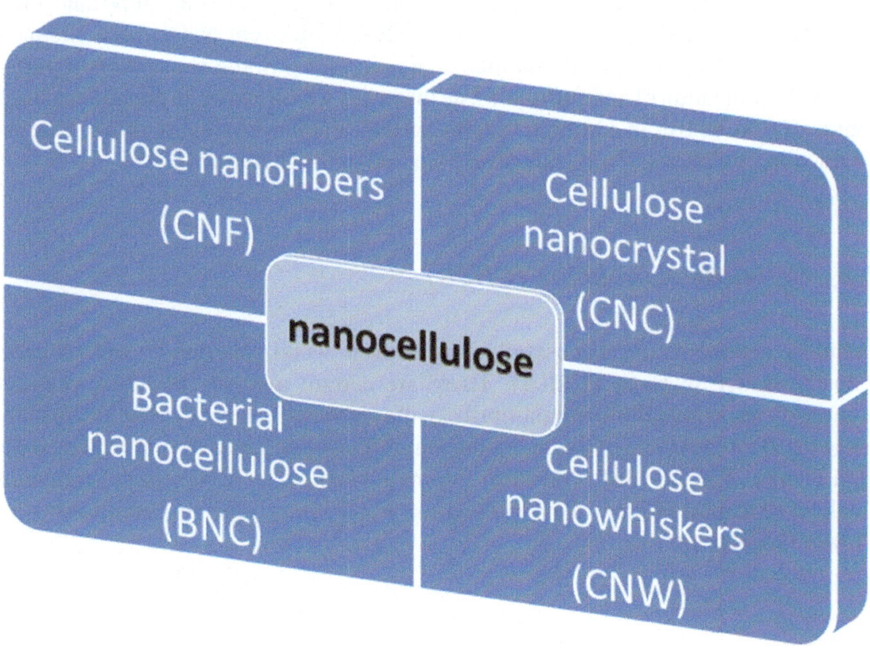


**Fig. 2** Nanocellulose broad classification

# 2 Nanocellulose Synthesis

The versatile cellulose has applications in various fields. In addition, materials scientists are still occupied to expand its usage, hence it is of paramount importance to produce a sustainable helpful process like pre-treatment. To yield pure nanocellulose from pre-treatment processes essentials steps are required. A lot of literature is available which stated the synthesis of nanocellulose by means of the methodology of pre-treatments. The preparation of nanocellulose is mainly carried out by mechanical pre-treatment or chemical pre-treatment methods [18]. Cryocrushing and refining mechanical pre-treatment technique involves important steps such as refining in the tank and crushing under inert N2 surrounding to neglect contamination [98].

On the basis of physical properties like shape and characteristics, the sources of cellulose synthesis are prescribed in the following sections.

## 2.1 *Cellulose Nanocrystals (CNCs)*

To date, various synthetic methods have been employed to produce the CNCs. Some of the methods are summarized: one of the methods reports the synthesis of acid-free cellulose nanocrystals by TEMPO oxidation followed by cavitation [110]. Microcrystalline cellulose, which is generally termed MCC, is also an important class of cellulose materials. For CNC synthesis, MCC underwent NaOH/urea chemical treatment followed by revival and then sonication [26]. Some researchers reported successful utilization of agricultural waste materials which can be an abundant natural source of CNCs [16]. In addition, various agricultural bio-wastes are utilized for the production of CNCs and are extensively studied in the literature [63].

## 2.2 *Cellulose Nanofibers (CNFs)*

Cellulose nanofibers are successfully derived from wood materials as well as paper industry sludge material [1, 108]. There are certain methods like high-pressure homogenization for cellulose nanofibrils synthesis which can prepare cellulose in the range of 10–50 nm scale of width but the major drawback for these methods includes high energy demand [38]. The most common method for CNFs is pre-treatment catalytic oxidation of plant-derived cellulose fibres under aqueous conditions using TEMPO reagent [21, 73, 79].

### 2.3 *Microfibrilillated Cellulose (MFC)*

Mechanical shearing with a combination of high-pressure homogenization is effectively used for MFCs and strong gel synthesis with enzyme treatment [39]. Wood is a commonly known source of cellulose synthesis. The composites of nanocellulose polypyrrole with MFCs have been successfully derived from wood sources [59]. Used tissue papers waste materials are considering to be a potential source to synthesize MFCs by way of acid hydrolysis (Kim and Kim). Cryocrushing and high shear refining are also reported methods to synthesize [84].

### 2.4 *Cellulose Nanowhiskers (CNWs)*

CNWs are a special class of cellulose-based materials. The production of CNWs has been explored using different agricultural residues [49], such as maize stalk residue [53], and other types of plant-based sources like bamboo fibres [67] and coconut husk [20].

### 2.5 *Bacterial Nanocellulose (BNC)*

The BNC can be prepared from economically viable sources, including agricultural-based industrial waste [11, 93]. The use of waste cotton fabrics is also reported for the preparation of BNC [29]. Gluconacetobacter xylinus used for the cellulosic saccharification of waste fabric via enzymatic pre-treatment route is also reported and found to be a viable source for the production of BNC [72].

## 3 Gas-Sensing Application of Cellulose-Based Composites

An extensive contribution in the field of designing sensor devices for gas sensing areas with respect to that thorough study on the material preparation and its structure and properties was also explored in the previous studies [4, 90, 92]. The rise of research interest for the development of sensor devices emerged from immediate issues in terms of environmental concerns and their sustainability. Gas sensors are mainly developed and fabricated using (a) semiconductor metal oxides and (b) organic conducting polymers due to their theoretically considered structure and properties.

The gas-sensing ability and the respective properties of metal oxides, mainly tin oxide [99, 103], titanium dioxide [8, 41], zinc oxide [33, 85], tungsten trioxide [48, 87] or ferrous oxide [12, 102] materials, are possibly raised by introducing metal entities, and the procedure generally termed to be doping [47, 96]. Similarly, in organic conducting polymers, polythiophene (PPT), polypyrrole (PPy) and polyaniline (PANI) [34, 62, 104] have some drawbacks associated with their structure, and the electronic properties can be resolved. Metal-oxide-based semiconductors suffer from instability due to the surface ionic conducting charge of metal oxides, while on the other side, high-density electrons in the backbone of conducting polymer make it unstable. High working temperature is a major shortcoming associated with semiconductor metal oxides while conducting polymer materials have low operating temperature as an advantage.

Other considerable major drawback polymer materials possess which are not good for sensor devices developments, and the drawback includes low response time for gas molecules and a very long recovery time to set to original conditions and it has low thermal stability. Research articles show materials scientists reported ample literature on the combination of polymers and metal oxides so as to get the synergetic effect of both and if its composites really work, it will surely go for engineering applications because of the properties it achieves [81, 88]. Cellulose biopolymer as described earlier is a versatile material, and it has a tendency to modify with other conductive polymers or simply with other metal oxides; in addition, cellulose can be effectively modified with chemical treatments called functionalization to be used in gas sensor devices. Because of the limited applications of inorganic particles, thus cellulose can work as a matrix material for holding inorganic particles in it so as to achieve desire sensor material properties like high flexible nature and helpful as a portable gas sensor. The hydroxyl group on cellulose structure makes it perfect to functionalize it for desire sensor device application [57, 58, 65]. The schematic representation shows applicability in a wide range of target gases (Fig. 3).

In the following sections, a detail explanation is given for gas sensors based on cellulose-based materials.

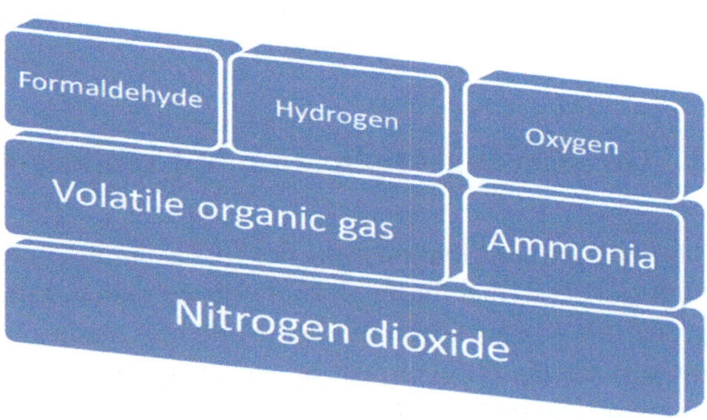


**Fig. 3** Applicability of cellulose-based sensors for a wide range of gases

### 3.1 Ammonia Sensor

An acquaintance of ammonia can lead to skin irritation, respiratory syndrome and headache, with long-term inhalation may be disastrous to individuals [2, 89, 95].

In one study, ammonia vapours were exposed in the presence of composite materials like cellulose-PANI and a hybrid system of cellulose-titanium dioxide and PANI nanofibers conveyed by Pang et al. [61]. The composite material cellulose/$TiO_2$/PANI displayed a better reaction to ammonia gas related to cellulose/PANI composite since the formation of p–n heterojunction amongst the PANI and titanium dioxide particles [61].

Carbon nanotube and cellulose paper-based bendable ammonia sensor described by Han et al. [25] have credited capability in the measure of 10–100 ppm with the detection limit of about 5 ppm. Additional greater property of this paper sensor encompassed consistency and consecutive repeatability likened to glass-built sensor materials, and thus it will be perhaps castoff in upcoming electronic tenders [25].

The multicomponent hybrid nanocomposite scheme of cellulose and with hydrothermally synthesized titanium dioxide and multiwalled carbon nanotube composites were synthesized by means of modest blending method by Mun S. et al. 2012 and consequently exploited as sensor material for room temperature ammonia sensing with the gas-observing range of 50–500 ppm. The projected sensor displays virtuous device characteristics in terms of sensitivity and repeatability [54].

Dai et al. [14] functioned on established an ammonia gas sensor by cholesteric CNC films with doping of copper salt. The highest red shift of 75 nm was detected at 225 mmol/g copper doping in cellulose nanocrystal composite because of the involvement of the steric space properties of copper ions. The assimilation of copper ions, the characteristic property of the calorimetric film sensor can regulate or moderate the colouration and morphology. The kinetic models of ammonia gas sensing to set up a quantitative correlation amid reflective wavelength shift and the entire ammonia initiation during the time were anticipated conferring to the real-time dependence response for $\lambda max$ [14].

Jia et al. [36] urbanized an innovative sensor by framing polyethylenimine and graphene-oxide-layered cellulose acetate nanofibers on a quartz crystal microbalance (CA/PEI/GO-based QCM) through electrospinning and LBL methods. This allows ammonia adsorption on the 3D permeable space of the CA nanofibrous membrane, and also the –COOH group on nanofiber provides necessary secondary bond interactions with $NH_3$ gas for steady adsorption. The resultant materials show impressive recognition ability and constancy of the QCM sensor. Besides, a minimal detection limit (1 ppm) and a rapid response ($<10$ s) were also noticed. Thus the robust features, being as reversible use, enhanced selectivity and reduced the unique ammonia gas sensor [36].

Jia et al. [37] magnificently formed a unique cellulose acetate/polyethylenimine/poly (acrylic acid) nanofibers/quartz crystal microbalance (CA/PEI/PAA nanofibers/QCM) ammonia gas sensor. In this work, the different ratios of PAA fillings are used to prepare nanofibrous CA/PEI/PAA membranes by electrospinning and a two-step LBL procedure. Finally, the prepared composite showed an ammonia gas disclosure with the exceptional 1 ppm level [37].

In another study, a smart vapochromic xerogel based on the microporous cellulose (CMC-BCM) was developed with a minimal detection of $9.0 \times 10^{-2}$ ppb for real-time ammonia contact by Khattab et al. [43]. As the straight detection technique to sense the ammonia is revocable, the porosive structure is additionally ready to use. Ammonia can be easily traced both in water and air samples with the auspicious CMC-BCM xerogel template. However, ammonia gas contact outcomes can be revealed in terms of colour change from yellow to purple. So CMC-BCP calorimetric indicator is a superficial technique for ammonia tracing by comparing with earlier reported approaches in which advanced electronic elements were used [43].

Jia et al. [35] developed composites based on cellulose for radiometric fluorescence, for monitoring the freshness of shrimp and crab in a real time using enhanced amine-response. The cellulose chains linked by reactive hydroxyl groups are exploited to covalently confine the fluorescein isothiocyanate (FITC) as display and protoporphyrin IX (PpIX) as interior reference onto cellulose acetate (CA), correspondingly. These composites show a sensitive, colour-receptive, quick and direct response to ammonia in a varied range of 5.0 ppm to $2.5 \times 10^4$ ppm [35].

Han et al. [25] prepared composites based on carbon nanotube (CNT) and cotton yarn for the detection of $NH_3$ gas for smart textile. The fabricated sensors showed a detection limit of around 8 ppm which follows OSHA standards. The material possesses high flexibility and adhesion, thanks to the hydrogen bonding between CNT and cotton yarn without losing their stability against chemicals. The cost-effective and high-capacity dispersed CNT-cotton yarns can be rendered as a promising agent [25].

Anju [3] used the solution method to prepare cellulose nanofibers/polyaniline/polyvinyl alcohol (PANI/CNF/PVA and PANI/NFC/PVA) composites with the help of ultra-sonication-assisted mixing of PANI-coated CNF/NFC fibres with the polyvinyl alcohol solution. The prepared material showed excellent sensing ability for the humidity as well as for $NH_3$ at room temperature due to the synergic effect of PANI/CNF and PVA [3].

Narwade et al. [56] designed cellulose nanofibrils (CNFs)/hydroxyapatite (HAp) composites with numerous weight percentages of nanoparticles structured into a dissimilar morphological composite. The composites were prepared using the freeze drying method and showed high sensitivity maximum of up to 575% for $NH_3$ gas with lowest detection limit around 5 ppm, response time 210 s and recovery time 30 s at room temperature [56].

### *3.2 Volatile Organic Vapour Sensor*

Prolonged exposure to volatile organic vapour (VOC) of alcohol is responsible for major health issues such as mental and physical diseases. Largely, the research is focused on the occupational exposure of alcohol and general use of alcohol-related products for hygiene maintenance such as hand sanitiser; but in recent literature new source was discovered of VOC alcohol exposure through e-cigarettes containing alcohol called e-liquid; in this case, the cellulose-based sensors VOC alcohol detection found to be an impressive tool for maintaining the exposure situations [51, 71, 76, 101].

Yun [106] developed cellulose-based composites using MWCNTs and imidazolides. These composites were developed using the formation of covalent bonds between MWCNTs and imidazolides. MWCNTs and cellulosic polymeric chains are aligned using mechanical stretching in the second part of preparation. The alignment helps in enhancing the mechanical stability of material as well as sensing ability via improving the electron pathways. The composite was tested with four polar solvent analytes which showed quick response stability against different solvents and response with flexibility while varying concentrations of analytes resulted in a differentiated response by the composite. A study revealed that linear response with changing concentration can prove MWCNTs/cellulose composite material as a promising tool for detection of 1-propanol [106].

In a recent study, Zhang [107] prepared a cost-effective and used simple procedure for the production of gas sensors to detect VOCs. The material is developed using graphene and ethyl cellulose (EC) in the ratio 20:40%, respectively. Graphene/EC materials possess high flexibility and showed excellent conductivity, thanks to graphene's electrical properties. The novel method used a gas-sensing device for reducing strain sensitivity by adopting a suspended structure and created a hole in PET substrate and also fixed graphene/EC film on electrodes that are prefabricated on its surfaces. The sensor shows relative resistance change within 0.3% in response to a minimum bending radius of 3.18 mm (corresponds to maximum strain) and around 0.2% relative resistance change during 400 bending cycle measurements with a bending radius of 5 mm, demonstrating its limited strain response. In contrast, the sensor possesses very large responses towards VOCs, with detection limits ranging from 37 to 167 ppm by considering the strain-induced noise level. The sensor can be used for wearable devices as it is stable against the normal hand gesture and movement and showed good response in ambient conditions [107].

Qi et al. [66] developed a cellulose-based aerogel for the detection of VOC. The material was prepared using CNT and cellulose, and this electrically conductive sensor possesses a simple design, efficient and high sensitivity, reversibility and reproducibility. The mechanism of sensing depends upon the change in resistance accompanied by the adsorption of solvent on the surface of composites which increase the distance between the separate nanotubes. The aerogels having a 3D porous structure and a high surface area for capturing the analytes are not possible in a common type of solid CNT-polymer composites. The rapid response of aerogel due to their high capacity of response and considerable change in resistance provides a wider scope of analyte for its detection [66].

In another study, $KMnO_4$ is used for doping in bacterial cellulose (BC). The different ratios of concentration of $KMnO_4$ are used and their optical and physical parameters are studied for the determination of ethylene in banana packaging by Pirsa

[64]. The fabricated $KMnO_4$-doped BC was used for the determination of ethylene concentration. The study is based on the change in clarity of the film as well as colour change, and also the absorbance of a film with varying ethylene concentration (10–2000 $\mu g\ mL^{-1}$) is studied using a spectrophotometer. Finally, the ethylene optical sensor was used for the detection and determination of ethylene concentration in the banana packages. The study concluded that temperature and ethylene gas production in the banana packaging are easily monitored using a spectrophotometer [64].

Aparicio-Martínez et al. [6] reported nanostructured material composed of a conductive polymeric matrix of PANI and WO3 doped with cellulose was applied for the detection of acetone at room temperature and low concentrations. The developed sensor showed a limit detection of 10 ppm after repeated exposure to acetone, which proves the suitability of the developed material for room temperature sensing of acetone without the major shortcomings of larger systems required by high-operating temperatures [6].

### 3.3 $NO_2$ Gas Sensor

Nitrogen dioxide is a vital air pollutant, and its exposure resulted in tragedies such as heavy breathing, cough, common cold, flu and bronchitis [86, 105, 109].

Sadasivuni et al. [77] intended room temperature $NO_2$ sensor by means of cellulose nanocrystals (CNCs) with the capability of sensing at a low level in less time. The work has demonstrated growth of iron oxide nanoparticles on the CNC surface which make dissimilar sensing material appropriate in a numeral of drives counting safekeeping, and leakage finding in laboratories. Iron oxide and cellulose nanocrystals were showed as ultrasensitive sensors for their great surface-to-volume ratio by way of excellent physical and electrical properties. So, these devices are appropriate for numerous tenders owing to their potentials like an elevated response, elasticity, room temperature procedure and cost-effectiveness.

The CNT/$WS_2$ composite gas sensor was developed on the cellulose substrate. Owing to its modest and speedy manufacturing procedure, the sensor can be used as a throwaway device designed by Lee [46]. The functionalization of $WS_2$ on the CNTs progresses the $NO_2$ sensing response, and by taking the benefit of the cellulose paper, the sensor is exceedingly bendable and twistable. These qualities can be protracted to other applications [46].

### 3.4 Formaldehyde Gas Sensor

Formaldehyde can cause a nuisance to human health upon contact [17, 55, 74]. Hu et al. [31] have invented a novel quartz crystal microbalance layered with nanofibrous polyethylenimine/bacterial cellulose membrane-coated (PEI/BC) gas sensor and exposed to formaldehyde at room temperature The technologically advanced sensors demonstrated decent linear affiliation for higher formaldehyde concentrations. Furthermore, the sensing properties were mostly pretentious by the content of PEI constituent in nanofibrous membranes, the concentration of formaldehyde and comparative humidity. The study correspondingly demonstrated that the sensor had worthy choosiness to formaldehyde further readily than former conservative volatile organic compounds. The applied method for fabricating of sensors was completed and is a cost-effective probable for mass production.

### *3.5 LPG Gas Sensor*

Polyaniline-carboxymethyl cellulose (PANI and PANI-CMC) composite has been manufactured by chemical polymerization technique. A significant consequence of the LPG sensing reveals that the sensitivity of the composite is maximum compared with pristine PANI. Investigation on its steadiness, retrieval and response time has demonstrated that the composite has the competence to advance a well-organized LPG sensing device that can be functioned at room temperature in addition to low-cost and easily operatable [70].

### *3.6 Hydrogen Gas Sensor*

A new method to modify the morphology of Pd nanocomposites by changing the counterparts of sodium carboxymethyl cellulose to Pd precursor, regarding the usage of a continuous-flow microfluidic platform, was testified by SwaminathanáIyer [91]. Hydrogen sensors commissioning numerous Pd nanostructures displayed the variation in present responses, reliant on the nature of the Pd nanocomposites. Meaningfully, the devices are tough in gaining improved firmness throughout testing [91].

### *3.7 Oxygen Gas Sensor*

Hu et al. [30] developed gold electrode arrays on porous cellulose membranes by using the inkjet printing method. This is trailed in succeeding PGEAs infatuated various sole characteristics for growing thin-film electrochemical gas sensors, e.g., excellent conductivity, outstanding flexibility, simple component and cost-effective. The ionic liquids that can be used as electrolytes for PGEAs created a paper-based solid-state electrochemical oxygen sensor, which seemed like a paper slice but assimilated analytical presentation [30].

## 4 Conclusion

Because of its multifunctional properties, cellulose can be employed in different application sectors. Herewith, we have demonstrated its usability in the gas sensor domain. Because of cellulose structure, it will provide more modification sites for desired application point of view. It leads to modification cellulose biopolymers causing synergetic effects through gas sensing and gas detection procedures. Furthermore, cellulose sensors reveal the following features: flexibility, cheap and easy to manufacture choice. Perilous gases like VOCs, $NO_2$, $NH_3$ etc. gases have been described as competently sensed using cellulose-based composite sensors. Thus, we can say that cellulose-based materials will be subjugated over all ritual sensor tools.

## References

1. Adu C, Berglund L, Oksman K et al (2018) Properties of cellulose nanofibre networks prepared from never-dried and dried paper mill sludge. J Clean Prod 197:765–771
2. Anjana NS, Amarnath A, Nair MVH (2018) Toxic hazards of ammonia release and population vulnerability assessment using geographical information system. J Environ Manag 210:201–209
3. Anju VP, Jithesh PR, Narayanankutty SK (2019) A novel humidity and ammonia sensor based on nanofibers/polyaniline/polyvinyl alcohol. Sens Actuators A Phys 285:35–44
4. Anjum SR, Narwade VN, Bogle KA, Khairnar RS (2018) Graphite doped Hydroxyapatite nanoceramic: selective alcohol sensor. Nano-Struct Nano-Objects 14:98–105
5. Ansell MP, Mwaikambo LY (2009) The structure of cotton and other plant fibres. In: Handbook of textile fibre structure. Elsevier, pp 62–94
6. Aparicio-Martínez E, Osuna V, Dominguez RB et al (2018) Room temperature detection of acetone by a PANI/cellulose/WO3 electrochemical sensor. J Nanomater 2018
7. Basatneh R, Najafi B, Armstrong DG (2018) Health sensors, smart home devices, and the internet of medical things: an opportunity for dramatic improvement in care for the lower extremity complications of diabetes. J Diabetes Sci Technol 12:577–586
8. Benkstein KD, Semancik S (2006) Mesoporous nanoparticle TiO2 thin films for conductometric gas sensing on microhotplate platforms. Sens Actuators B Chem 113:445–453
9. Briciu-Burghina C, Heery B, Duffy G et al (2019) Demonstration of an optical biosensor for the detection of faecal indicator bacteria in freshwater and coastal bathing areas. Anal Bioanal Chem 411:7637–7643
10. Brown RM Jr (2004) Cellulose structure and biosynthesis: what is in store for the 21st century? J Polym Sci Part A Polym Chem 42:487–495
11. Castro C, Zuluaga R, Putaux J-L et al (2011) Structural characterization of bacterial cellulose produced by Gluconacetobacter swingsii sp. from Colombian agroindustrial wastes. Carbohydr Polym 84:96–102
12. Chen J, Xu L, Li W, Gou X (2005) α-Fe2O3 nanotubes in gas sensor and lithium-ion battery applications. Adv Mater 17:582–586
13. Coulet PR, Blum LJ (2019) Biosensor principles and applications. CRC Press
14. Dai S, Prempeh N, Liu D et al (2017) Cholesteric film of Cu (II)-doped cellulose nanocrystals for colorimetric sensing of ammonia gas. Carbohydr Polym 174:531–539
15. de Amorim JDP, de Souza KC, Duarte CR et al (2020) Plant and bacterial nanocellulose: Production, properties and applications in medicine, food, cosmetics, electronics and engineering. A review. Environ Chem Lett 18:851–869

16. de Oliveira FB, Bras J, Pimenta MTB et al (2016) Production of cellulose nanocrystals from sugarcane bagasse fibers and pith. Ind Crops Prod 93:48–57
17. Delikhoon M, Fazlzadeh M, Sorooshian A et al (2018) Characteristics and health effects of formaldehyde and acetaldehyde in an urban area in Iran. Environ Pollut 242:938–951
18. Dufresne A (2013) Nanocellulose: a new ageless bionanomaterial. Mater Today 16:220–227
19. Eichhorn SJ, Rahatekar SS, Vignolini S, Windle AH (2018) New horizons for cellulose nanotechnology. Philos Trans R Soc A Math Phys Eng Sci 376:20170200
20. Fahma F, Iwamoto S, Hori N et al (2011) Effect of pre-acid-hydrolysis treatment on morphology and properties of cellulose nanowhiskers from coconut husk. Cellulose 18:443–450
21. Fukuzumi H, Saito T, Iwata T et al (2009) Transparent and high gas barrier films of cellulose nanofibers prepared by TEMPO-mediated oxidation. Biomacromol 10:162–165
22. García JA, Monzón-Hernández D, Bustos E (2017) Fiber-optic sensor networks: environmental applications. Undergr Sens Monit Hazard Detect Environ Infrastruct 322
23. George J, Sabapathi SN (2015) Cellulose nanocrystals: synthesis, functional properties, and applications. Nanotechnol Sci Appl 8:45
24. Gupta PK, Raghunath SS, Prasanna DV et al (2019) An update on overview of cellulose, its structure and applications. Cellulose 846–1297
25. Han J-W, Kim B, Li J, Meyyappan M (2013) A carbon nanotube based ammonia sensor on cotton textile. Appl Phys Lett 102:193104
26. Hashaikeh R, Abushammala H (2011) Acid mediated networked cellulose: preparation and characterization. Carbohydr Polym 83:1088–1094
27. Hassan HS, Elkady MF (2020) Semiconductor nanomaterials for gas sensor applications. In: Environmental nanotechnology, vol 3. Springer, pp 305–355
28. Hazra A, Samane N, Basu S (2020) A Review on metal oxide-graphene derivative nanocomposite thin film gas sensors. Multilayer Thin Film Versatile Appl Mater Eng 1
29. Hong F, Guo X, Zhang S et al (2012) Bacterial cellulose production from cotton-based waste textiles: enzymatic saccharification enhanced by ionic liquid pretreatment. Bioresour Technol 104:503–508
30. Hu C, Bai X, Wang Y et al (2012) Inkjet printing of nanoporous gold electrode arrays on cellulose membranes for high-sensitive paper-like electrochemical oxygen sensors using ionic liquid electrolytes. Anal Chem 84:3745–3750
31. Hu W, Chen S, Liu L et al (2011) Formaldehyde sensors based on nanofibrous polyethyleneimine/bacterial cellulose membranes coated quartz crystal microbalance. Sens Actuators B Chem 157:554–559
32. Idrees Z, Zheng L (2020) Low cost air pollution monitoring systems: a review of protocols and enabling technologies. J Ind Inf Integr 17:100123
33. Islam MR, Podder J (2009) Optical properties of ZnO nano fiber thin films grown by spray pyrolysis of zinc acetate precursor. Cryst Res Technol J Exp Ind Crystallogr 44:286–292
34. Jain S, Karmakar N, Shah A et al (2017) Ammonia detection of 1-D ZnO/polypyrrole nanocomposite: effect of CSA doping and their structural, chemical, thermal and gas sensing behavior. Appl Surf Sci 396:1317–1325
35. Jia R, Tian W, Bai H et al (2019) Amine-responsive cellulose-based ratiometric fluorescent materials for real-time and visual detection of shrimp and crab freshness. Nat Commun 10:1–8
36. Jia Y, Yu H, Zhang Y et al (2016) Cellulose acetate nanofibers coated layer-by-layer with polyethylenimine and graphene oxide on a quartz crystal microbalance for use as a highly sensitive ammonia sensor. Colloids Surf B Biointerfaces 148:263–269
37. Jia Y, Zhang Y, Yu H et al (2016) Fabrication of novel cellulose acetate/Polyethylenimine/poly (acrylic acid) nanofibers/quartz crystal microbalance sensor for ammonia gas detection. J Nanosci Nanotechnol 16:12351–12355
38. Jonoobi M, Oladi R, Davoudpour Y et al (2015) Different preparation methods and properties of nanostructured cellulose from various natural resources and residues: a review. Cellulose 22:935–969

39. Kangas H, Lahtinen P, Sneck A et al (2014) Characterization of fibrillated celluloses. A short review and evaluation of characteristics with a combination of methods. Nord Pulp Pap Res J 29:129–143
40. Kanmani P, Aravind J, Kamaraj M et al (2017) Environmental applications of chitosan and cellulosic biopolymers: a comprehensive outlook. Bioresour Technol 242:295–303
41. Karunagaran B, Uthirakumar P, Chung SJ et al (2007) TiO2 thin film gas sensor for monitoring ammonia. Mater Charact 58:680–684
42. Kaushal S, Priyadarshi N, Pinnaka AK et al (2019) Glycoconjugates coated gold nanorods based novel biosensor for optical detection and photothermal ablation of food borne bacteria. Sens Actuators B Chem 289:207–215
43. Khattab TA, Dacrory S, Abou-Yousef H, Kamel S (2019) Development of microporous cellulose-based smart xerogel reversible sensor via freeze drying for naked-eye detection of ammonia gas. Carbohydr Polym 210:196–203
44. Kim H-C, Kim J Waste paper: a potential source for cellulose nanofiber and bio-nanocomposite applications
45. Kumar S, Pavelyev V, Mishra P et al (2020) A review on 2D transition metal di-chalcogenides and metal oxide nanostructures based NO2 gas sensors. Mater Sci Semicond Process 107:104865
46. Lee WS, Choi J (2019) Highly sensitive, foldable, and twistable gas sensor based on tungsten disulfide-functionalized carbon nanotubes on cellulose paper. In: 2019 IEEE 32nd international conference on micro electro mechanical systems (MEMS). IEEE, pp 311–314
47. Lee YC, Huang H, Tan OK, Tse MS (2008) Semiconductor gas sensor based on Pd-doped SnO2 nanorod thin films. Sens Actuators B Chem 132:239–242
48. Liu Z, Yamazaki T, Shen Y et al (2007) Influence of annealing on microstructure and NO2-sensing properties of sputtered WO3 thin films. Sens Actuators B Chem 128:173–178
49. Ludueña LN, Vecchio A, Stefani PM, Alvarez VA (2013) Extraction of cellulose nanowhiskers from natural fibers and agricultural byproducts. Fibers Polym 14:1118–1127
50. Meng K, Zhao S, Zhou Y et al (2020) A wireless textile-based sensor system for self-powered personalized health care. Matter 2:896–907
51. Mirzaei A, Leonardi SG, Neri G (2016) Detection of hazardous volatile organic compounds (VOCs) by metal oxide nanostructures-based gas sensors: a review. Ceram Int 42:15119–15141
52. Moon RJ, Martini A, Nairn J et al (2011) Cellulose nanomaterials review: structure, properties and nanocomposites. Chem Soc Rev 40:3941–3994
53. Motaung TE, Mtibe A (2015) Alkali treatment and cellulose nanowhiskers extracted from maize stalk residues. Mater Sci Appl 6:1022
54. Mun S, Chen Y, Kim J (2012) Cellulose–titanium dioxide–multiwalled carbon nanotube hybrid nanocomposite and its ammonia gas sensing properties at room temperature. Sens Actuators B Chem 171:1186–1191
55. Mundt KA, Gentry PR, Dell LD et al (2018) Six years after the NRC review of EPA's draft IRIS toxicological review of formaldehyde: regulatory implications of new science in evaluating formaldehyde leukemogenicity. Regul Toxicol Pharmacol 92:472–490
56. Narwade VN, Anjum SR, Kokol V, Khairnar RS (2019) Ammonia-sensing ability of differently structured hydroxyapatite blended cellulose nanofibril composite films. Cellulose 26:3325–3337
57. Narwade VN, Khairnar RS, Kokol V (2018) In situ synthesized hydroxyapatite—cellulose nanofibrils as biosorbents for heavy metal ions removal. J Polym Environ 26:2130–2141
58. Narwade VN, Khairnar RS, Kokol V (2017) In-situ synthesised hydroxyapatite-loaded films based on cellulose nanofibrils for phenol removal from wastewater. Cellulose 24:4911–4925
59. Narwade VN, Pottathara YB, Begum S et al (2020) Cellulose based bio polymers: synthesis, functionalization and applications in heavy metal adsorption. In: Biofibers and biopolymers for biocomposites. Springer, pp 247–257
60. Nevárez LM, Casarrubias LB, Canto OS et al (2011) Biopolymers-based nanocomposites: membranes from propionated lignin and cellulose for water purification. Carbohydr Polym 86:732–741

61. Pang Z, Yang Z, Chen Y et al (2016) A room temperature ammonia gas sensor based on cellulose/TiO2/PANI composite nanofibers. Colloids Surf A Physicochem Eng Asp 494:248–255
62. Park SJ, Kwon OS, Lee JE et al (2014) Conducting polymer-based nanohybrid transducers: a potential route to high sensitivity and selectivity sensors. Sensors 14:3604–3630
63. Pereira PHF, Waldron KW, Wilson DR et al (2017) Wheat straw hemicelluloses added with cellulose nanocrystals and citric acid. Effect on film physical properties. Carbohydr Polym 164:317–324
64. Pirsa S, Chavoshizadeh S (2018) Design of an optical sensor for ethylene based on nanofiber bacterial cellulose film and its application for determination of banana storage time. Polym Adv Technol 29:1385–1393
65. Pottathara YB, Narwade VN, Bogle KA, Kokol V (2019) TEMPO-oxidized cellulose nanofibrils–graphene oxide composite films with improved dye adsorption properties. Polym Bull 1–15
66. Qi H, Liu J, Pionteck J et al (2015) Carbon nanotube–cellulose composite aerogels for vapour sensing. Sens Actuators B Chem 213:20–26
67. Qian S, Sheng K (2017) PLA toughened by bamboo cellulose nanowhiskers: role of silane compatibilization on the PLA bionanocomposite properties. Compos Sci Technol 148:59–69
68. Qiao X, Shao Z, Bao W, Rong Q (2017) Fiber Bragg grating sensors for the oil industry. Sensors 17:429
69. Qing Y, Sabo R, Zhu JY et al (2013) A comparative study of cellulose nanofibrils disintegrated via multiple processing approaches. Carbohydr Polym 97:226–234
70. Ravikiran YT, Kotresh S, Vijayakumari SC, Thomas S (2014) Liquid petroleum gas sensing performance of polyaniline-carboxymethyl cellulose composite at room temperature. Curr Appl Phys 14:960–964
71. Reid M, Reid RD, Oswal P et al (2018) NaDos: A real-time, wearable, personal exposure monitor for hazardous organic vapors. Sens Actuators B Chem 255:2996–3003
72. Reiniati I, Hrymak AN, Margaritis A (2017) Recent developments in the production and applications of bacterial cellulose fibers and nanocrystals. Crit Rev Biotechnol 37:510–524
73. Rodionova G, Saito T, Lenes M et al (2013) TEMPO-mediated oxidation of Norway spruce and eucalyptus pulps: preparation and characterization of nanofibers and nanofiber dispersions. J Polym Environ 21:207–214
74. Rovira J, Roig N, Nadal M et al (2016) Human health risks of formaldehyde indoor levels: an issue of concern. J Environ Sci Heal Part A 51:357–363
75. Rudnik E (2019) Compostable polymer materials. Newnes
76. Ryu G-S, Nketia-Yawson B, Choi E-Y, Noh Y-Y (2017) Diketopyrrolopyrrole-based polymer transistors for hazardous volatile organic compound detection. Org Electron 51:264–268
77. Sadasivuni KK, Ponnamma D, Ko H-U et al (2016) Flexible NO2 sensors from renewable cellulose nanocrystals/iron oxide composites. Sens Actuators B Chem 233:633–638
78. Sadeghifar H, Venditti R, Jur J et al (2017) Cellulose-lignin biodegradable and flexible UV protection film. ACS Sustain Chem Eng 5:625–631
79. Saito T, Kimura S, Nishiyama Y, Isogai A (2007) Cellulose nanofibers prepared by TEMPO-mediated oxidation of native cellulose. Biomacromol 8:2485–2491
80. Sanabria V, Gall S, Gensch F et al (2019) Backward rod extrusion of bimetallic aluminum-copper alloys at room temperature. In: AIP conference proceedings. AIP Publishing LLC, p 30001
81. Sevinchan Y, Hopkinson PE, Bakulin AA et al (2016) Improving charge separation across a hybrid oxide/polymer interface by Cs doping of the metal oxide. Adv Mater Interfaces 3:1500616
82. Shi P, Wang W, Liu D et al (2019) Structural and electrical properties of flexible ITO/In2O3 thermocouples on PI substrates under tensile stretching. ACS Appl Electron Mater 1:1105–1111
83. Sindhu KA, Prasanth R, Thakur VK (2014) Medical applications of cellulose and its derivatives: present and future. Nanocellul Polym Nanocompos 437–478

84. Spence KL, Venditti RA, Rojas OJ et al (2011) A comparative study of energy consumption and physical properties of microfibrillated cellulose produced by different processing methods. Cellulose 18:1097–1111
85. Srinivasan G, Kumar J (2008) Effect of Mn doping on the microstructures and optical properties of sol–gel derived ZnO thin films. J Cryst Growth 310:1841–1846
86. Srivastava V, Jain K (2016) At room temperature graphene/SnO2 is better than MWCNT/SnO2 as NO2 gas sensor. Mater Lett 169:28–32
87. Stankova M, Vilanova X, Llobet E et al (2005) Influence of the annealing and operating temperatures on the gas-sensing properties of rf sputtered WO3 thin-film sensors. Sens Actuators B Chem 105:271–277
88. Stubhan T, Krantz J, Li N et al (2012) High fill factor polymer solar cells comprising a transparent, low temperature solution processed doped metal oxide/metal nanowire composite electrode. Sol Energy Mater Sol Cells 107:248–251
89. Sun L, Semenova Y, Wu Q et al (2017) High sensitivity ammonia gas sensor based on a silica-gel-coated microfiber coupler. J Light Technol 35:2864–2870
90. Sureshkumar N, Dutta A (2020) Environmental gas sensors based on nanostructured thin films. Multilayer Thin Film Versatile Appl Mater Eng 63
91. Swaminathanályer K (2012) Pd–sodium carboxymethyl cellulose nanocomposites display a morphology dependent response to hydrogen gas. Green Chem 14:906–908
92. Taha S, Begum S, Narwade VN et al (2020) Development of alcohol sensor using TiO2-Hydroxyapatite nano-composites. Mater Chem Phys 240:122228
93. Tsouko E, Kourmentza C, Ladakis D et al (2015) Bacterial cellulose production from industrial waste and by-product streams. Int J Mol Sci 16:14832–14849
94. Ummartyotin S, Manuspiya H (2015) A critical review on cellulose: from fundamental to an approach on sensor technology. Renew Sustain Energy Rev 41:402–412
95. Vaghela C, Kulkarni M, Haram S et al (2016) Biopolymer-polyaniline composite for a wide range ammonia gas sensor. IEEE Sens J 16:4318–4325
96. Van Hieu N, Duc NAP, Trung T et al (2010) Gas-sensing properties of tin oxide doped with metal oxides and carbon nanotubes: a competitive sensor for ethanol and liquid petroleum gas. Sens Actuators B Chem 144:450–456
97. Walker JM, Akbar SA, Morris PA (2019) Synergistic effects in gas sensing semiconducting oxide nano-heterostructures: review. Sens Actuators B Chem 286:624–640
98. Wang B, Sain M, Oksman K (2007) Study of structural morphology of hemp fiber from the micro to the nanoscale. Appl Compos Mater 14:89–103
99. Wang B, Zhu LF, Yang YH et al (2008) Fabrication of a SnO2 nanowire gas sensor and sensor performance for hydrogen. J Phys Chem C 112:6643–6647
100. Wang H, Li D, Yano H, Abe K (2014) Preparation of tough cellulose II nanofibers with high thermal stability from wood. Cellulose 21:1505–1515
101. Wang H, Lustig WP, Li J (2018) Sensing and capture of toxic and hazardous gases and vapors by metal–organic frameworks. Chem Soc Rev 47:4729–4756
102. Wang Y, Wang S, Zhao Y et al (2007) H2S sensing characteristics of Pt-doped α-Fe2O3 thick film sensors. Sens Actuators B Chem 125:79–84
103. Wu Q-H, Li J, Sun S-G (2010) Nano SnO2 gas sensors. Curr Nanosci 6:525–538
104. Yoon H (2013) Current trends in sensors based on conducting polymer nanomaterials. Nanomaterials 3:524–549
105. Yu L, Guo F, Liu S et al (2016) Both oxygen vacancies defects and porosity facilitated NO2 gas sensing response in 2D ZnO nanowalls at room temperature. J Alloys Compd 682:352–356
106. Yun S, Kim J (2010) Multi-walled carbon nanotubes–cellulose paper for a chemical vapor sensor. Sens Actuators B Chem 150:308–313
107. Zhang Q, An C, Fan S et al (2018) Flexible gas sensor based on graphene/ethyl cellulose nanocomposite with ultra-low strain response for volatile organic compounds rapid detection. Nanotechnology 29:285501
108. Zhao Y, Moser C, Lindström ME et al (2017) Cellulose nanofibers from softwood, hardwood, and tunicate: preparation–structure–film performance interrelation. ACS Appl Mater Interfaces 9:13508–13519

109. Zhou Y, Gao C, Guo Y (2018) UV assisted ultrasensitive trace NO2 gas sensing based on few-layer MoS 2 nanosheet–ZnO nanowire heterojunctions at room temperature. J Mater Chem A 6:10286–10296
110. Zhou Y, Saito T, Bergström L, Isogai A (2018) Acid-free preparation of cellulose nanocrystals by TEMPO oxidation and subsequent cavitation. Biomacromol 19:633–639

# Chapter 10
# Nanotechnology for Defence Applications

**Mayyadah S. Abed and Zeinab Abbas Jawad**

## 1 History of Defence Technology

There is a strong link between civilization and the advancement of humanity, material science, and defense technology. Arms are developed from wood, stones, as natural weapons or even bones in prehistoric era, to the recent defense technology, which is dynamic and involved different substances for offensive and defensive purposes, different war fields (sea, land, aerospace, and space), different threat's level (lethal weapons), and threat perceptions due to the availability of accurate detection technology [6]. The continuing development and evolution of defense technology can be classified into five distinct forms depending on contemporary substance ages. They are (i) Primitive arms (Stone Age), involves eventually polished, of raw stones, wood, and bones. (ii) Old-production Bronze-Age arms-German, British, Scandinavian, Gallic, and Celtic. (iii) Iron Age Arms, which are believed to be the end of the old age and the beginning of the medieval period. (iv) Medieval armaments of the XVII and XVIII centuries and (v) pre-World War and post-World War weaponry [14].

The trajectory is long, and it covers the production of bows and arrows that achieve targets within a few hundred meters of current ICBMs (guided intercontinental ballistic missiles) that have struck a hypothetical target thousands of kilometers away from targets for directed energy weapons (DEW), as illustrated in Fig. 1. This path of weapon production and associated defense technology has been a source for material development over thousands of years [14].

---

M. S. Abed (✉)
Department of Materials Engineering, University of Technology-Iraq, Baghdad, Iraq
e-mail: mayyadah.s.abed@uotechnology.edu.iq

Z. A. Jawad
Department of Chemical Engineering, College of Engineering, Qatar University, P.O. Box: 2713, Doha, Qatar

 
M. Mujawar et al. (eds.), *Nanotechnology for Electronic Applications*, Materials Horizons: From Nature to Nanomaterials, https://doi.org/10.1007/978-981-16-6022-1_10

**Fig. 1** Projectile weapons with the respective periods, from bows and arrows to guided missiles

## 2 Nanotechnology for Defence

The nanomaterial is any material with at least 1D in the nanoscale (nanometer equals a billionth of a meter), which may include few molecules or atoms. Nanomaterial has unusual and tunable physical and chemical characteristics that allow the production of high strength, lightweight, optical, electromagnetic, and chemical adjustable properties. The handling of materials in nano size, which range from agriculture and control of water pollution, energy storage and generation, healthcare, defense, and defense, to a wide range of applications, using a so-called nanosize technology, is now a matter of fact, as shown in Fig. 2. Nanotechnology has broken the boundaries between various science and technical fields, making the area unique and multidisciplinary. Thus, nanotechnology is recognized as one of the main innovations of the twenty-first century [2].

Similarly, the company promises quality improvements for a wide variety of protection and safety applications by offering lightweight vehicles and platforms, increased reliability, range, the lethality of weapons, stealthy component, stable aerodynamics, sensing and mitigating chemistry, higher agility, biology, radiology, or nude. Additionally, nanotechnology is advantageous in the information fields

**Fig. 2** Nanotechnology applications in defense

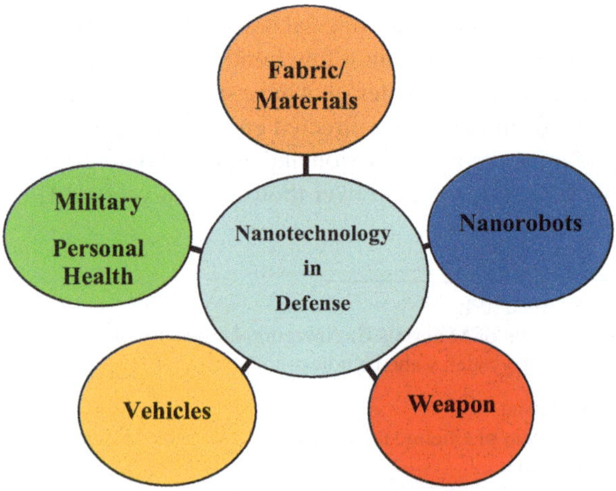

and signal processing, intelligence, and autonomy. This is due to concerning information technology, particularly the capacity of threats, modern electronic displays, and interface systems, an essential position in developing miniature combat vehicles and robotics to obtain significant benefits. In particular, this will include threat identification [2].

Furthermore, nanomaterials' increased chemical reactivity makes them good candidates for high energy and explosive propellants due to a high percentage of atoms/molecules in the nanoparticle surface (NP). Nanomaterials can be utilized either as functional (for thin films, coatings, textiles, etc.) or as structural (for reinforcing composite). Therefore, nanotechnologies use as a pad, either in structural form or as a functional coating surface or fabrics such as nanocomposites (nanoparticles, nanofibers, or nanotubes as reinforcement for polymer or metal matrix), provide fascinating opportunities for radical modifications in composite function. Nanoscale enhanced polymer enhances matrix properties to the degree that compressive strength is increased; tangled nanotube networks can also enhance the material, like carbon nanomaterials, for instance, resulting in an improvement in tensile strength and elastic modulus, components that have high temperature-resistance of ceramic nanomaterials compounds [2, 14].

# 3 The Potential Application of Nanotechnology in Offense-Defense

## *3.1 Protection*

### 3.1.1 Liquid Armor

Body armors can be classified into soft and hard categories according to flexibility. For higher impact protection levels, hard body armors are used, so metal, ceramic, or composite plates are embedded. For the moderate ballistic protection level, soft body armors are made from large numbers of 20–50 high-performance fabric panels. This large number of layers makes the body armor uncomfortable to wear due to its' inflexibility and heavyweight. Therefore, developing hybrid composite materials to fabricate soft, lightweight, flexible body armor has been investigated [7, 21].

Material requirements of anybody's armor have to be flexible, low bulk, lightweight, and protective. Kevlar (p-aramid) and high molecular weight polyethylene (UHMWPE) are typically utilized as high-performance fabrics to prepare for flexible soft body armors. To impede the pullet from penetrating the armor, many fabrics and layers are needed to lead to nonflexible, heavy, and uncomfortable use. So, it is important to focus on enhance fabrics' performance and reducing the number of layers without influence the comfortability or increasing weight in body armor studies [15]. Conventional body armor involves many layers of untreated Kevlar. Pieces of ceramics may add for high-threat situations in the armor's front to absorb

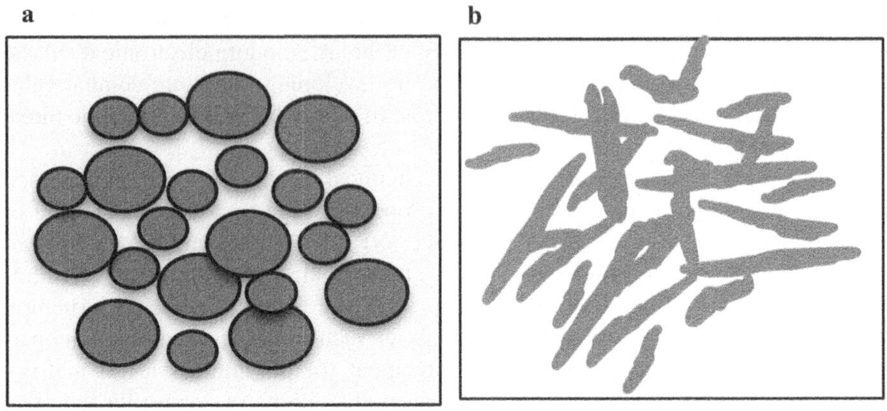

**Fig. 3** **a** isotropic silica and **b** anisotropic calcium carbonates with particle size approximately 400–600 nm used in STF

more energy by fractures. Up to date, there is no armor for extremities like arms, legs, and neck. Battlefield statistics from WWII, Korea that ~16% of deaths in war were due to extremities trauma, whereas ~70% of extremities injures were non-fatal [26].

Shear thickening fluid STF has the main attention in these aspects to increase the energy dissipation, make the armor more flexible, lighter, and less cost. Shear Thickening Fluid (STF) is a liquid phase filled with rigid particles with a high volume fraction of 55–65%. Its viscosity increased at a high shear rate. These particles may become isotropic like silica or anisotropic as calcium carbonates with particle size approximately 400–600 nm, as demonstrated in Fig. 3. The ideal carrier fluid is polyethylene glycol (PEG) due to its' eco-friendly and stable properties. When applying high shear rates, repulsive forces occur among particles and are rearranged as hydro-clusters. This collide becomes rigid enough to impede the impact force. Impregnation of Kevlar fabric with STF is a potential application to body armor. Thus, STF behaves like a liquid with high flexibility, little or no impediment to motion at normal motion (low shear rates). While at ballistic impact (high shear rates), fibers' motion within fabric deforms STF at a high rate. Therefore, STF transit to rigid phase and enhances ballistic protection of fabric. Shear thickening transition is at a shear rate of ~101 to 103 $s^{-1}$. The shear rate during ballistic experiments is approximately 104–105 $s^{-1}$. Therefore, high-velocity impact (ballistic impact) should transition fluid to a rigid phase. Stab test, drop weight impact with different impactor tips, and the ballistic test are different experimental tests used to determine the required protection level [26], as presented in Figs. 4 and 5.

Furthermore, conventional fabric armors' energy absorption mechanisms are yarn pull-out, plastic deformation of fiber, and fiber fracture. Meanwhile, the failure mechanism in fewer Kevlar layers with STF was fewer pull-outs in STF composite and more fiber fracture in STF composite, as illustrated in Fig. 6. Hence, STF restricts

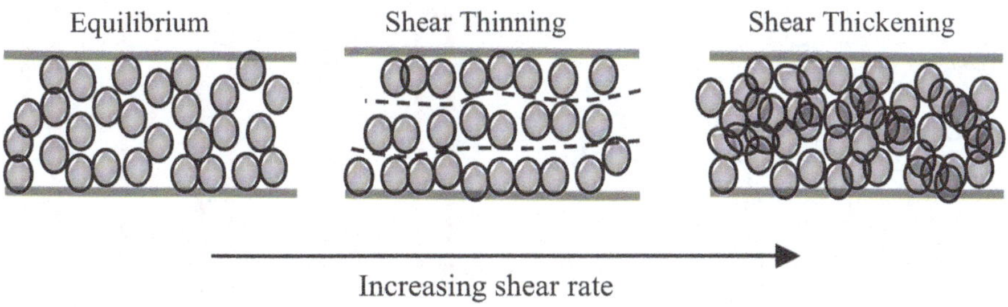


**Fig. 4** Shear-thinning, shear thickening mechanism

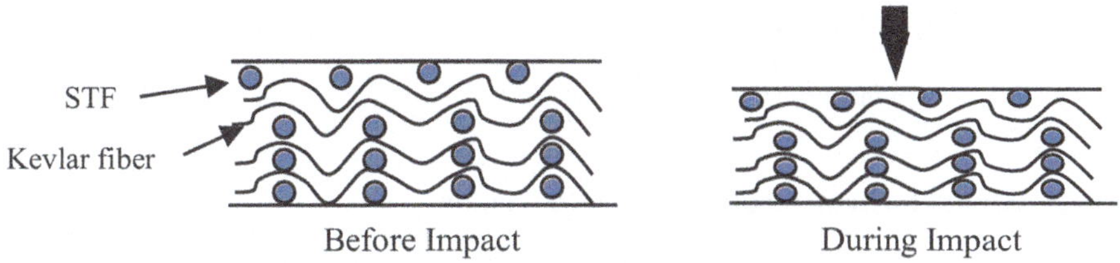


**Fig. 5** The effect of impact force on STF/ fabric composite

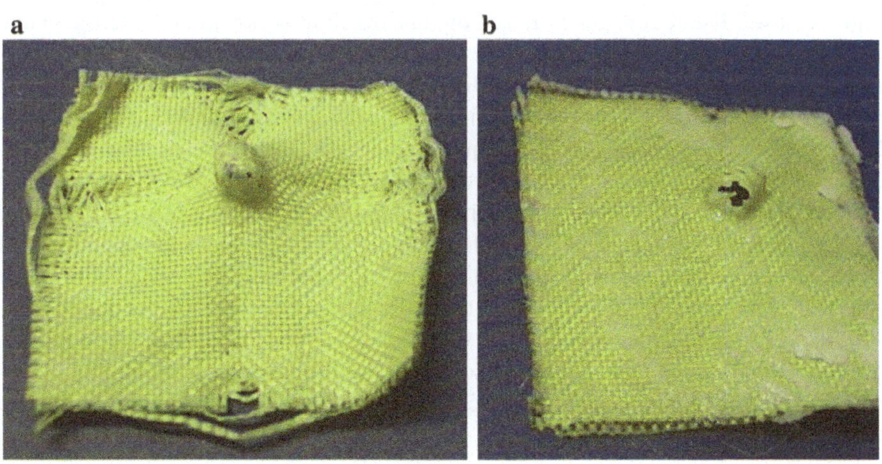


**Fig. 6** **a** First layer of Kevlar (back three layers show comparable pull-out), and **b** first layer of Kevlar (back three layers show little pull-out, no fracture) [26]

yarn motion and allows yarns to be loaded to failure energy absorbed by fiber fractures. STF increases pull-out energy; less pull-out is required to achieve high energy absorption [26].

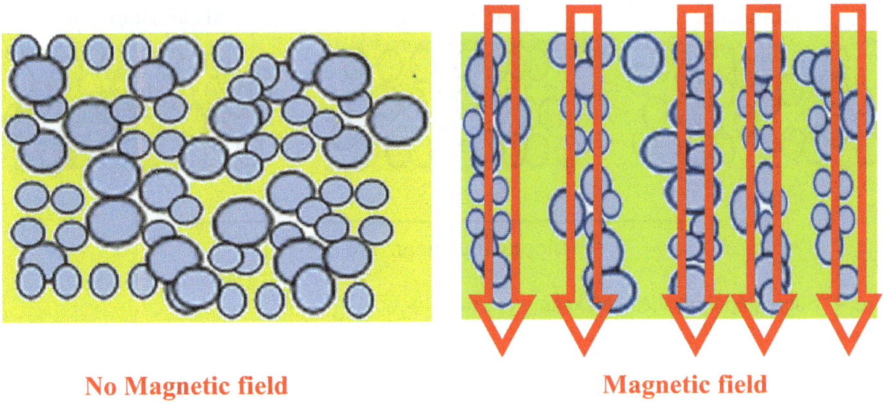


**Fig. 7** The effect of magnetic field on MR fluid

### 3.1.2 Magnetorheological Fluid-Based Armor

The magnetorheological fluid (MR fluid) is a colloidal of ferrous particles that impede a body penetration on the entire material proportionally with magnetic field strength [4], as illustrated in Fig. 7. This change leads the fluid status to change from a liquid to a solid-state reversibly. Based on this, the fluid is considered intelligent and controllable by the magnetic field. The magnetorheological fluid consists of ferrous particles typically from 20 to 50 μm in diameter, basic fluid, and stabilizers [19]. In the magnetic field's presence, the nanoparticles align linearly and parallel to the field [1]. Response times are established that require remarkably low voltages. Ahmadian and Norris [1] have proved magnetorheological fluid's ability to be used as bulletproof and structural membranes. The weapon's strength depends on the composition of the fluid for military applications. Weaving small circuits of the armor had been recommended by [1]. When power is applied through the cables, the armor stiffness and the armor returns to its liquid, fluid state while the current is turned off. Armor technology can be designed to protect soldiers in various combat environments, depending on the of particles' type. The agility of soldiers could increase nanotechnology. This can be achieved by increasing the mechanical properties and versatility of warfare technology [1].

### 3.1.3 Nano Coat to Withstand Extreme Conditions

The soldier has used fabric made from nanopolymers as a breakdown bio, chemical warfare agents, strain resistance, self-cleaning, fireproof warfare [2]. In 2007, Karen Gleason et al. developed a new superhydrophobic film using chemical vapor deposition (CVD). They deposited Teflon nanolayers on Kevlar fiber yarns, making ballistic vests [11].

Primitive nuclear, biological, chemical suit (NBC) is hard to pass air. Hence, it becomes unbreathable. Nanocomposite fabrics can be 25% lighter and stronger than regular fabrics used for military uniforms. Additionally, the air permeability or breathability can be improved over three times than the current military uniform. Besides, it offers enhanced abrasion resistance, fireproof fabric, and offer combination biological/chemical resistant, waterproof uniform. Nano-silver-coated socks and underwear remain hygienic during days in a war situation in which underwear cannot be change [24].

## *3.2 Nano Robotics*

Nanorobotics is a new pathway in a nanotechnology engineering discipline that deals with robotic components and machines built at the nanoscale or close to, which are expected to be utilized in space and military applications [23]. Nanorobotics also involves a system in the micro range or below and build from nanoscale components [25]. In general, nanorobotics includes two partitions. The first one deals with robot design and control. In contrast, the second one deals with the creation, manipulation, and assembly of nano components with each other and or with macro manipulators [17], states that nanorobots should have three characteristic abilities: swarm intelligence, self-assemble ability, and replication at, and ability to have a nano or macro interface architecture with control and maintenance [17].

On the battlefield, nanorobots move in water, air, and ground utilizing propulsion basics known for larger systems, including tracks, wheels, rotor blades, jets, and wings [3]. Besides, self-assembly robots can be designed and act with each other in high numbers for specific tasks such as communication, reconnaissance, sensing, target destination [3].

In fact, these robots have been designed for specific military tasks such as reconnaissance, communication, target destination, and sensing capabilities. Self-assembling nanorobots could act together in high numbers, blocking windows, putting abrasives into motors and other machines, and other unique tasks [23].

## *3.3 Weapons*

### 3.3.1 Non-lethal Weapons

Poisons are considered lethal weapons, while nanotechnology provides a completely new menu of poison options with its ability to activate complex brain functions. For example, a liar poisons stopping any person from telling the truth. The poison Akleptomaniac can prevent a person from stealing items. A poison of alcoholic drinks can prohibit a person from consuming alcohol. The poison of obesity will kill a person and prefer "the poison of the lobotomy," which renders a person unable to be angry or mildly. Although, Nano Mind Erasers can be more effective than killing people to neutralize their memories. Micro fields flashing up in an unnoticeable number of small brain bursts can wipe out parts of an unnoticed brain [2].

In summary, nano-diameter needles are shot like a bullet from large distances, invisible to the human eye, shot like "pin" the human body from a wall, or freeze its physical activity. Nano needles are the ultimate non-lethal weapon because of their small diameter. They seldom have scares and damage. Hence, it can be used as an attacker for arresting without harm or death [2].

### 3.3.2 Lethal Weapons

Nano-aluminum powder ammunition enhanced by nano-aluminum powder has been developed to boost the efficiency and mortality of mini-compact munitions required to minimize the amount of advanced aircraft munitions bays and arms for unmanned air vehicles (UAVs). Processing explosive formulations Nano-aluminum powder ammunition provides better mortality through improved air blast and shrapnel speed than standard formulations. Nanoparticle surfaces are much higher and thus more reactive than traditional powders for a specific weight [8].

The reaction rate can be balanced for optimum air blast and fragment formation and acceleration by the nanoparticle diameter. In conventional formulations for munitions, the development, production, and the characterization of advanced munitions by substantial additions to nano-aluminum powders were only barred by only a handful percent in volume nano-aluminum powder. A solvent dilutes explosives conventionally to be able to incorporate a nano-aluminum powder due to the honey-like viscosity. The solvent also desensitizes the formulation to be treated safer. The solvent then vaporizes during mixing, allowing a granular solution to be easily formed into ammunition containers, with brown sugar consistency [8].

Currently, nano-aluminum powder can be applied up to 50% by volume compared with just a few percentages using previous techniques. An increase of the order-of-magnitude of density is achieved, and the blast characteristics are better reproducible. Industrial partners, including ATK Thiokol and the Savannah River National Laboratory (SRNL), explore the extension of this processing method to introduce nano-aluminum powders to traditional ammunition formulations. Quality evaluations are in progress [8].

## *3.4 Sensing and Detection*

A significant impediment to using CNTs in a real application is the absence of solutions for the mass processing and organization of high quality and high uniformity carbon nanotube into useful architectures. The IR detector can make with a controlled Carbon Nanotube array. This work extends the template-dependent chemical vapor deposition method for producing high quality, regulated diameter, length, and spacing highly ordered range of carbon nanotubes in a silicon substratum [8].

Nanodevices and Photon-Plasmon-Electron Conversion allow for a new class in imaging cameras (as the sound waves are formed from oscillating air molecules, Plasmons are calculated by electron gas oscillations in a metal), a new capturing method has been developed and demonstrated. The Plasmon waves are generated in a metal film that is normally less than 100 nm wide and is bound to the substratum of a semiconductor. The Plasmon waves are slitted in the metal film, in which their energy is centered in the semiconductor for the creation of electron/hole couples. The pairs of the electron/hole, in turn, generate an electrical signal that is observable. Different slit orientations can produce different polarization signals from the same chip, and multi-spectral signals can be produced from a multi-slit chip [8].

## 3.5 Automotive Industry and Fuel Economy

To minimize engine emissions, healthy driving, quiet cars, self-healing bodies, and windscreens, nanomaterials for automotive applications are planned [20]. A long time ago, the concept of using nanofluids to boost coolant efficiency was introduced [5]. The outcome of that proposal was a 100% increase in liquid thermal conductivity, resulting in several more experimental research ventures in this field utilizing nanometer-scale particles [13]. However, improvements in coolants' performance provide drive and inspiration for further improvements in engine efficiency and reduced cooling systems' size and weight. There is also another study focusing on enhancing lubricants' thermal and rheological properties by using nanoparticles [16].

An improvement in the weight reduction of vehicles is a significant problem for motor vehicles, as the key way to boost fuel economy is to reduce weight. The urgent need for new low-cost, high-performance, lightweight metal replacement materials is being generated by global demands for fuel efficiency and low manufacturing and transport emissions. The nanocomposite is the name used for the final solid matrix containing nano-objects (for example, a new class of polymer-based materials with superior mechanical, thermal, and processing properties, ideal for automotive and other metal replacement applications) [10]. The nanoparticles' potency is only 0.5–5% by weight of the amount of material added. This nanocomposite has characteristics superior to traditional microscale composites that can be synthesized using economic techniques. To improve production speed, environmental and thermal stability, recycling, and weight, nanocomposites in vehicle parts and systems are planned. Applying this technology to non-critical structural components such as front and rear parts, cowl vent grills, valve/timing address, and truck beds could save several billion kilograms of weight per year [6, 10].

Furthermore, nano-energy materials (NEMs) have much higher energy densities, quicker release rates, higher stability, and greater protection. The batteries used for combat devices with NEMs have much lighter weight and relief for soldiers.

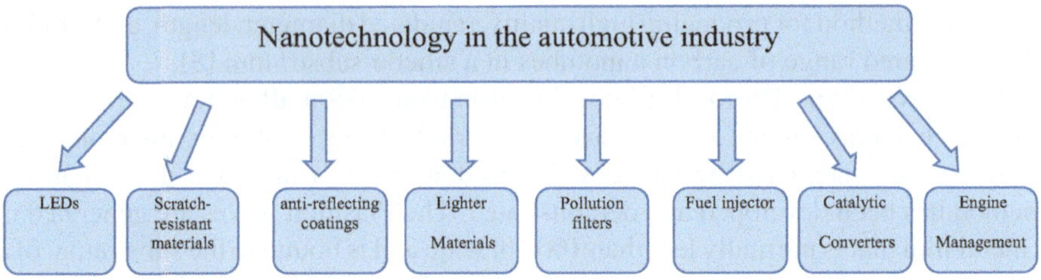


**Fig. 8** Examples of application fields of nanotechnology in the automotive industry

Nano-inspired batteries provide reliable power to military weaponry while needing less volume. The drop-in batteries would cause the size of devices to decrease. NEMs' battery weights are reduced by 20–30%, ensuring that the military keeps fewer batteries in combat [24]. Figure 8 summarizes the role of nanotechnology in the automotive industry [6].

Fortunately, the American Navy is having an all-electric warship of the next decade. This new technology regulates the ship and results in significant performance, survival, and cost reduction changes. The energy production space needed for warships is being dramatically reduced. This freezes room abroad to be used for other functions [12].

## *3.6 Military Personnel Health*

Nanofood, filter, medications, diagnosis, gas, and biological sensing are the main nanotechnology applications in military personnel health. They are classified as discussed below [8, 12, 24].

### 3.6.1 Nanofood

NT would make it possible for food to last longer without rotting, perfect for a war where it is difficult to supply food. Nano-Silver food packed resist to damage-free days. These foods can be consumed in emergency recovery teams in disaster and war, such as high altitudes with fresh food issues.

### 3.6.2 Nano Filter

Filter nanomaterials for removing selected impurities from fluids could result in meager costs and, therefore, omnipresent and several small but distinct concentrations of potentially toxic impurities. The filters "Argonide Nanomaterials" can be utilized for a war where the greatest hazard is water pollution.

### 3.6.3 Nanomedicine

If a soldier is injured in the arm or leg, nanofibers will constrict into a tourniquet. This would be a real lifesaver because half of all battlefield deaths are due to massive blood loss before wounded soldiers can be treated [12].

### 3.6.4 Diagnosis

Anti-Toxin Guard is a nanotechnological device that uses nanotechnology to detect micro-organisms such as disease-related pathogens. Sensors would also provide the soldier's vital signs and location to medics via radio. When the new nanosensor garment has been developed, soldiers are expected to wear an adhesive chest patch fitted with sensors and a tiny radio. Biosensors may be used for the health monitoring of a soldier. Physical safety nano-sized silicon carbide particles [12]. Theoretical studies have developed a class of amplifying fluorescent polymers that detect susceptible hazardous substances. Additionally, a family of self-segregating nano-engineered additives was produced to transport active materials to the polymer/air interface to self-detoxify contaminates [8].

### 3.6.5 Biosensing

NT allows small sensors known as Nano units, which have single types such as "smart materials" available that shift in light and heat reaction. "Nanobots" are lightweight mobile robots, theoretically feasible, that have been constructed and assembled into larger equipment. Artificial "electronic nose" for the identification of biochemical and nuclear agent toxic gases and vapors, "Nanowires," made out of sub-micrometer layers which can be used as 'barcode' for detecting various pathogens, like smallpox, ricin, anthrax, and botulinum toxin, including gold, silver, and nickel. At the same time, the method could recognize several pathogens with their specific fluorescent features. Intelligent RFID NT Alloy is proposed as a war tag for nanosensors, and Indian Army soldiers are currently using a metallic element with a graved name, number, and religion. Nanosensors embedded in the battle suit can relay signals to search and rescue operators in the miniaturization of gadgets such as RFID (Radio-frequency identification)/search and rescue missions [24].

### 3.6.6 Artificial Muscles

It improves soldiers' abilities with artificial muscles, enabling soldiers to jump high walls with nanotechnology and electricity if wounded or needed. As power cuts, it flexes and relieves itself when energy is disconnected. However, its responses have been much too sluggish to date. Nano allowed soldier medicine for and healing field-less crack bones. Armed forces field ambulance units hold this instant NT that enables war medicine. If a soldier's arm or leg is wounded, nanofibres uniform, the tissue may tighten into a tourniquet. This is a rescue since half the deaths in the war area are attributed to a significant blood loss before treating wounded soldiers [24].

The development of "environment-intelligent networks" (hereafter AIN) is one of the first applications of nanotechnology solutions for soldiers. A test of AIN is the monitoring of human condition and health, in which small devices like cellular phones or receptors are made accessible for different heart rate and blood monitoring systems or calorie counters. AIN has the big benefit of focusing on non-invasive technologies. If the device is not functioning, soldiers could feel the wearer's condition and can produce rapid reactions by the release of medications or using small materials to compress the injuries. If the fighting atmosphere is contaminated by bacterial, bacteriological, or chemical agents, invasive nano-agents can also help the military advance on the ground. Nanostructured warning systems may cause rapid drug and other medication delivery before searches and rescue equipment pull wounded troops from the battlefield infected area [18].

### 3.7 *Invisibility by Nanotechnology: Stealth Movement*

Scientists are manipulating light, so soldiers seem to vanish. Scientists are also working on electrochromic camouflage, a cloth that immediately changes colors to fit in with the environment [22]. Electrochromic camouflage with NT electrochromic camouflage leads to the disappearance of soldiers in the neck. Electrochromic camouflage changes colors to fuse immediately into the environment. In fact, nano has allowed painting to create a special paint that disappears, or, to be more accurate, makes drones, missiles, or avions very hard to detect. The object's envelope is formed by NT, absorbs the radar emission waves, and releases them through space as heat energy. This makes it hard to be recognized by radar the material disguises the object [24].

Furthermore, science fiction no longer plays with invisibility. The system shows the same dispersive properties as the vacuum to render an object transparent and perfectly invisible. Objects would absorb, diffract, or refract light from the atmosphere, which would make the object visible. Bending the light around an object, certain materials to be invisible, based on this, the unit should not reflect any light and cast no shadows along with the object to be camouflaged. Fortunately, the artificial metamaterials emerging have made it possible to manipulate electromagnetic waves and create new features with exceptional versatility and have taken the ancient dream one step closer to the fact that an entity is invisible. The metamaterials display the electromagnetic waves' negative refractive index material (NIM) behavior. Nanotechnology can also generate metamaterials, not just microwaves, which can bend visible light [22].

Based on [22], NIM objects in the atmosphere that absorb, reflect, or refract light from the environment that makes the object visible. NIM is not a refractive index material. Figure 9a shows Light refraction at the crossroads. As mentioned

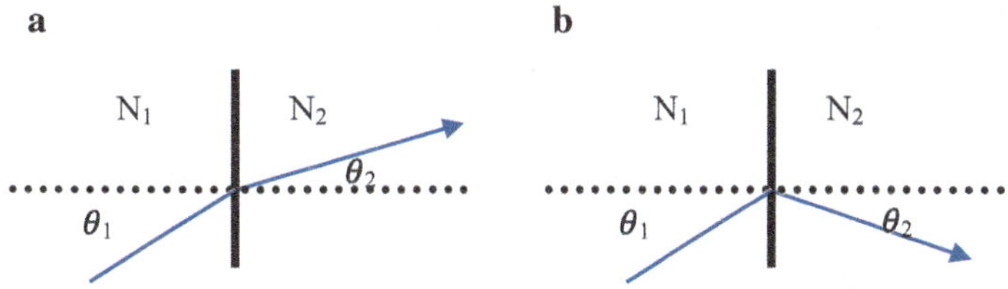


**Fig. 9** **a** Refraction of light at an interface between materials, and **b** refraction of light at a NIM interface

previously, bending the light around the object would change these materials to be transparent. However, this problem is minimal. Thus, there is no such bending of light by any traditional material. Instead, the refractive index of the material should be negative to bend the light as it turns out. Figure 9b shows the light refraction at a NIM interface with the positive refractive, which indices all traditional materials.

In general, metamaterials are known as metamaterials in this specific category of material. They are human-made materials that overtake the optical characteristics of traditional materials by their same description. Contrary to the traditional materials, these metamaterials would bend the light and cause a curve around the invisible object. They would be incredibly abnormal in the straw within the fluid, as shown in Fig. 10 [22].

However, it is challenging to synthesize a metamaterial capable of bending electromagnetic radiation into the visible range. This is achieved by using suitable geometries to play the electromagnetic waves and light. Their wavefronts and propagation directions relatively small. However, complex micro- and nano-structuring assemblies require complex production technology, which remains a challenge [22].

In another study by the University of Central Florida [9], they achieve a wide area of visible light using a multi-layer 3D. They used a nano transfer printing technique to produce the material, creating composites of met-al/dielectric films. These are arranged in a 3D architecture to achieve nanoscale models in the visible spectral range for operation. The regulation by the structural handling of electromagnetic

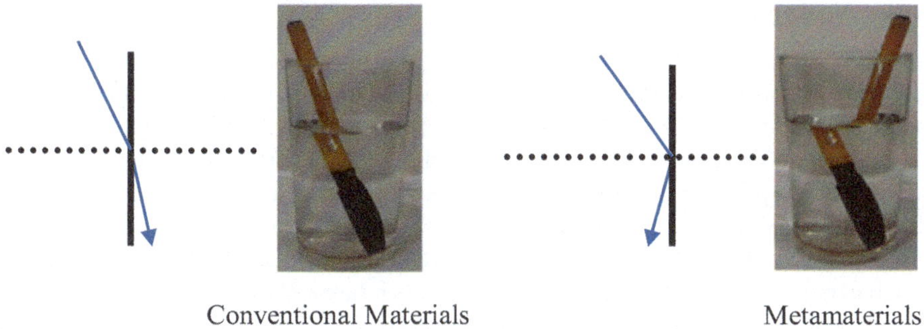


**Fig. 10** Metamaterial cloak that works for visible light

resonances over the 3D space allows precise control of light propagation. The metamaterial made of three dimensions in a UCF laboratory allowed electromagnetic radiation to be regulated in the visible area [9].

### 3.8 Unmanned Vehicle

Currently, missile-armed unmanned aerial vehicles (UAVs) are a commonly recognized characteristic of modern military organizations. Unmanned devices have often been involved in crisis and military activities since the end of the Cold War. Surprisingly, materials that are lighter and stronger among the first applications of nanotechnology. They would have different properties than today's available materials. NT would first give rise to changes in current device configurations in that context. It must be stressed that such reforms would be of primary importance to the defense community. Nanostructured materials would provide the defense industry with the means required to develop smaller, more mobile, more agile and more reliable military platforms, such as light armoured vehicles, tanks, fighter jets, man-transportable, and micro-unmanned aircraft. Developments in the NT would ensure the development of platforms capable of adapting to all types of climatic and environmental conditions. For example, the aerial platforms could be operated from land bases or sea-based carriers more easily and feasibly in the future, regardless of the environmental constraints (maritime spindrifts, desert sand, tropical vegetation, and intense wetness). It is anticipated that conventional weapons would benefit from mid-term and long-term nanotechnology advances. Additionally, on-board computers in missile systems would be smaller, allowing for a substantial reduction in missile signatures and an improved destructiveness of payload inclusion. Machines can also aid a dismounted soldier. It is believed that advanced and discrete sensors (some of which are invasive), near-real, or real-time data on how troops progress on the battlefield would prove technically feasible [18].

### 3.9 Nanocomposites

Enhanced nano-engineered fiber-coated nano polymer composite has been developed to improve the composite polymer matrix's interface fiber/matrix properties. The demonstration has started at the laboratory research specimen level with improved energy absorption while enhancing and preserving other composite properties. The nanoparticles of zinc and carbon composites can be used to manufacture Air-Activated Heaters. Electrochemical oxidation has been developed to heat self when exposed to air. The heater has also been synthesized to heat the typical ready-to-eat (MRE) operating meal. The Nanocomposite Space Device Insulation (AF) is another application of nanocomposite. The nanostructured silicate molecules for rocket motor isolation are produced and marketed. The major advantage of these materials is that they have lightweight motor and good corrosion resistance. Furthermore, nanotechnical applicators and actuator design projects in various areas, specifically Australia, are being developed to change their physical properties (strength, conductivity, etc.), in which nanoparticles are applied to the material as composites. Meanwhile, French/German/Canadian cooperative to build a new generation of ceramic, aeronautical, catalytic, and "new energy" nanostructured bulk composites and clothing [8].

Moreover, the nanocomposite is used for Armour Vehicle applications. Armour Vehicle uses $Al_2O_3$, which has a high-quality standard, 2–10 μ grain size; Vicker hardness (HV 10); and 410 GPa elastic modulus. Alumina ($Al_2O_3$), silicon carbide (SiC) nanocrystalline variant with 50 nm and 20 nm of grain respectively, with 70/840 MPa and young modules of 680–900 GPa, this gives a hardener value of 3,000. "Nanodyne" is made of a composite tungsten-carbide powder (grain size less than 15 nm) used to create a sintered alloy as hard as diamond and which in turn is used to make cutting devices, drills, armor, and jet engine parts "Nanodyne" [24].

Composites also minimize vibrations in aircraft. The challenge is mainly in the aircraft, especially in rotator wing vibration suppression. Composite NT decreases vibrations. NT enabled Un-Manned Combat Aerial Vehicle UCAVs as very small with low RCS, as most of the composite parts are made of composites. Aircraft composites also lower vibration levels. Aircraft is a typical design challenge, in particular in rotary-wing vibration suppression. Composite NT decreases vibration. In contrast to traditional weapons, the NT allows weapons to be tiny and lethal. These unpiloted UCAVs can be deployed on hostile land and unfavorable weather in impassable fighters. Nano-consistent high-force plastics are required to replace metal and reduce military weapons platforms' weight by the Radar Signature (RCS). Intelligent components, with integrated conditions and load control sensors, improve monitoring of key parameters. Auto sufficiency or self-healing materials are also possible in memory alloys that form a memory with NT. Shape memory alloys in fighter aircraft sleeves are used. Aircraft composites further at lower vibration levels. Aircraft design is a challenge, in particular in rotary-wing vibration suppression [24].

### 3.9.1 Stealth Nanotechnology

Coatings, camouflage, and composites for radar absorption minimize RCS. NT is used for the production of anti-ballistic structures, reactive armor nanoparticles, and nanotubes absorbing shock. Nanoparticles are used to develop stronger, smoother, and/or stealthier as surface coverings. The composite IR picture is also reduced [24].

### 3.9.2 Nano Fiber

Primitive nuclear/bio/chemical (NBC) reduces the solid ability because the air is hard to pass through. NT can be fabricated 25% stronger and lighter in nonwoven composite cloths than existing military uniform woven textiles. Moreover, the new weaved military uniform textiles' air permeability or breathability can be enhanced over three times. Uniform fabric provides increased abrasion resistance, improves the chemical/biological resisting and fire resistance, and offers the combination of water repellent composite fabric. In the war scenario, nano-silver socks/undergarments remain hygienic when undergarments cannot be changed.

Furthermore, carbon nanotubes (CNTs)/ polymer composites have a high young module, high thermal efficiency, and crush resistance. CNTs/Polyimide, CNTs epoxy, and CNTs/PP are some of the CNTs applications for the aircraft structure. SiC nanoparticles improve the Aeros engine's performance, and SiC nanocrystallites are used for wear-resistant coating, incorporated in amorphous $Si_3N_4$.SiC nanoparticles [24].

## *3.10 Navy Nanomaterials*

### 3.10.1 Electronics

In many technologically key areas, the detection and control of fundamental magnetic spin phenomena would have far-reaching effects, especially quantum calculations. For imprinting of knowledge in Si spin systems, the Navy demonstrated the ability to produce, modulate, and detect pure electron spin currents. A spin-polarized bias current has been demonstrated to regulate the spin of individual quantum shell status of self-assembled quantum dots [8].

A quantum dot thermometry technique has been developed to quantify temperature gradients over minimal distances and differentiate between electronic and grid vibrations in thermal transportation. The quantum dot thermometry achievements are relevant in many fields, like most fundamental scientific science, but have originated from thermoelectric fundamental studies of nanofilms [8].

### 3.10.2 Nanostructure Enhanced Thermoelectric Properties

Experimentally established in nanostructured nanowires, thin films and bulk materials can partly separate the transport of phonons and electrons utilizing the disengaging of composite semiconductor structures containing epitaxial nanoparticles. Theoretical models support the size-dependent decrease in thermal conductivity by nanoparticles with minimal impact on electric conductivity, leading to record thermoelectrically efficiency. Higher performance, sound-state applications allowing fuel-flexible, mobile power generation, vehicle waste heat recovery/fuel efficiency, and electronic appliance/system cooling to be included in these developments [8].

### 3.10.3 Energy Storage

The internal structures of the Ultraporous Carbon Nanofoam can be protected using low-cost scalable electro-less deposition methods by the ultra-thin (<50 nm) coatings of a metal oxide ($MnO_2$ or $FeO_x$). The nanoscale metal oxide sheath functions as a load-storage bank enabled by ion–electron reactions with electrolyte cations and nano-structure supporting electrons. This is important to achieve high discharge rates in high energy–density batteries [8].

### 3.10.4 Magnetics

Nanocrystallites are rich in cobalt alloy and have been formed with considerably improved fractures over traditional iron-based magnetic alloys. Complex magnetic part manufacturing rates can be as low as 20%. Damage-resistant magnetic materials offer extended engine design possibilities with considerably reduced production costs [8].

### 3.10.5 Semiconductors

AgPbmSbTem + 2 (m = 1 – 18) is the first discrete quaternary nanocrystal synthesized using a low-temperature solution route that demonstrates a new approach to achieving uniform solid-solution alloys that do not possess large volumes of size and composition-controlled semiconductor properties. This semiconductor class of nanocrystals can be used for thermoelectric energy transfer, IR sensing, photovoltaic memory, and phase change [8].

## 4 Conclusion

Nanotechnology has significant impact applications in the military which help defense in the following: Protection (armor-like fabric body armor, and vehicle shields), temperature tolerance clothes, improve NBC suit, waterproof uniform, self-cleaning, anti-degradants military uniform, diagnoses chips for drug doses auto control, invisibility ware, increase surveillance probabilities for better protection, smaller cameras and robots, cheap, small, and more effective weapons, anti-scratching surfaces, stronger, thinner, and cheaper glass, faster intensive medical help, lighter, faster, and stronger aircraft, vehicle, ships which use less fuel, radar undetected planes and submarines by electromagnetic camouflage.

## References


1. Ahmadian M, Norris JA (2008) Experimental analysis of magnetorheological dampers when subjected to impact and shock loading. Commun Nonlinear Sci Numer Simul 13:1978–1985
2. Altmann J (2006) Military Nanotechnology: potential applications and preventive arms control. Routledge, London and New York
3. Altmann J, Gubrud MA (2004) Military, arms control, and security aspects of nanotechnology. Discov Nanoscale: 269–277
4. Ashour O, Rogers CA, Kordonsky W (1996) Magnetorheological fluids: materials, charcterization, and devices. J Intell Mater Syst Struct 7:123–130
5. Choi SUS (1995) Enhancing thermal conductivity of fluids with nanoparticles. Am Soc Mech Eng Fluids Eng Div FED 231:99–105
6. Coelho MC, Torräo G, Emami N, Grácio J (2012) Nanotechnology in automotive industry: research strategy and trends for the future-small objects, big impacts. J Nanosci Nanotechnol 12:6621–6630
7. Dehkordi M, Nosraty H, Shokrieh M et al (2013) The influence of hybridization on impact damage behavior and residual compression strength of intraply basalt/nylon hybrid composites. Mater Des 43:283–290
8. Department of Defense Director (2009) Defense nanotechnology research and development program. Defense Research & Engineering USA
9. Gao L, Kim Y, Vazquez-Guardado A et al (2014) Negative index materials: materials selections and growth conditions for large-area, multilayered, visible negative index metamaterials formed by nanotransfer printing. Adv Opt Mater 2:255
10. Garcés JM, Moll DJ, Bicerano J et al (2000) Polymeric nanocomposites for automotive applications. Adv Mater 12:1835–1839
11. Gleason KK, Rutledge GC, Gupta M et al (2007) Superhydrophobic fibers produced by electrospinning and chemical vapor deposition. US Pat Appl Publ 26pp
12. Jones A (2008) Nanotechnology in the military, pp 1–16
13. Keblinski P, Eastman JA, Cahill DG (2005) Nanofluids for thermal transport. Mater Today 8:36–44
14. Kumar N, Dixit A (2019) Nanotechnology for defence applications. Springer Nature Switzerland AG
15. Majumdar A, Butola BS, Srivastava A (2013) An analysis of deformation and energy absorption modes of shear thickening fluid treated Kevlar fabrics as soft body armour materials. Mater Des 51:148–153
16. Marquis FDS, Chibante LPF (2005) Improving the Heat transfer of nanofluids and nanolubricants with carbon nanotubes. JOM 57:32–43
17. Mavroidis C, Ferreira A (2013) Nanorobotics: past, present, and future. In: Mavroidis C, Ferreira A (ed) Nanorobotics. Springer, New York, NY, pp 3–27
18. Neve AD (2014) Military uses of nanotechnology and converging technologies: trends and future impacts
19. Olabi AG, Grunwald A (2007) Design and application of magneto-rheological fluid. Mater Des 28:2658–2664
20. Presting H, König U (2003) Future nanotechnology developments for automotive applications. Mater Sci Eng C 23:737–741
21. Sarasini F, Tirillò J, Valente M et al (2013) Hybrid composites based on aramid and basalt woven fabrics: impact damage modes and residual flexural properties. Mater Des 49:290–302
22. Srinivas K (2017) The role of nanotechnology in making metamaterials for object invisibility. Int J Emerg Res Manag Technol 6:172–176
23. Tate JS, Espinoza S, Habbit D et al (2015) Military and national security implications of nanotechnology. J Technol Stud XLI:22–30

24. Tiwari A (2007) Military nanotechnology. Int J Eng Sci Adv Technol Adv Technol 2:825–830
25. Weir NA, Sierra DP, Jones JF (2005) A review of research in the field of nanorobotics. SAND2005-6808, 1–50
26. Wetzel ED, Wagner NJ (2002) Advanced body armor utilizing shear thickening fluids. In: 23rd army science conference Orlando, FL December 2002

# Chapter 11
# Nanotechnology for Biomedical Devices: Cancer Treatment

**Andrew Cappuccitti, Benjamin Daniels, Christina Galloro, Kevin Kung, Kevin Ly, Abdul Malik Mohammad, and Yaser Dahman**

## 1 Introduction

The human body is composed of trillions of cells that actively reproduce, grow, differentiate, and die. In an event mainly caused by a mutation, this cell cycle of a group of cells becomes abnormal, resulting in uncontrolled cell growth and/or inadequate cell apoptosis, giving rise to a tumour. When this tumour is contained and localized, it is called benign and can be safely removed from the body. However, when this tumour spreads to other cells and tissues of the body, it is called malignant tumour, which results in cancer. According to the Canadian Cancer Society, in the year 2010, 7.98 million people died of cancer worldwide making 14% of all deaths cancer related. In Canada, 196,900 people were diagnosed with cancer in 2015 out of which 78,000 died (Canadian Cancer Society 2015). This makes cancer one of the most deadly diseases on the planet.

Hanahan and Weinberg published a monumental study on cancer in 2000 titled "The Hallmarks of Cancer." In this study, the authors gave a list of six hallmarks that cause and propagate cancer (Hanahan and Weinberg 2000):

1. Self-sufficiency in growth signals, where cancerous cells stimulate their own cell growth
2. Insensitivity to anti-growth signals, where the cells resist the inhibitory signals to stop cell growth
3. Evading apoptosis, where the cells resist and evade the normal programmed cell death
4. Limitless reciprocating potential, in which the cells reproduce uncontrollably
5. Sustained angiogenesis, where the cells stimulate the growth of blood vessels and other channels for the supply of nutrients to the tumour

A. Cappuccitti · B. Daniels · C. Galloro · K. Kung · K. Ly · A. M. Mohammad · Y. Dahman (✉)
Department of Chemical Engineering, Ryerson University, Toronto, ON M5B 2K3, Canada
e-mail: ydahman@ryerson.ca

 
M. Mujawar et al. (eds.), *Nanotechnology for Electronic Applications*, Materials Horizons: From Nature to Nanomaterials, https://doi.org/10.1007/978-981-16-6022-1_11

6. Tissue invasion and metastasis, where cells invade local tissue and spread to distant sites.

These six hallmarks make cancer treatment extremely difficult and multifaceted. The current modalities in cancer treatment include local therapy, which includes surgery and radiation; systematic treatment, which includes chemotherapy, hormonal therapy, monoclonal antibodies and radioactive material; supportive care, which deals with symptomatic treatment; and nonconventional therapy including the alternative medicine techniques (Rajkumar 2014). Perhaps the most well-known and effective treatment option listed above is chemotherapy. However, side effects of chemotherapy including immunosuppression and myelosuppression, typhlitis, gastrointestinal disease, anaemia, fatigue, nausea and vomiting, hair loss, secondary neoplasm, infertility, Tumour Lysis Syndrome, organ damage, cognitive impairment, teratogenicity, peripheral neuropathy, and others, are too serious to overlook (Rajkumar 2014). Therefore, there is an urgent need for novel cancer treatment options that can stop the disease with less harmful effects.

The much-needed novel cancer treatment and diagnosis options can be created using different fields in nanotechnology. The diameter of nanoparticles (NPs) is significantly smaller than the animal cells, and can be used in NP mediated tumour targeting, cancer imaging, drug delivery, controlled drug release, gene therapy, photodynamic therapy, and multimodality treatment. This review will explain the use of nanoparticles in cancer studies while highlighting the pros, cons, and scope for future development in the field.

## 2 Cancer Cell Targeting Using Nanoparticles

The targeting of cancer cells and tumours using nanoparticles is done primarily using two methods, which are passive and active targeting. Both methods use nanoparticles to deliver the cancer drugs to the cancer tumours but they differ in terms of how the nanoparticles interact and target individual cancer cells and in the way the drugs are distributed throughout the entire tumour.

### *2.1 Passive Targeting*

#### 2.1.1 Exploiting Nanotechnology to Passively Target Cancer

The early detection of cancer is essential to increasing the odds of successful chemotherapy. Current detection of cancerous lesions often occurs through clinical biopsy of a progressed mass, or vague microscale marking detectors such as prostate serum antigens that can exhibit misleading behaviours [23]. By reducing the markers to the nanoscale, higher detection sensitivity can be achieved with an

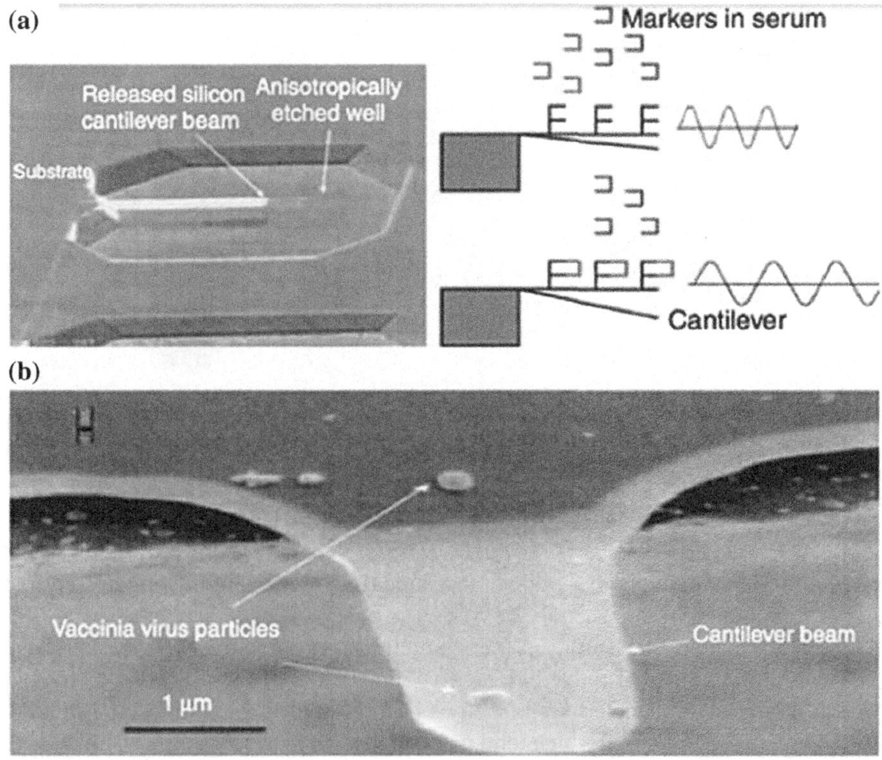


**Fig. 1** **a** Cantilever. **b** Cantilever bound to cancer [77]

increase in specificity and accuracy. One approach on this scale is the dip-pen nanolithography technique, which can be used with nanocantilevers to detect markers at extremely low concentrations [35] (Fig. 1).

The premise of the technology is to "print a nanoscale oligonucleotide chip on metallic and insulating substrates, which exhibits sequence-specific binding properties of the DNA" [77]. The nanocantilevers are deflected by the binding events. This changes their resonating frequencies, which can be interpreted as a positive detection of the marker. The extreme sensitivity and specificity of these cantilevers at the nanoscale not only makes them a more accurate indicator of cancerous legions but also gives medical practitioners advanced warning and ultimately the ability to isolate and treat cancer before it further develops and spreads throughout the host.

The effective delivery of chemotherapeutic drugs is essential for the treatment of cancer and the overall health of the patient in care [77]. Current delivery methods are somewhat crude and expose the entire body to the drug, taking a heavy toll on the body of the patient as the immune system responds to the foreign matter. By applying chemotherapy on the nanoscale, a high level of precision can be achieved with each treatment, and the drugs can be masked from the immune system (reducing the heavy

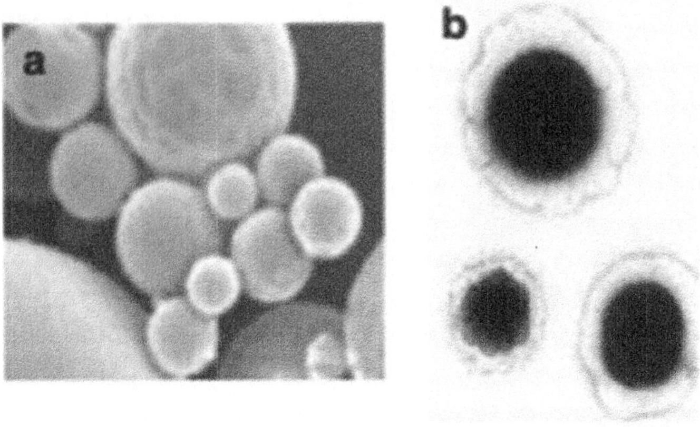

**Fig. 2** **a** Polymer-coated nano-cell. **b** Cross-section of a polymer-coated nano-cell. As the tumour exhibits hypoxic properties, the nano-cell degrades and releases the liposome to bind to the tumour, effectively treating it [77]

collateral impact on the body) while at the same time increasing the effectiveness of tumour treatment (Moghimi et al. 2005).

One approach of this nature is the synthetic lipid derivatives of polyethylene. The hydrophilic nature of polyethylene glycol allows the coated liposomes to pass through the tissue undetected by the body's immune system. The liposomes tend to accumulate in the tumours (which tend to have pore sizes as large as 600 nm and are easily permeated by the liposomes). The tumours also exhibit retentive capacities that prevent the liposomes from escaping once they permeate. This drug delivery—although not specific "passive targeting"—is extremely effective, with 3–10 times more drug delivered to solid tumours on average than conventional chemotherapy. Tumour response to this form of treatment has been observed to be as high as 80% (Allen et al. 1991) (Fig. 2).

### 2.1.2 Iron Oxide and Quantum Dots in Vivo

Iron oxide nanocrystals are used during MRI due to their super-paramagnetic properties, which alter spin–spin relaxation times of nearby $H_2O$ molecules (Moghimi et al. 2005). Lymph-node metastases that are normally undetectable due to the inability to permeate the lymph nodes effectively have been observed using these super-paramagnetic nanocrystals. The nanocrystals move through the lymph nodes by interstitial-lymphatic fluid transport due to their lymphotropic properties, where their effects on the surrounding bodies are observed under high-resolution MRI (Harisinghani et al. 2003).

Another nanotechnology used for tumour imaging and metastases detection are Quantum dots (Q-dots or QDs). Q-dots are "particles with physical dimensions smaller than the exciton radius" (Chan et al. 2002). Similar to the Iron oxide nanocrystals, Q-dots are used for tumour imaging as a less destructive alternative to radiocontrast dyes. Q-dots can be manufactured to emit fluorescent emissions in the visible and infrared light spectrums. Q-dots are manufactured with a polyethylene glycol chain (PEG) covalently bonded in order to avoid being consumed by the body's reticuloendothelial system [71]. PEG also renders the surface of the Q-dot to be hydrophilic—easily passing through the body. Additionally, peptide coatings on the Q-dots may target specific structures in the body, such as tumour vasculature or the lungs, increasing the overall effectiveness of the Q-dots passing into suspected metastases (Akerman et al. 2002).

The biocompatibility of both nanocrystals and Q-dots still needs to be studied in order to declare these technologies safe for use in humans. Q-dots generally employ heavy metals such as Cadmium, and although these heavy metals are coated in organic shells and present in amounts far below known toxic levels, their health effects are unknown. In the same way, the nanoparticles are more effective at targeting tumours and passing through the body undetected, there may be a potential for negative interactions throughout the body that are just as effective as the intended benefits (Akerman et al. 2002). Q-dots are discussed more in detail in Sect. 3.2.2.

### 2.1.3 Passively Utilizing Enhanced Permeation and Retention

One of the general characteristics of a cancerous tumour is its inability to drain interstitial fluid (a common defect in the lymphatic function of tissue drainage). This deformity can be taken advantage of, as macromolecules tend to accumulate in tumours [57], and inevitably will cause a buildup of nanoparticles (that are > 4 nm diameter). The nanoparticles are able to perforate the tumour through fenestrations in imperfect tumour blood vessels. As the tumours grow, the blood supply to the tumour eventually needs to increase substantially. Select cells die, and release growth factors that promote angiogenesis. The capillaries formed during this period are riddled with fenestrations that are typically between 200 and 2000 nm in diameter. Most nanoparticles (Q-dots, carbon nanotubes, etc.) are of small enough size that they enter the fenestrations easily, and become entrapped in the tumour [7].

Nanoparticles between 2 and 3 nm in diameter accumulate faster in tumour vasculature, and tend to penetrate the tumour more deeply and thoroughly than larger particles. As easy as it is for nanoparticles of this size range to diffuse into the tumour, it is just as easy for them to diffuse back out of the tumour. A more appropriate size range for nanoparticles that are intended to accumulate in a tumour is between 7 and 25 nm in diameter. Nanoparticles should not be greater than 70 nm in diameter as they do not accumulate effectively in tumours that are poorly permeable [7]. One technique for effectively delivering a large quantity of nanoparticles to a tumour while maintaining effective permeation is to manufacture a nanoparticle ~90 nm in diameter by weakly bonding a number of smaller nanoparticles together. The weak bonds ensure that the large particle is easily sheared apart into the smaller nanoparticles (~10 nm in diameter) so that they are all delivered to the same location, but are able to still permeate the tumour with the effectiveness of a smaller nanoparticle [81].

The general characteristic of a cancerous tumour's inability to drain interstitial fluid can be taken advantage of as macromolecules tend to accumulate in tumours [57], and inevitably will cause a buildup of nanoparticles (that are > 4 nm diameter). The shape of these nanoparticles can be adjusted so that they penetrate tumour vasculature effectively, but still exhibit important properties such as a high surface area (especially important in the application of drug delivery). Single wall carbon nanotubes with a length to diameter ratio between 100:1 and 500:1 have been found to clear the kidneys better due to their elongated shape. Still effective at penetrating and accumulating inside tumours, the carbon nanotubes will not accumulate as easily in other benign, yet poorly permeable areas in the host. This makes them an excellent candidate for their application in tumour penetration and treatment [7].

These nanoparticles must not only be able to permeate the tumour's vasculature, but also be able to bypass or escape the host's defence mechanisms (such as phagocytes which are often achieved by coating the nanoparticle in a water-soluble neutrally charged substance such as polyethylene glycol). The surface charge of the nanoparticle can limit diffusion and make the nanoparticle 'sticky' while inside the tumour. Alternatively, an undesirable charge (positive or negative) can promote reactions of the coating with ions in the host prior to tumour entry. It is extremely important that there are more nanoparticles in the bloodstream than the desired amount inside a tumour, to encourage diffusion into the tumour vasculature and prevent back-diffusion of the nanoparticles into the bloodstream [7].

### 2.1.4 Improving Passive Targeting by Altering Nanoparticle Shape

Proof-of-principle studies on the use of carbon nanotubes to deliver cancer therapy have been promising in the past few years. Carbon nanotubes have the advantage of the capability to carry and discharge therapeutic agents on an intracellular level. The study of carbon nanotubes is one focus of a broad effort to develop nanotechnologies that can combat cancerous tumours actively and passively. Different nanostructures present unique behaviours 'in vivo', and must be understood in order to fully exploit their capabilities, and to understand when to use which type of nanostructure on a case-by-case basis.

In the study conducted by Fabbro, carbon nanotubes displayed three distinct capabilities outlined as follows: "they deliver biologically active molecules cytoplasmatically and by-pass many biological barriers (acting as a cellular needle), they have a large surface area and internal cavity that can be decorated with targeting ligands and filled with therapeutic or diagnostic agents, and finally they have unprecedented electrical and thermal conductivity properties." To refine this technology, Fabbro states that carbon nanotubes need to be further experimented on to produce better-characterized structures to combat and target specific biological activity (Fig. 3).

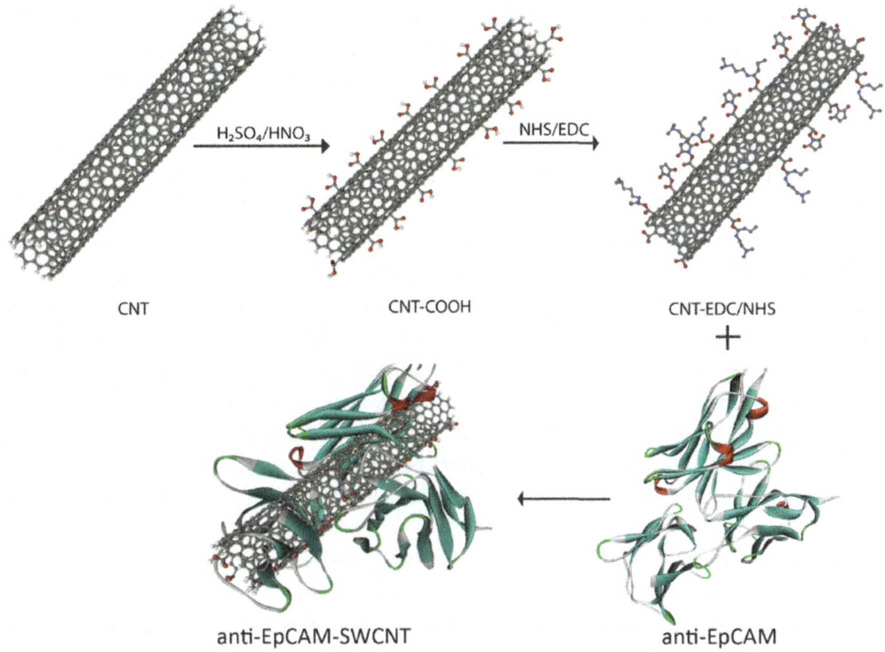


**Fig. 3** Carbon nanotube variants [60]

As cancer metastasizes, it becomes increasingly difficult to treat effectively—impossible in many circumstances. The best that medical practitioners can do once cancer has metastasized is to try and treat as many parts of the metastases as they can. This often leads to a prolonged life of the patient, but ultimately cannot stop cancer entirely.

Tests of a new nanotechnology were performed by Mauro Ferrari (Houston Methodist Research Institute, Texas) on mice that were host to breast cancer that had spread to the lungs. Fifty percent of the test subjects were cured after 8 months of follow up (estimated to be similar to 24 human years of long-term survival). Nanoscale silicon (porous) is filled with doxorubicin (anti-cancer drug) in what is referred to as an injectable nanoparticle generator (iNPG). Being composed of a biodegradable polymer, the iNPG holds strands of silicon monomers containing the drug. The silicon monomers are able to collapse and enter cancer tumours where the drug is released and the tumour is killed (Conner 2016).

The first trial on human patients is expected to be in early 2017. Ferrari stated, "To my very best understanding, this is the first case we've ever seen of a therapy with a well understood mechanism that can provide long-term, disease-free survival of our preclincal animal populations" (Ferrari 2016). Ferrari also stated it would be the first demonstration of a cure of metastatic lung cancer. Ferrari attributed the successful result of the nanotechnology due to the nature of the drug delivery mechanism that concentrates the nanoparticles in the tumour and avoids healthy cells.

This is a win–win as both the tumour becomes treated more effectively due to the higher concentration of nanoparticles, and the toxic side effects are minimized due to the specific targeting of cancer cells only (Connor 2016).

### 2.1.5 Exploiting Nanotechnology to Penetrate the Blood–Brain Barrier

In animals and humans, the blood–brain barrier is, "a neurobiological frontier that isolates brain tissues from the blood vascular system" [32]. This barrier is a membrane that keeps the central nervous system and the brain protected from the variations in blood composition (hormones, nutrients, drugs), yet allows water and lipophilic molecules to pass through. Diffusion is the main driving force across the blood–brain barrier, of which molecule size, lipid solubility, and concentration gradient play a decisive role in what passes through the membrane. The membrane poses a unique challenge to the effective delivery of cancer-combatant therapeutic drugs, as most drugs are not suitable to pass through the dense network of capillary endothelium cells [32].

The use of nanotechnology to bypass the blood–brain barrier has proven an effective solution. Micelles, liposomes, dendrimers, microcapsules, and polymeric nanoparticles have all exhibited beneficial characteristics in conquering this plight. Their exceptional stability, biocompatibility, and effective ability to deliver and distribute drugs make them competitive solutions to permeating the blood–brain barrier. Future research will maintain a focus on safety and the neurobiological processes involved in regulating the blood–brain barrier as potential side effects may include neurological disorders [32].

## *2.2 Active Targeting*

Nanoparticle active targeting is an extension of passive targeting with the difference being that instead of using nanoparticles to target a specific tumour area and releasing the drugs outside the cell, you are now specifically targeting the cancer cells themselves through molecular recognition and releasing the drugs from within. Molecular recognition of cancer cells can be done either through ligand-receptor or antibody-antigen recognition although the former is more commonly used and researched [83]. Ligands are biomolecules that bind to a target protein or receptor located on the cell membrane. Receptor ligand interactions cause specific cellular responses ranging from chemical signals to initiating and facilitating transport through the cell membrane which is of most importance to nanoparticle uptake. The type of ligands used for the nanoparticles are selective ligands which only bind or have an affinity to specific receptors. Antibodies are protein molecules that are used by the immune system to actively target and attach to specific antigens, which are also proteins that are located on the surface of cells, in order to trigger a larger immune response. The ligands and antibodies are surface functionalized or attached to the surface of nanoparticles which then gives the nanoparticles an affinity for the corresponding receptor or antigen. However, molecular affinity is not enough for the specific targeting of cancer cells since healthy cells also may have the same receptors. The phenomenon that is specific only to cancerous cells that allows for specific targeting of these cells is called overexpression [83]. Overexpression is the over replication of proteins due to cell membrane receptors and tumour antigens. With a much higher number of receptors and antigens, the surface functionalized nanoparticles will have a much higher affinity for these cells instead of the healthy cells. Once attached to the surface, the nanoparticle can be internalized through the process of receptor-mediated endocytosis or phagocytosis [7]. Receptor-mediated endocytosis is the active transport across the cell membrane in which both the ligand and receptor diffuse across the cell membrane and into the cell while phagocytosis is the encapsulation of the ligand by the cell membrane. Once inside the cell, the cytotoxic drugs can be released at once or slowly over time depending on the nature of the nanoparticle. The entire process of active targeting can be seen in a step-by-step diagram in Fig. 4.

From Fig. 4, it can be seen that the nanoparticles are first administered into the bloodstream by injection. They then travel along the bloodstream towards the tumour micro circulatory system. The nanoparticles enter the tumour vesicle network and remain in that network due to EPR effects. Through ligand-receptor interactions,

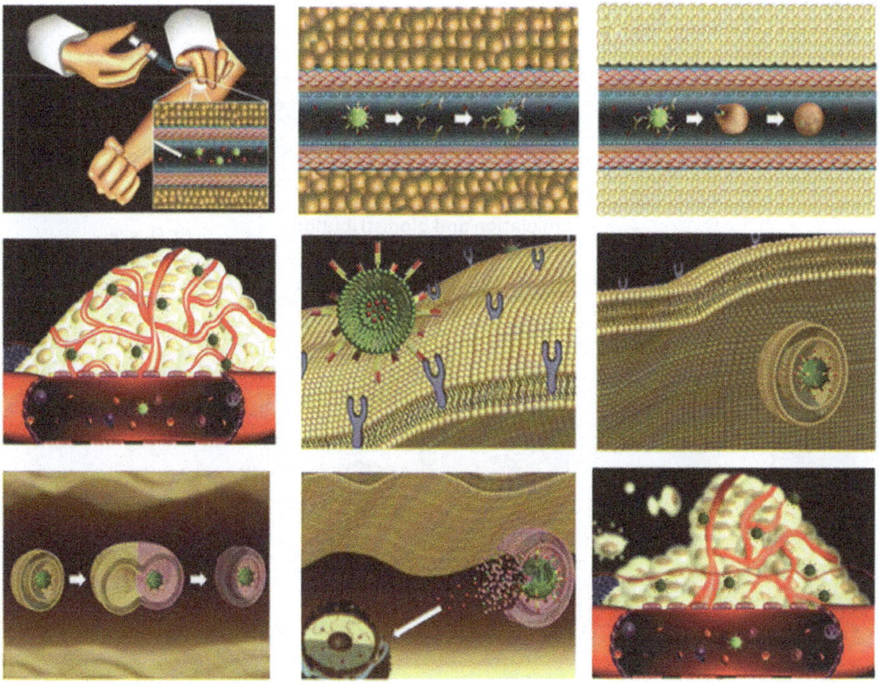

**Fig. 4** The process of active targeting [38]

the nanoparticles attach to the cell membrane and are then internalized by receptor-mediated endocytosis forming an early endosome. The endosome then fuses with a lysosome in which the low pH environment causes the release of the drugs from the nanoparticle if the drug release mechanism is due to a pH difference. However, there could be other drug release mechanisms like thermal release mechanisms.

In addition to targeting the specific cells, active targeting can also target the surrounding neo-vasculature system of the solid tumours. These neo-vasculature systems also contain many receptors that can be targeted to administer drugs that reduce the size of these micro-circulatory networks needed for tumour sustainability. A decrease circulatory blood network would decrease the amount of nutrients available and would ultimately lead to tumour reduction as the cells start to die from a lack of nutrients.

### 2.2.1 Factors that Affect Nanoparticle Active Targeting

There are many physiological and physical properties that can affect the ability of the nanoparticle from targeting cancer cells. These range from blood stream circulation retention to the properties of the nanoparticle itself and its construction. The following sections will describe several of the factors that will impact the performance of the nanoparticle for active targeting. Figure 5 shows a diagram of the many factors that will affect nanoparticle performance.

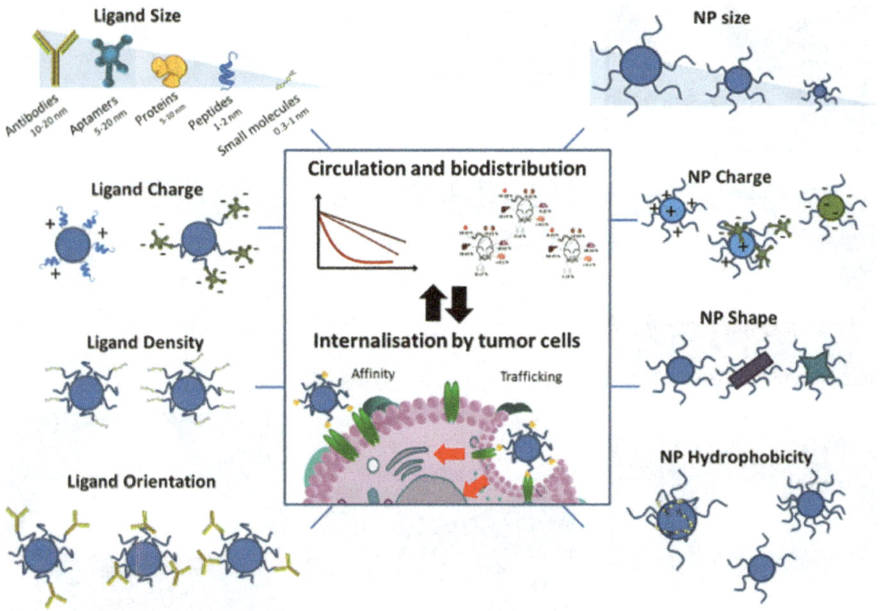


**Fig. 5** Factors that affect nanoparticle active targeting [7]

### Nanoparticle Ligand Density

The density of the ligands on the surface of the nanoparticle has a large impact on the ability of the ligands to attach to their substrate receptors. Increased ligand density on the surface results in multiple interactions with the cell membrane receptors. This forces the clustering of the receptors around the nanoparticle causing the cell membrane to wrap itself around the nanoparticle which prevents the nanoparticle from detaching [7]. This wrapping phenomena allows for the use of weak ligand-receptor interactions as high ligand density will prevent detachment. However, if the density is too high, steric hindrances may occur in which the ligands will block one another from attaching to a receptor [7]. In some cases, increasing ligand density can result in increased recognition by the immune system resulting in clearance out of the circulatory bloodstream.

### Nanoparticle Size

The size of a nanoparticle will affect the retention times in the blood circulatory system as well as the specific targeting of the cancer cells. Retention time in the bloodstream is mainly determined by the ability of the immune system to detect and remove the nanoparticle from circulation. Nanoparticles less than 5.5 nm cannot be used since they are readily removed from the bloodstream by the kidneys through glomerular filtration [38]. The larger the particle, the higher chance of detection from the immune system. In some cases, optimal cellular uptake of the nanoparticles is a strong function of size with a very narrow size range. For example, the size range for optimal uptake from breast adenocarcinoma cells is a 20–25 nm range [7]. Lastly, the size will affect the lateral mobility of the nanoparticle in the bloodstream. In order for a nanoparticle to target a cell, it needs to leave the bloodstream. Thus it needs to move laterally towards the cell membranes. This concept of lateral movement is called margination and the factors that affect margination are diffusive, buoyancy, and gravitational forces [85]. In terms of size, since the diffusion coefficient is inversely proportional to the size, smaller nanoparticles will result in higher diffusive forces compared to any other momentum forces thus will have higher margination and a higher degree of deposition to the cell membranes. For example, from a test between 25 and 60 nm particles, the 25 nm particle showed a higher degree of deposition and internalization by cancer cells [7].

### Nanoparticle Shape

The amount of lateral movement in the bloodstream is also dependant on the amount of torque acting on the nanoparticle. The more oblique the shape, the more torque and lateral movement the nanoparticle will have in the bloodstream. Therefore, nanospheres will have the smallest margination rates while nanorods will have one of the highest rates. In a test comparing gold nanospheres and gold nanorods, the

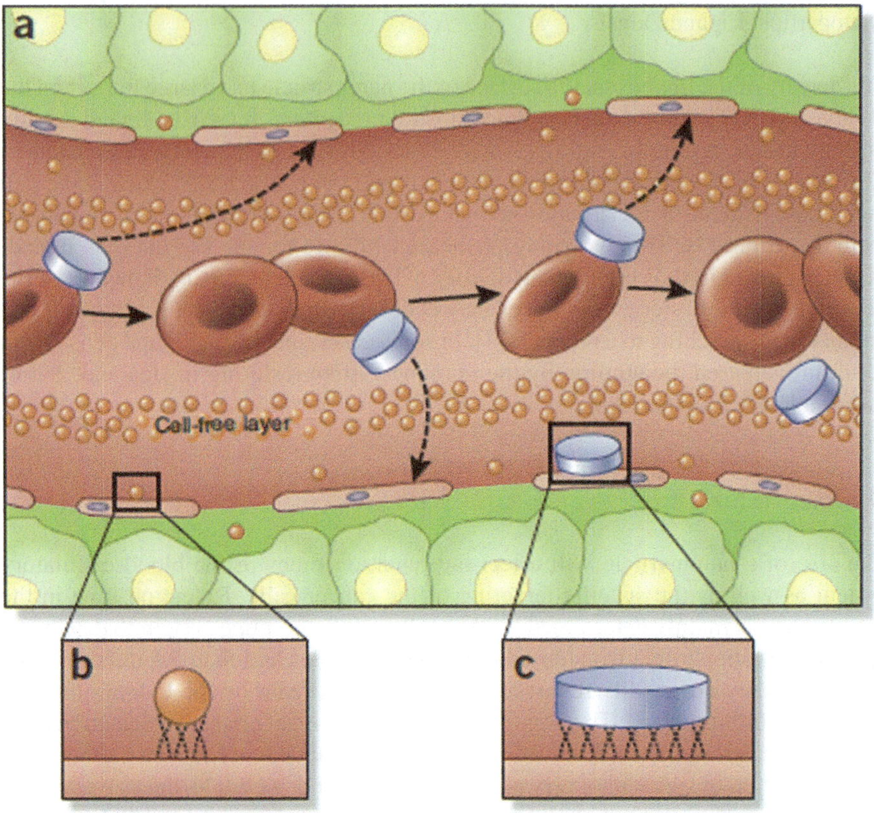


**Fig. 6** Effect of size and shape on margination [10]

nanorods had movement 8 times greater than the nanospheres [85]. The concept of margination can be seen in Fig. 6.

As seen in Fig. 6, the cylindrical-shaped particles tumble through the blood stream allowing for more lateral movement whereas the spherical particles travel in a laminar fashion. Also, spherical particles present the lowest amount of contact points thus cylindrical particles will have better binding efficiency than spherical particles [10].

Nanoparticle Density

The density of a nanoparticle has a large effect on the momentum it has in the bloodstream thus will affect margination towards the cell membranes. Lighter nanoparticles like liposomes will have higher margination rates than heavier nanoparticles like gold. In fact, when comparing liposomes to gold nanoparticles in an experiment to study the effects of density on margination, liposomes resulted in 57 times greater deposition than gold nanoparticles [85]. However, gold is still a very viable material for nanoparticles due to its physiological effects and advantageous physical properties. Compared to the size and shape of the nanoparticle, the density has the highest effects on the margination rates.

Surface Charge of Nanoparticles

The surface charge of a nanoparticle impacts the amount of cellular binding and uptake into the cancer cells. Since the cell membrane is negatively charged, the overall surface charge of the nanoparticle should be positively charged in order to maximize uptake into the cell. The overall surface charges depend on the surface coatings, ligand densities and charges, nanoparticle materials, and the type of formulation strategies [7]. Various surface coatings can be applied to the nanoparticle to make it positively charged. These coatings can include chitosan, dextran, folic acid polyethylene imide, and many others [62].

### 2.2.2 Active Targeting Current Research

Compared to passive targeting, active targeting is a very new concept that has only seen extensive research in the past few years. As of 2014, there are only 5 targeted liposomes and 2 targeted polymeric nanoparticle treatments that have gone to clinical development [7]. One of the main drawbacks of actively targeted nanoparticles is the fact that each nanoparticle formulation targets only one specific receptor. Given the fact that many different types of cancer exist overexpressing different receptors, the treatments may be very limited to only a few types of cancer. Therefore, current research and clinical trials are now looking into targeting receptors that are common to many different types of cancers. One such receptor is the transferrin receptor, which is the receptor used to transport iron into the cell [7]. This receptor is overexpressed in many types of cancer since increased iron intake is necessary to sustain the cell. Table 1 shows a vast majority of the currently researched nanoparticle treatments

**Table 1** Nanoparticle active targeting current research [7]

| Formulation | Nanoparticle type | Type of receptor | Drug type |
|---|---|---|---|
| BIND-014 | Polymeric | PSMA (small molecule) | Docetaxel |
| MM-302 | Liposome | Her2 | Doxorubicin |
| MBP-426 | Liposome | Tf-Receptor | Oxaliplatin |
| CALAA-01 | Polymeric | Tf-Receptor | siRNA |
| SGT53 | Liposome | Tf-Receptor | p53 plasmid DNA |
| Rexin-G | Retroviral Vector | Collagen | Human cyclin G1 gene |

including the ligand and receptor types used, the cancer drugs used, and the clinical phases already performed.

## *2.3 Nanoparticle Opsonization Protection*

One of the major challenges in using nanoparticles for drug delivery is the ability to avoid detection and clearance by macrophages or white blood cells especially in the liver and spleen where the majority are located. In fact, this is one of the primary reasons why many cancer drugs and early developed unprotected nanocarriers fail. In almost all cases, unprotected nanoparticles can be cleared from the circulatory system within minutes upon introduction in the bloodstream [15]. The danger of this is that the intercepted nanoparticles can release their drug in other parts of the body like the spleen or liver leading to organ damage. The way the immune system removes the nanoparticles from circulation is a process called opsonization which is the marking of a foreign substance by an opsonin molecule which are plasma proteins that attach themselves to any foreign objects chemically enhancing the interactions of the nanoparticles with white blood cell receptors. The factors that affect this process depend on the size, surface charge, and the hydrophilicity of the nanoparticles [15]. To protect the nanoparticle from being marked by an opsonin molecule, the surface has to be surface functionalized with a highly hydrophilic polymer molecule that will prevent attachment of the opsonin molecule and ultimately provide stealth to the nanoparticle from the immune system. Currently, the polymer that has proved the most effective against opsonization is polyethylene glycol (PEG). The hydrophilicity of PEG provides steric hindrance to the opsonin molecules preventing adsorption to the surface by creating a hydrating layer around the nanoparticle due to the strong interactions with the surrounding water molecules. However, PEG on the surface of nanoparticles can prevent ligands from attaching to receptors. Since only a very small amount of PEG is needed to provide adequate stealth, the amount of PEG on the surface should be minimized. Experiments have shown that as little as 0.5 wt % of PEG is all that is needed to significantly reduce opsonization [3]. Other types of opsonization protection can be seen in Fig. 7.

The figure shows that self-peptides like CD47 can be attached to the surface of nanoparticles, which will cause the phagocytes to recognize the nanoparticles as one of its own preventing phagocytosis. The nanoparticles can also be coated with extracted cell membranes providing biomimetic properties to the nanoparticle causing recognition as a cell [10].

# 3 Nanoparticle Materials and Formulations

The type of material used for cancer nanoparticles is very important in terms of their eventual targeting capabilities and bioavailability or circulatory retention. Safety is

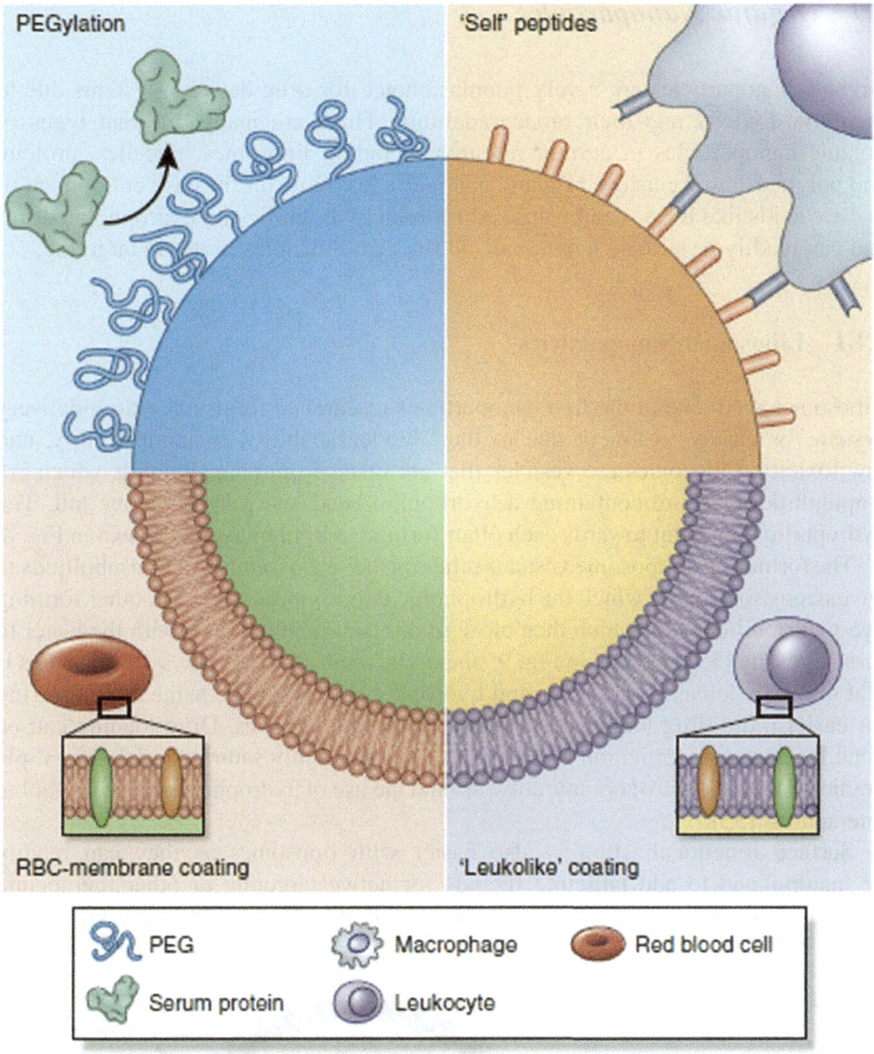


**Fig. 7** Opsonization protection [10]

also a major concern when it comes to materials as the material must be biocompatible, biodegradable, and low in toxicity. Once the nanoparticle has unloaded the drug, it should be able to degrade and be metabolized into nontoxic substances which are then cleared through recirculation. There are two main categories of nanoparticles researched currently which are the organic nanoparticles and the inorganic nanoparticles. The organic nanoparticles will be described first followed by the inorganic carriers.

## 3.1 *Organic Nanoparticles*

Organic nanoparticles are a very popular choice for drug delivery systems due to their low toxicity and their biodegradability. There are many different types of organic nanoparticles in current research including liposomes, micelles, protein, and polymer-based carriers. Organic nanoparticles allow for the easy entrapment of nucleic acids like RNA, can be utilized for both hydrophilic and hydrophobic drugs and can readily be surface functionalized for stealth features or active targeting.

### 3.1.1 Liposomal Nanoparticles

Liposomes were one of the first nanoparticles researched for use as a drug delivery system for cancer treatment due to their biodegradability, biocompatibility, and low toxicity. Liposomes are vesicles that are made from phospholipids which are amphiphilic in nature containing a hydrophilic head and a hydrophobic tail. The hydrophilic tails point towards each other forming a lipid bilayer as shown in Fig. 8.

The formation of liposome vesicles relies on the self-assembly of phospholipids in an aqueous solution in which the hydrophobic tails connect with each other forming two hydrophilic layers which then close in on itself as it interacts with the water to form a vesicle [55]. As seen in Fig. 9, one of the main characteristics of liposomes is that they can deliver hydrophilic and hydrophobic drugs at the same time allowing for easier multi-drug loading than many other nanoparticles. Drug loading can be done by liposome formation in aqueous solutions highly saturated with drugs, pH gradient methods, use of organic solvents and the use of hydrophilic and hydrophobic interactions [55].

Surface functionalization is also easier with liposomes as they can readily be manipulated to add targeting ligands for active targeting or other therapeutic molecules to the surface. Due to the size of liposome particles, they are highly

**Fig. 8** Liposome vesicle (Horiba Scientific 2016)

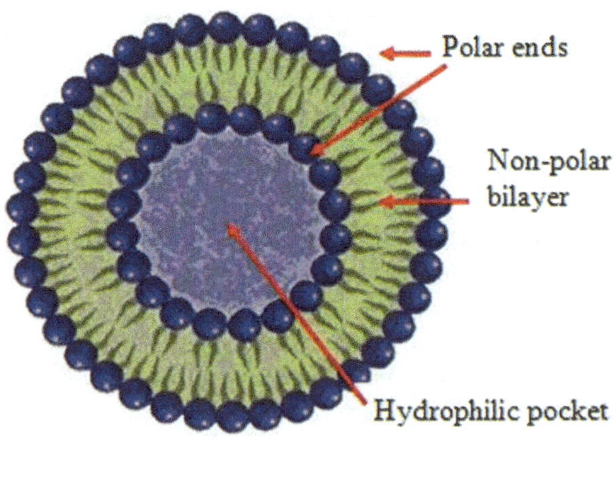

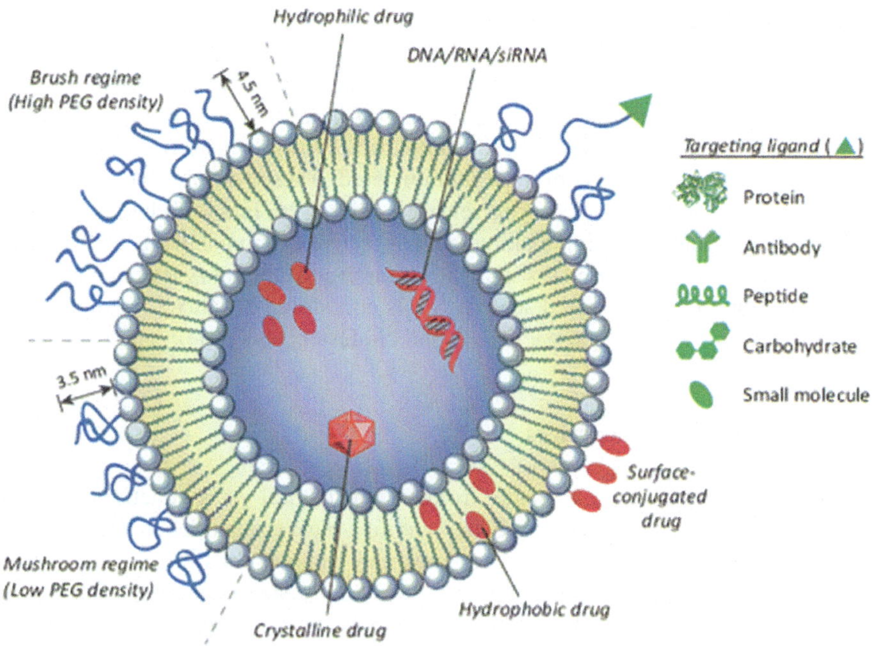


**Fig. 9** Liposome vesicle drug loading [14]

susceptible to detection by the immune system thus PEG surface functionalization is mandatory for added stealth.

Liposome nanoparticles are already approved for human use with many in the clinical trial phases and some that are commercially available. The most popular liposomal drug is called Doxil. This was the first nanoparticle drug approved by the FDA and is currently commercially available. Doxil is a 75–100 nm liposome nanoparticle that encapsulates the anti-cancer drug doxorubicin which is then used to target breast cancer, ovarian cancer, and Kaposi's sarcoma. It uses passive targeting to release the drug into the tumour tissues. Compared to free doxorubicin, tests have shown that Doxil reduces serious toxicities like cardiotoxicity and neutropenia [6]. Other approved liposomal formulations include Myocet, which is used to treat breast cancer, DaunoXome, and Depocyt. Currently, only passive targeting liposomes are approved but a lot of research is being done on active targeting carriers. Further research is looking into thermo-responsive liposomal particles which will release the drugs upon application of heat and magnetic liposome particles which can be guided to the tumour using magnets.

### 3.1.2 Solid Lipid Nanoparticles (SLN)

Solid lipid nanoparticles are a relatively new type of nanocarrier technology. It is a spherical nanoparticle that consists of a solid lipid core matrix through which lipophilic drugs can be loaded. These nanoparticles must be stabilized by a surfactant emulsion in order for them to be administered. These particles have extremely low toxicities, good controlled drug release, and can be mass produced easily by high-pressure homogenization making them attractive economically as well as functionality [55]. This technology is still in the research and laboratory testing phase with many formulations being tested such as SLNs containing docetaxel for melanoma, doxorubicin for colon cancer, paclitaxel, and cholesteryl butyrate.

### 3.1.3 Polymeric Nanocarriers

Polymeric nanoparticles are made from either natural or synthetic polymers. Compared to liposomes, they offer better stability, have a stealthier surface, and can be prepared with diverse polymer structures and compositions for increased functionality to meet any requirement necessary for drug stability and administration [90]. Therefore, its versatility is what makes polymeric nanoparticles very attractive. Many different polymers are FDA approved for consumption such as aliphatic polyesters, polypeptides, PEG, and dextran. There are many different types of polymeric nanoparticle structures due to their versatility such as micelles, nanogels, polymersome, etc. However, currently, micelles are the most commonly researched polymeric nanoparticles. Micelles are vesicles that are made from block copolymers that contain hydrophilic and hydrophobic units. The hydrophilic and hydrophobic units self-assemble in an aqueous solution forming a structure containing a hydrophobic core stabilized by a hydrophilic shell as seen in Fig. 10. The copolymers can be arranged in A-B type diblock copolymers, A-B-A triblock copolymers, or grafted copolymers [61].

The main advantage for polymeric micelles is that it allows for increased apparent solubility for hydrophobic drugs in the bloodstream which is mostly made up of water. Current micelle formulations in research can be made from polyethylene glycol, poly(D,L-lactide), poly (caprolactone), poly(propylene oxide), poly(L-aspartate) and polyamers [61]. There is currently one approved polymeric micelle formulation of poly(ethylene glycol) and poly(D,L-lactide) copolymers for commercial use named Genexol-PM which delivers the drug paclitaxel. Other formulations in clinical trials can be seen in Table 2.

In terms of active targeting, Bind-014 polymeric nanoparticle is one of the leading nanoparticle technologies for active targeting currently in clinical research used to target prostate cancer.

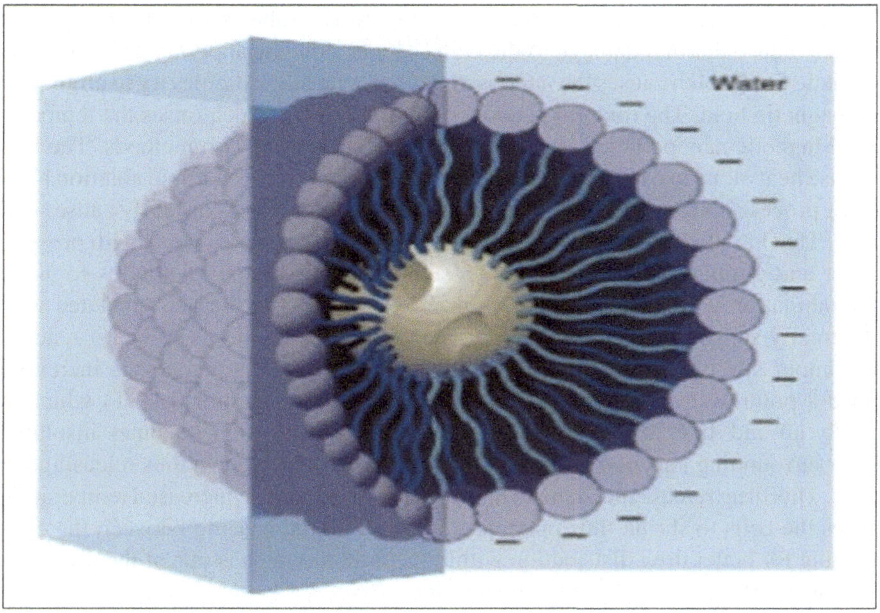


**Fig. 10** Micelle nanoparticle (JRH Co 2011)

**Table 2** Current micelle nanoparticle formulations [61]

| Micelle formulation | Copolymer | Drug | Diameter (nm) | Cancer type |
|---|---|---|---|---|
| NK012 | PEG-Pglu | SN-38 | 20 | Breast cancer |
| NK105 | PEG-P(aspartate | Paclitaxel | 85 | Stomach cancer |
| SP1049C | Pluronic L61 and F127 | Doxorubicin | 22–27 | Adenocarcinoma of oesophagus |
| NC-6004 | PEG-Pglu | Cisplatin | 30 | Solid tumours |
| Genexol-PM | PEG-P(D,L-lactide) | Paclitaxel | 20–50 | Breast, pancreatic, lung and ovarian cancer |

## 3.2 *Inorganic Nanoparticles*

### 3.2.1 Magnetic Iron Oxide Nanoparticles

One area of nanoparticle cancer research is to use magnetic nanoparticles to enhance drug accumulation in tumour and to induce cell destruction. Magnetic iron oxide nanoparticles are at the forefront of this research as iron oxide has the lowest toxicity and has the best bio distribution out of many other magnetic compounds [28]. Magnetic iron oxide nanoparticles are primarily used to induce hyperthermia within the tumour which is the raising of the cell temperatures to above normal body temperatures to cause thermal damage to the cells. This is done by applying an alternating magnetic field which causes the nanoparticles to rotate as the dipoles try to align with the magnetic field. The rotation generates friction or heat which raises the temperature. Magnetic nanoparticles are used to treat tumours using two methods. The first is to use heat alone to cause cell necrosis in a process known as thermal ablation [28]. In this process, the temperature is raised to around 60 °C which instantly causes cell death. This has already been tested in phase 1 clinical trials on patients with prostate cancer and glioblastoma with positive results [28]. The second method is to use it in combination with drug delivery. An iron oxide core is further encapsulated with a thermally sensitive polymer which also encapsulates the drugs. When it reaches the tumour site by either active or passive targeting, the temperature is increased above a point known as the lower critical solution temperature (LCST) which is usually around 40–45 °C [46]. Above the LCST, the polymer becomes insoluble in the surrounding aqueous environment and starts to break down thus releasing the drugs. This drug release mechanism has the advantage that the increased temperature causes the cells to shrink thus opening up the interstitial spacing between the cells allowing for better drug distribution within the tumour which is one of the problems with nanoparticles and cancer treatment [46].

### 3.2.2 Quantum Dots

Quantum dots (QDs) attracted the attention of many within the biological community when first introduced, as they "showed the potential to revolutionize biological imaging" [65] and be applied to various fields ranging from the energy to the medical industries [13]. However, their uses have been extensively researched in the field of medicine, particularly cancer research. QDs are one of the many novel utilizations of nanotechnology in the research of cancer and the fight against it. They exhibit their greatest ability in enhancing the detection and diagnosis processes—with a focus on cellular and in vivo molecular imaging.

QDs are nano-sized, semiconductor crystals. They are known for having unique optical, chemical, and electronic characteristics such as size- and composition-tunable light emissions, high luminescence, and resistance to photo-bleaching. The high luminescent of QDs is due to their high quantum yield, while their high molar-extinction coefficient allows for variable emission of wavelengths from the visible to infrared spectrums [8, 9]. High resistance to photo-bleaching, or fading, allows for imaging techniques to be observed and tracked over extended periods of time. In addition, QD can also be excited with a single light source with minimal spectral overlapping, producing multiple wavelengths and colours simultaneously [65].

The general structure of a quantum dot consists of three main components: the core–shell architecture, a polymer coating, and attached biomolecules. A simplified and more detailed diagram is seen below in Figs. 11 and 12, respectively.

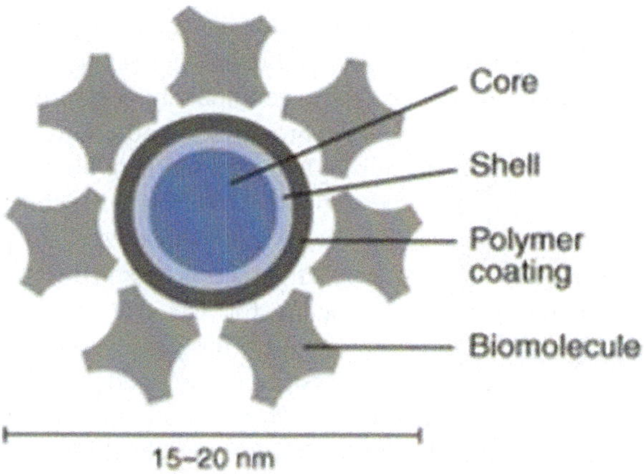


**Fig. 11** The basic structure of the QD is composed of the 3 main components: the core–shell architecture, a polymer (or organic) coating, and conjugated bio-molecules [8, 9]

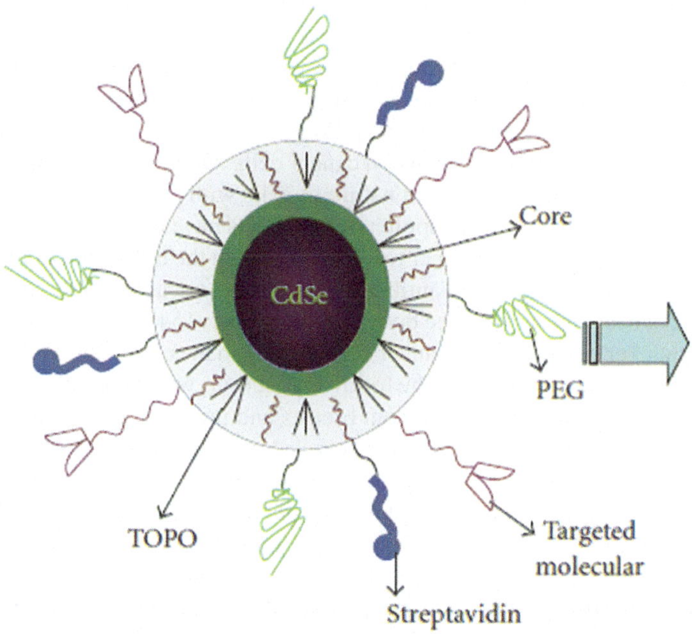


**Fig. 12** A detailed QD, composing of a CdSe core–shell architecture, trioctylphosphine oxide (TOPO) coating, and attached polyethylene glycol (PEG) and streptavidin conjugates [65]

The core–shell architecture is composed of semiconducting materials, with the semiconductor core compound being encased in another semiconductor shell material, such as CdSc and Zns, respectively. This component generally ranges from 2 to 10 nm in diameter [65]. This architecture provides a medium to engineer different photophysical properties of the QD, through manipulation of the fluorophores, at the single-particle level. As well, it can be augmented to change the collective behaviour of the particles [13]. The fine-tuning of the size, thickness, and particles of the core–shell system changes the inter-particle interaction to emit different wavelengths.

The polymer coating provides a means for the QD to combat its natural hydrophobic attribute, making the QD compatible in aqueous conditions. The coating is also a protective layer to help weather harsh conditions and prevent the heavy metal ions, existing in the core–shell architecture, from dissolving into the environment [13]. Though the coating may also be made of organic and inorganic material and may have multiple layers, which would effectively increase the hydrodynamic radii to 15 nm or more [13]. It provides a platform for functional biomolecules to bond with the targeted probe molecules such as cancer cells.

Bio-molecules are the last main component of the QD. The polymer coating provides a large surface area for conjugation with bio-molecules that function mainly for the active targeting of their specified tumour cells. Additionally, as QDs are naturally hydrophobic, solubilizing QDs is also achieved through surface modification with these bio-functional molecules [65]. Popularly utilized are the attachments of antibodies that target certain antigens on cancer cells. Once attached, the QD acts as a bio-mark to reveal the location and nature of cells with high specificity [65].

Current research and possible applications of QDs have been focused on the area of imaging and diagnostics of cancer research. Through the utilization of multicolour QDs conjugated with a designed antibody, researchers are able to identify multiple, and specific cancer cells around the body simultaneously [54]. This would encourage better and quicker diagnostics of cancer cells and aid in the elimination of such due to the earlier detection of these cells. The following sections will focus on the application of QDs in imaging and diagnostics from breasts and prostate cancer studies.

One application of QDs in cancer research has been in the profiling of breast cancer cells. Breast cancer is the most common cancer among Canadian women, and one of the main causes is due to the amplification of the HER2 gene. This gene generally helps control how healthy breast cells grow, divide, and repair themselves; however, in 25–30% of breast cancers, the HER2 gene fails to operator regularly and creates too many HER2 genes. This leads to an overexpression of HER2 proteins and eventually the manifestation of the breast cancer tumour (Breastcancer.org 2015).

Yezhelyev et al. took advantage of this knowledge and combined it with the functionality of QDs to demonstrate their capabilities of detecting both cultured and human breast cancer cells. However, instead of detecting the HER2 proteins alone, they expanded their research by profiling five significant tumour marks: HER2, ER, PR, GFR, and mTOR proteins. These tumour marks were detected using multiple bio-conjugated QDs set at different frequencies and the resulting images were also compared to conventional imaging methods [54].

The study demonstrated the strong and effective possibilities of simultaneous detecting of multiple target proteins, or 'multiplexing' capabilities of the QDs in cancer research. This ability of detecting multiple target proteins can offer a clearer view of cancer tissues that would offer more effective therapeutic decisions [54]. The resulting spectra collected from the emission of excited QDs can quantify the number of breast cancer cells at each body site—providing a more detailed representation of their status or growth. Yezhelyev et al. studies also reveal a more vibrant and distinct imaging of breast cancer cells compared to conventional methods. Both the resulting images of the multiple proteins, their spectra, and a comparison to conventional imaging methods can be seen in Figs. 13, 14, and 15.

Further studies of QD-based imaging of breast cancer cells by Peng & Li achieved similar results. The study was conducted on 700 patients with invasive breast cancer, and using similar QD methods as Yezhelyev et al., they conducted their imaging experiments solely targeting HER2 proteins. The results were compared with conventional immunohistochemistry (IHC) analysis and concluded that a QD-based approach to imaging was more sensitive, accurate, and possibly economic. This indicated that QD-based methods have the potential "for clinical application, especially in developing countries" (Peng and Li 2009). Figure 16 shows the comparison of the two methods from their study.

QD technology may play a critical role in the early diagnostics of all cancers, including prostate cancer. Prostate cancer is the most common cancer among men

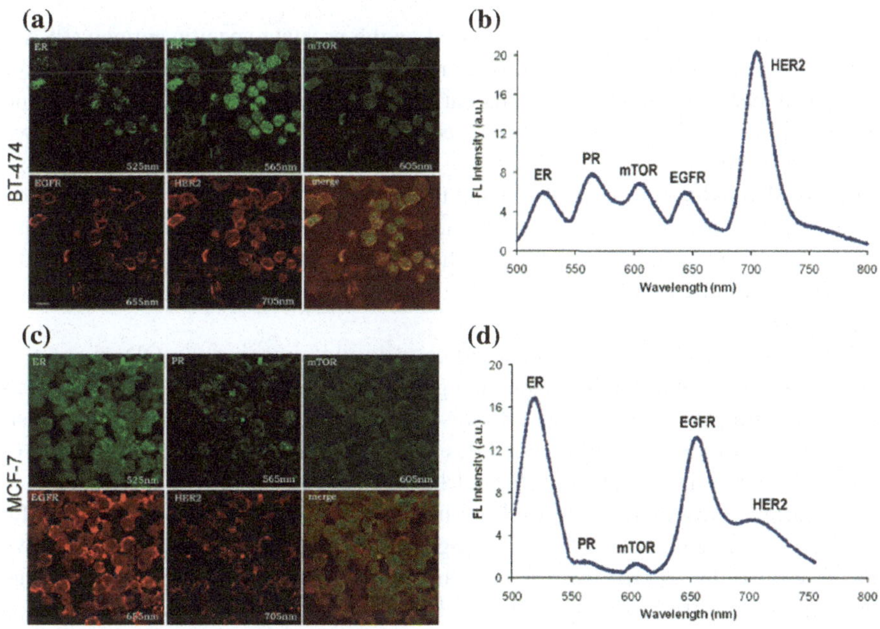


**Fig. 13** The resulting QD-based imaging of the five breast cancer tumour marks of cell lines BT-474 and MCF-7 and their respective spectra [54]

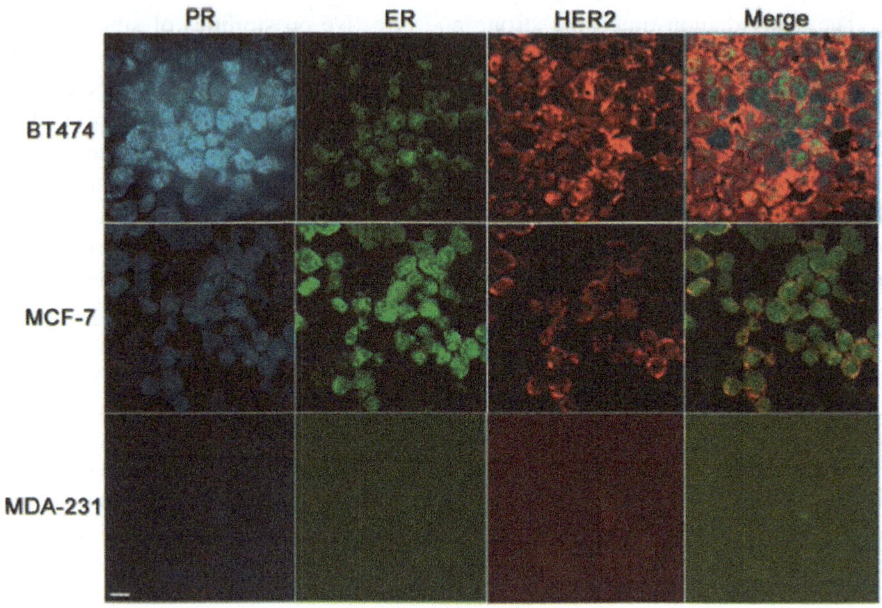


**Fig. 14** A juxtaposition of a more vibrant QD-based cancer cell imaging (left) compared to conventional imaging methods (right) [54]

in the West and its early diagnosis is "based on the prostate-specific antigen (PSA)" [65]. The detection of PSA and androgen receptors (AR) is an important marker for prostate cancer cells, and therefore its early detection. Research utilizing QDs and the detection of both PSA and AR has been conducted since 2004, and has shown promising results.

Gao et al. conducted a study regarding the potential of using QDs in the detection and analysis of prostate cancer cells in mice. QD probes conjugated with PSA antibodies were used to identify and target the prostate cancer cells. Figure 17 reveals the QD-based imaging of a mouse with a growing prostate tumour, compared to a healthy mouse.

The sensitivity and multicolour abilities of in vivo imaging via QDs were also put to the test in this study, again using mice. QD-based in vivo imaging was compared with conventional green fluorescent proteins (GFP), as well as the sensitivity and brightness of multicolour QDs in a separate experiment. The results of both experiments are seen below (Figs. 18 and 19).

Concerning the experiments resulting in the above figure, it was shown that QD-based in vivo imaging and GFP based imaging resulted in similar brightness (the two images to the right of (**a**)); however, only the QD signal was observed in vivo (orange dot in the bottom right of (**a**)). Gao et al. comment that an intensity comparison between QD and conventional GFP imaging could not be achieved due to optical variables such as tissue scattering. On the other hand, this qualitative result does

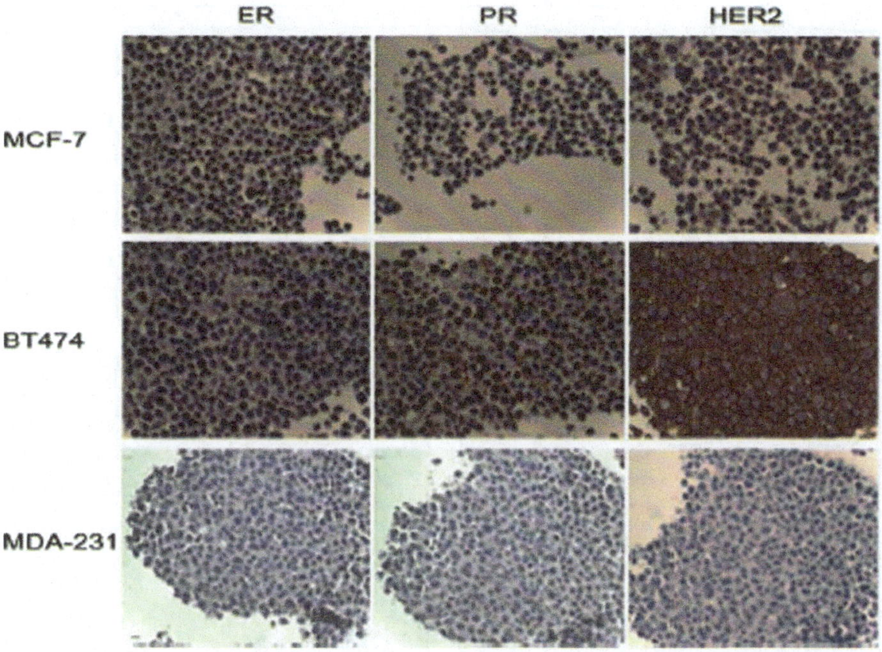


**Fig. 15** A juxtaposition of a more vibrant QD based cancer cell imaging (left) compared to conventional imaging methods (right) [54]

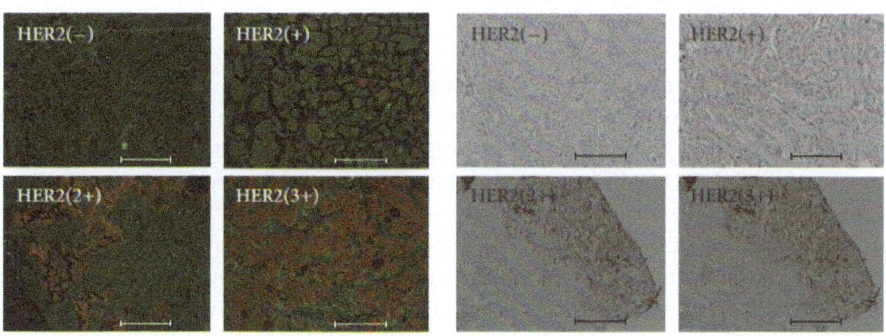


**Fig. 16** The juxtaposition of QD-based imaging and conventional IHC analysis of breast cancer cells; with a focus on targeting the HER2 protein (Peng and Li 2009)

demonstrate that the "emission spectra of QDs can be shifted away from the autofluorescene" [36]. Therefore, allowing the detection of QDs at lower signal intensities and providing more qualitative information regarding the prostate tumour.

As for the multicolour experiment in (**b**) of the same figure, it was observed that the QDs of all three colours were observed: "simultaneously in the same mouse and with a single light source" [36]. This can be attributed to the QDs feature of

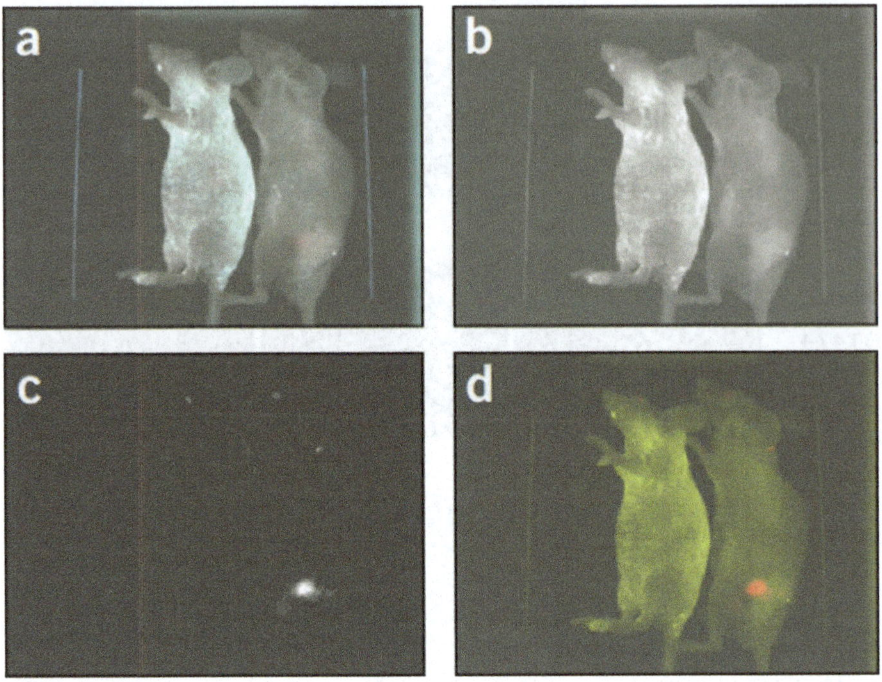


**Fig. 17** Images of a mouse with a growing prostate tumour (right) and a healthy mouse with no tumour (left) injected with QDs. **a** Original image; **b** unmixed autofluorescene image; **c** unmixed QD image; and **d** super-imposed image [36]

having a high molar extinction coefficient in comparison to organic dyes (about 10–50 times larger) as well as high quantum yields. As such, the absorption rates, of QDs compared to organic dyes will be proportionally faster. As a result, the in vivo imaging utilizing QDs will be much brightness and vivid than images using organic dyes [36].

The study conducted by Gao et al. concluded in achieving sensitive and multi-colour florescence imaging of the prostate cancer cells. This study showed the early potential of the unique optical properties of QDs as they "provide new opportunities of multicolour imaging and multiplexing" [36] due to their demonstration of highlighting a large number of genes, proteins, or small molecules. Furthermore, the study also reveals that the QDs have a longer excited-state lifetime (20–50 ns) than conventional organic dyes (2–5 ns)—nearly one order of magnitude [36]. In turn, QD probes would be practical for the "fluorescence lifetime imaging of cells, tissue specimens and living animals" [36].

In 2008, a study conducted by Shi et al. compared the use of QD-based imaging to conventional IHC based imaging of prostate cancer cells. They constructed QDs conjugated with bio-molecules to detect both AR and PSA, Fig. 20 shows the resulting images.

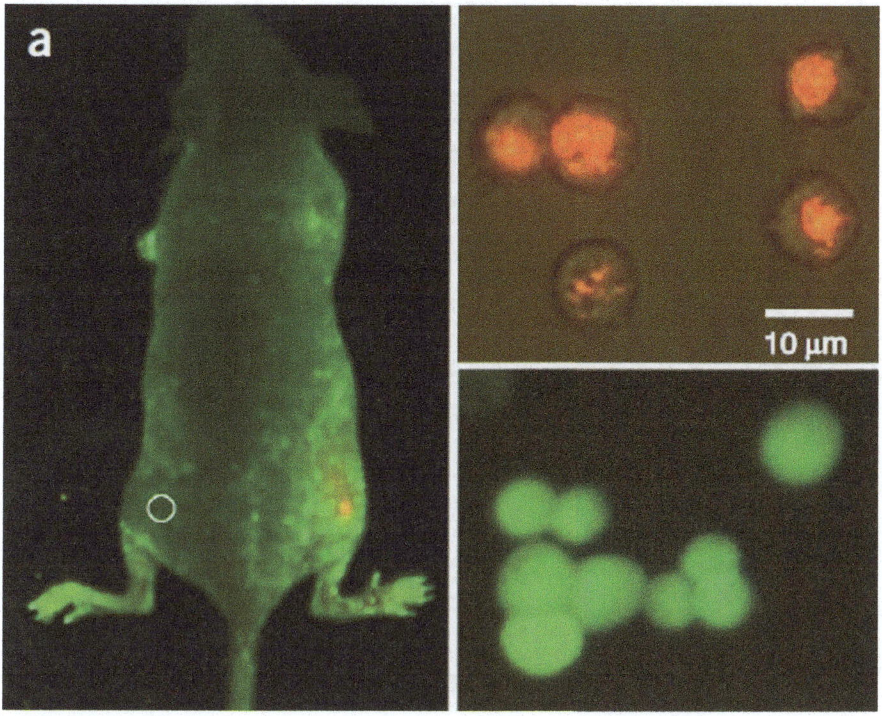


**Fig. 18** The sensitivity and multicolour capabilities of QDs in living mice [36]

The results of their study confirmed that QD-based imaging of AR and PSA is "superior to conventional IHC" [78]. As such, this was directly correlated with the QDs having the ability for multiplexing and also their simultaneous detection under one excitation source [78].

Overall, the QD plays an important role in locating and understanding the condition of the cancer cells for a specific area of the body in much detail as seen in various experiments and studies. QDs have a superior brightness factor compared to organic dyes due to their high quantum yield and capabilities to fine-tune their core–shell architecture [36]. This allows QDs to produce more vibrant and distinct imaging of the cancer cells and tumours. As per the size and composition of the core–shell architecture, the wavelength emitted by the QDs can be tuned, in turn, affecting the overall sensitivity of the analysis and spectra produced for the targeted cancer cell. In addition, the broad excitation spectra that QDs may produce highlight their multiplexing capabilities under a single excitation source. The simultaneous detection of multiple signals and colours, and therefore cells, may lead to quicker and more distinct diagnosis of cancers [8, 9]. Lastly, their high resistance to photo-bleaching, or fading, can be attributed to the coating that protects the core–shell design [65]. This factor is important to real-time imaging tracking of cancer cells in the long term. All of such factors allow QDs to significantly aid in the early detection, diagnosis,

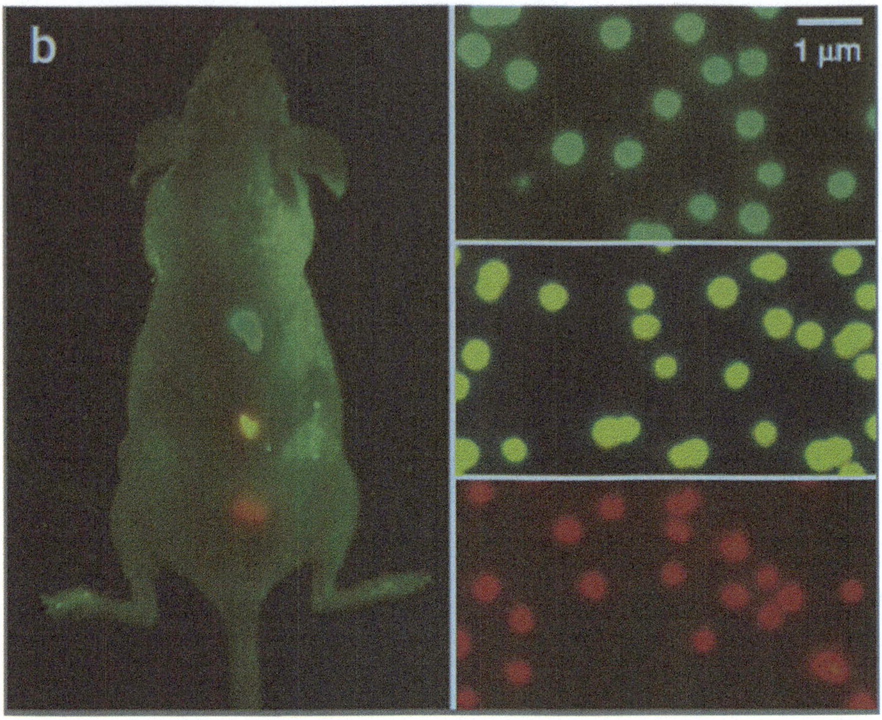


**Fig. 19** Green-, yellow-, and red-coloured QDs under one excitation source in a mouse [36]

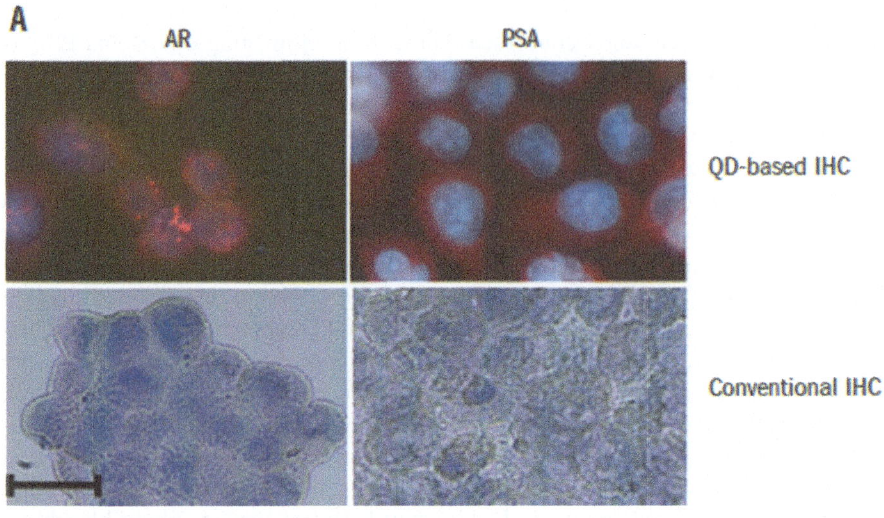


**Fig. 20** Comparative imaging of conventional IHC and QD-based IHC of prostate cancer cells [78]

and treatment of cancerous tumours and may revolutionize current imaging methods used in practice.

### 3.2.3 Gold Nanoparticles

Gold nanoparticles (Au-NPs) were initially proposed to be used for early cancer detection, largely due to their "light absorption and emission characteristics" [8, 9]. They are capable of emitting 200 times more light than QDs, but typically take up a volume approximately 60 times greater. Methods for cancer detection include fixing an antibody with an affinity towards cancer cells to an Au-NP platform. Au-NPs can be used through injection, exposure to a tissue sample, or mixed with a blood sample. The samples are checked using a microscope under white light. Depending on the type of surface proteins exhibited by certain cancer cells, the Au-NPs will align in various positions. When alignment changes, the light scattering of incident light also changes. The type of cancer can be identified by the light scattering pattern, which is characteristic of the alignment of Au NPs on the cell surface. Au-NPs can withstand greater exposure to light without burning out when compared to QDs. They also do not require chemicals such as the contrast agents used in QDs to yield equivalent data. Currently, Au-NPs are found to be inherently biocompatible and as well as non-malignant [8, 9]. Figure 21 is an image of gold nanoparticles being used in cancer applications.

In the image above, the illuminated tissues are cancerous while the surrounding area consists of healthy cells. The Au-NPs are the small bright spots visible throughout the sample.

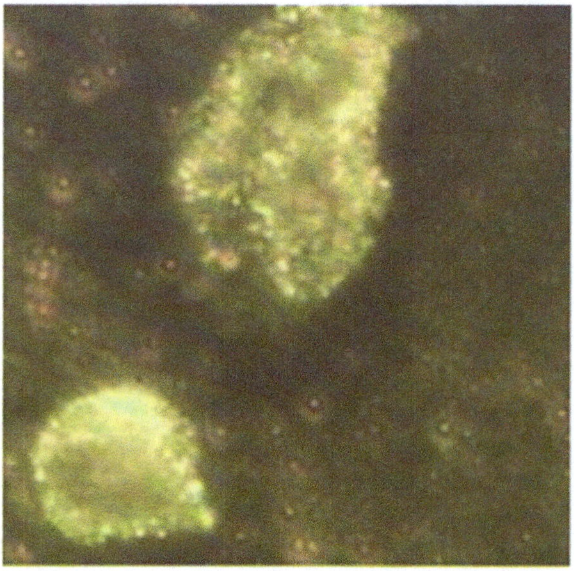

**Fig. 21** Gold nanoparticles detecting cancer cells [8, 9]

Au-NPs can be modified to allow for different types of cancer treatment and therapy. They can be constructed in geometries such as multipods, stars, rods, shells, boxes, and cages. The latter three geometries are all types of hollow Au-NPs which provide novel location-controlled drug delivery methods. The following image shows the types of treatment a single administration of hollow Au-NPs is capable of [1].

Figure 22 illustrates the concepts that in hollow geometries such as nanocages, Au NPs are capable of imaging, drug delivery, and photothermal therapy. Theranostic treatment is the term coined for the ability of a single treatment to have both therapeutic and diagnostic functionality. Gold nanocages, which are hollow Au-NPs with a porous shell, were invented by Sun et al. and reported in 2002 [82]. These nanocages excel when applied for theranostic treatment for cancer for a multitude of reasons. Gold nanocages have flat planar areas for functionalization, good mechanical properties, are easily mass produced, can have shell thicknesses of 2–10 nm, and can produce variable wavelengths (600–1200 nm) and different absorption/scattering patterns. The porous shell can transport a payload and also control the dosage of drugs released. The biodistribution of these nanocages for different cellular interactions can be modified by controlling the size from 20–500 nm. All these factors sum up the promising application of hollow Au-NPs in theranostic cancer applications [1].

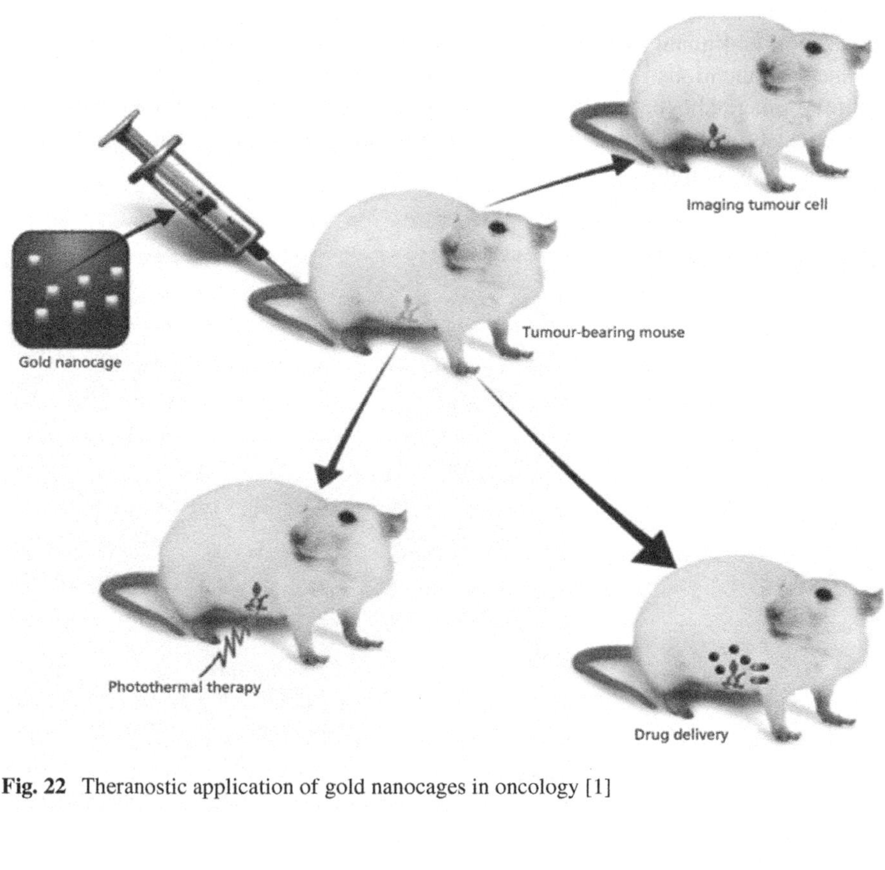


**Fig. 22** Theranostic application of gold nanocages in oncology [1]

Experiments have been conducted with antibody-functionalized, polymer-coated, 5 nm Au-NPs. Plasma-polymerized allylamine (PPAA), illustrated as the green shell around a yellow Au-NP core below, was used as the polymer coating. The process uses 1-ethyl-3(3-dimethylaminopropyl)carbodiimide hydrochloride (EDC), N-Hydroxysulfosuccinimide sodium salt (NHSS), and 2-(N-morpholino)ethanesulphonic acid (MES), to bio-conjugate the monoclonal antibody Cetuximab to the polymer shell in an amide-bond formation reaction [56] (Fig. 23).

In vitro and in vivo testing has proved the antibody-functionalized Au-NP is successful in targeting cancer cells. It was also discovered that the biodistribution of the monoclonal antibody is the same when conjugated with the Au-NP, making for an ideal platform. Further research is being conducted for the addition of a radioactive Au-NP core to provide therapeutic treatment as well [56].

Au-NPs have been studied for direct targeting of cancer cell nuclei to induce cell death and prevent cell division. The Au-NPs used in these experiments were 30 nm, PEG coated, and conjugated with arginylglycylaspartic acid (RGD) and an amino acid structured nuclear localization signal (NLS). RGD helped with cancer cell targeting and particle attachment to cells. NLS was used to allow the particle to enter the nucleus of the cell. Figure 24 shows the results of the administration of

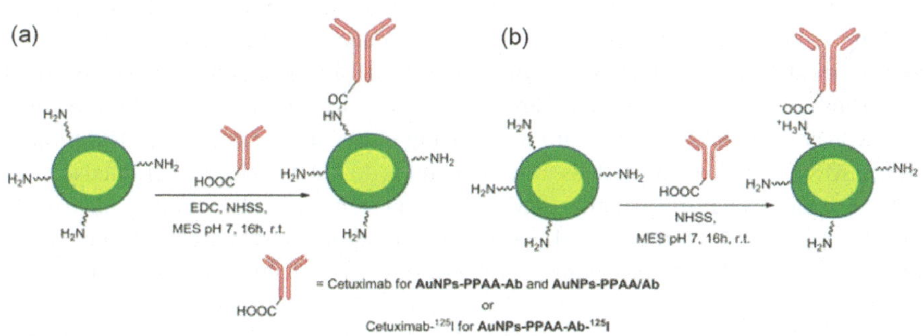


**Fig. 23** Synthetic pathway for the synthesis of covalent AuNPs-PPAA-Ab (**a**) and physiadsorbed AuNPs-PPAA/Ab bioconjugates (**b**) [56]

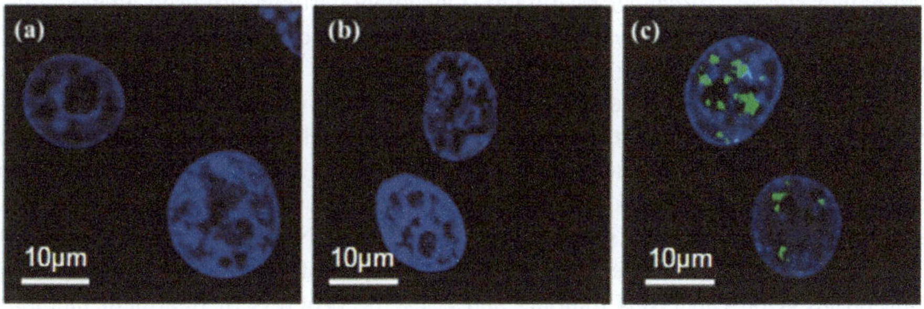


**Fig. 24** Au-NPs in 3 densities: **a** no nanoparticles, **b** lower density, and **c** high concentration [47]

RGD/NLS-AuNPs in three concentrations [47].

In the above figure, the nuclei of the cancer cell are stained blue using DAPI. After a certain concentration is reached, a point is observed when double-strand breaks (DSBs) occur with the cancer cell DNA. The green spots in the previous figure show where DSBs occurred with high concentrations. Lower concentrations of Au-NPs were not able to transport into the nuclei of the cancer cell, similar to if the Au-NPs were not conjugated with NLS. Targeting the DNA within cancer cells shows promise as one method to prevent division by stopping cytokinesis, leading to apoptosis. Results showed that 0.1 nM Au-NP concentrations only stopped 10% of cancer cells from progressing to the G1 phase, compared to 35% for 0.4 nM. During these experiments, the normal, healthy cells surrounding the cancer cells were not hindered during mitosis [47].

## 4 Cancer Imaging and Diagnosis

In the diagnosis and treatment of cancer, imaging and screening are major factors that must be optimized. Imaging is a tool that is used to guide decisions about cancer treatments as well as in monitoring the efficacy of the administered therapies. If imaging is not done correctly, or cannot detect cancer in a patient, a wrong diagnosis or treatment plan may be administered, therefore increasing the possibility of cancerous cells accumulating into larger and more harmful tumours. Currently, only non-invasive imaging methods are used, using conventional screening techniques, such as MRI, X-ray, and CT scans. These methods do not use tumour-targeted agents, and consequently can only readily detect cancerous tumour/cells once they have made a visible change to the tissue [86]. Even when a visible change to the tissue is seen through these methods, the nature of the tumour (malignant or cancerous) and the characteristics are unknown and must be determined using invasive procedures, such as biopsy. Nanotechnology can be applied as an aid in cancer detection by having tumour-targeted contrast agents/imaging probes, which increase the sensitivity and specificity of tumour imaging [86]. These nanoparticles can be used in the diagnosis in both cancer patients and healthy individuals that can lead to early detection, and higher success of cancer treatments. Once nanoparticles are used, they can tag cancer cells to then be easily detected by conventional screening techniques, as well as newer molecular imaging methods.

Two main factors that need to be met in the development of nanoparticles for cancer imaging are to: identify the cancerous cells, and enable the cells to be seen when screened [86]. Nanoparticles are able to provide rapid and sensitive detection of cancer-related molecules. This allows for the detection of molecular changes, even when they occur in a small percentage of cells. Nanoparticles when used in imaging applications can be used in the early stages of cancer, and can give real-time progress on cancerous tumours and track the progress of treatment methods [86]. For any nanoparticle to be approved by the FDA for clinical use, rigorous pre-clinical testing, which includes safety and toxicity studies, must be completed.

Nanoparticle imaging agents must undergo more safety considerations before being used, compared to nanoparticle therapeutics, as they can be administered to healthy individuals [84].

Nanoparticle cancer imaging can be divided into two main areas of study [40]:

1. Nanodetection for sensing protein in cancer cells
2. Nanoparticle or nanovector formation for high-contrast imaging.

Within these categories, nanoparticles have been used to tag a range of medically important targets such as bacteria, biomarkers, and molecules including proteins and DNA [40]. One of the main areas of interest in the use of nanoparticles for cancer imaging is the capture of circulating tumour cells, which generally only account for 1–2 cancer cells per milliliter of blood [40]. The nanoparticles in use are conjugated with cancer-specific antibodies or ligands that improve the capture of these cancerous cells. When able to detect cancerous cells in the blood, diagnosis can be made without a tumour formation. Easier detection is seen when a cancerous tumour is present. Figure 25 shows an example of how nanoparticles can be used in imaging cancerous tumours. In part **a** of the figure, the tumour is circled before the injection of nanoparticles, and in part **b** the same tumour is seen to appear dark after injection, increasing the contrast of the cancerous cells [39]. Both parts of the image were screened using magnetic resonance imaging (MRI).

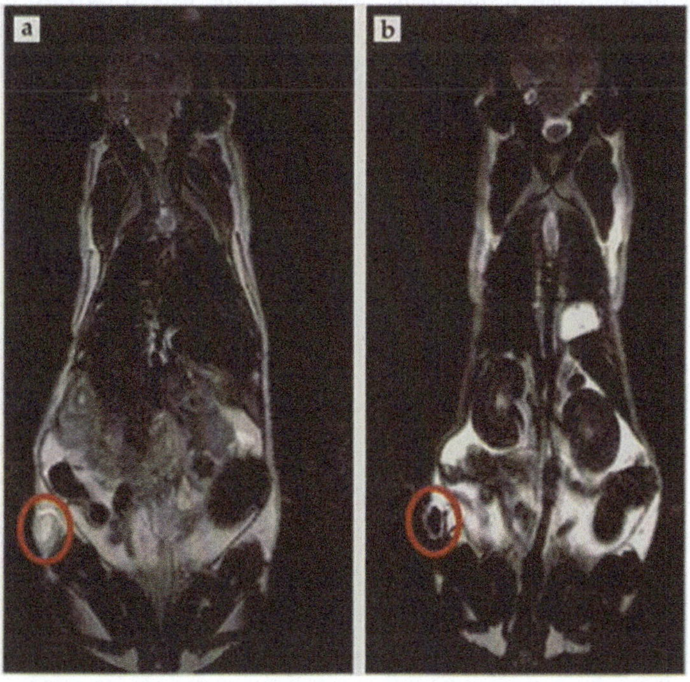


**Fig. 25** Iron oxide nanoparticles injected into a mouse implanted with a colon cancer tumour [39]

When nanoparticles are used for cancer imaging nanoparticles facilitate the association of hundreds of thousands of imaging moiety per construct [84]. This can allow for a signal amplification of up to one million times, enabling even very small amounts of cancer cells to be detected. A new field of molecular imaging has been developed that focuses on the visualization of biological events and processes in living systems [86]. Present approaches in molecular imaging include positron emission tomography (PET), single-photon emission tomography, and optical imaging including near-infrared fluorescence reflectance (NIRF) imaging and fluorescence-mediated tomography [86]. Figure 26 shows the use of quantum dots and optical imaging techniques in the localization of tumours in a mouse. After injection, bioluminescence imaging was first used to locate the tumours (shown with arrows), which were then confirmed using optical (NIRF) imaging [86].

Another field in the imaging of cancer using nanoparticles includes fluorescent nanoplatforms that can have targeting agents such as antibodies or ligands conjugated to them [40]. The goal of this technique is used to advance nanoparticle tumour targeting is to provide enhanced sensitivity and specificity for tumour imaging. This can lead to earlier detection, but has the risk of leading to overdiagnosis and overtreatment [40]. This is a potentially harmful effect of using nanoparticles for cancer detection and imaging, and therefore needs to be considered before treatment begins. As these techniques are currently only a research tool, it is important to recognize the risks, and ensure that surgeons do not remove an excess amount of tissue, which can potentially increase patient morbidity, with very little benefit [40]. At the current time, there is a lot of experimental work being done to find possible cancer imaging techniques using nanotechnology, with very few reaching clinical trials or clinical

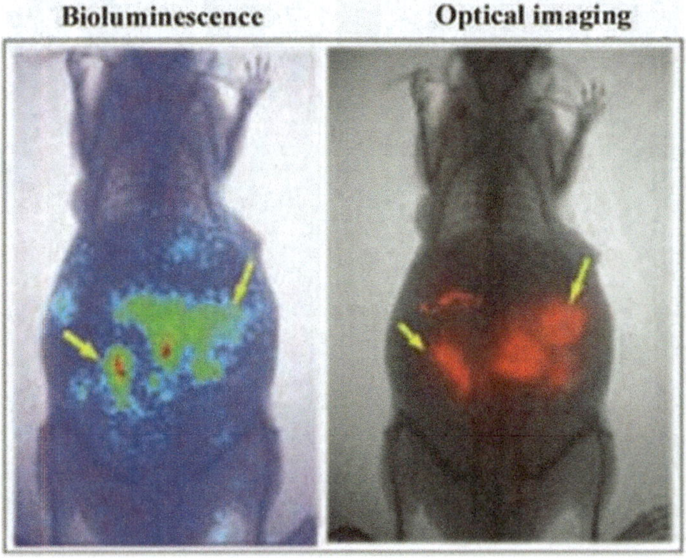


**Fig. 26** Optical imaging of tumour using quantum dots

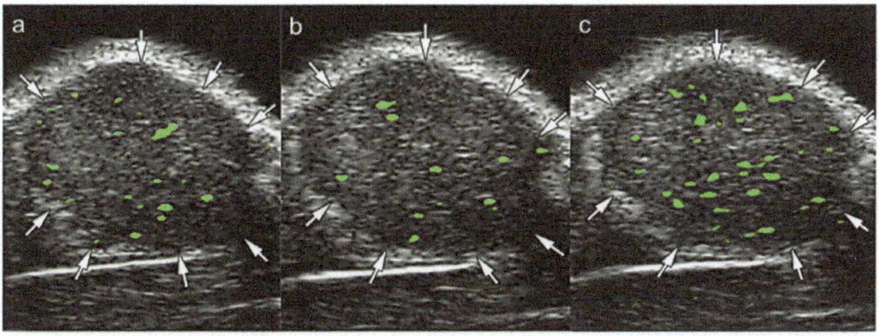


**Fig. 27** Nanoparticles used as contrast agents in cancer imaging and tumour targeting [84]

use. Figure 27 shows the use of fluorescent contrast agents used in cancer imaging and tumour targeting over time.

When fabricating nanoparticles for the use as contrast agents for tumour targeting, the following three characteristics are ideal [84]:

1. High rate of margination
2. Strong binding activity to tumour sites
3. Rapid internalization of targeted cells.

It is very challenging to produce a nanoparticle with all three of these characteristics because oftentimes in enhancing one design aspect, another will be minimized. Table 3 highlights the effects of nanoparticle design on pharmacokinetics, margination, and binding avidity.

The area of targeted nanoparticle imaging agents provides many new benefits to the accurate diagnosis and treatment of cancer [84]. To begin to maximize the

**Table 3** Effect of nanoparticle design on pharmacokinetics, margination, and binding avidity [84]

| Design parameter | Pharmacokinetics | Margination | Binding avidity |
|---|---|---|---|
| Size | Small particles (<10 nm) are cleared by end-stage renal, larger particles are cleared by the liver and spleen | More difficult for larger particles to escape the blood flow | Larger particles preferred—increased # of binding interactions |
| Shape | Oblate shapes preferred—more difficult to internalize by macrophages | Oblate shapes preferred—higher margination due to variable torques and drag forces | Oblate and discoidal shapes offer higher numbers of surface ligands available for interactions with receptor bed |
| Surface chemistry | Appropriate polymeric coating extends blood circulation | The length of the polymeric coating influences the hydrodynamic size | Optimum # of ligands maximizes binding avidity |

diagnostic accuracy of a nanoparticle imaging agent, it must be determined which design will maximize the transport of the nanoparticle to the cancerous tumour. One major advantage of targeted nanoparticle technology is the flexibility of the design in terms of material and platform. Materials that can be used include lipids, metals, and polymers [84]. The common nanoparticle platforms currently in use and testing are shown in Fig. 28. These nanoparticles displayed are commonly fabricated using top-down approaches, which allow for various sizes, shapes, and structures to be formed out of various materials [84]. Most commonly the top-down fabrication approaches such as the particle replication in non-wetting templates (PRINT) method and the dip-pen nanolithography method are used in the fabrication of nanoparticles for cancer imaging, as they allow for a large variety of shapes and structures [84]. Based on the type of technology used in the imaging process, most commonly X-ray and its derivatives, different nanoparticle platforms are used [84].

Multifunctional nanodevices hold the possibility of changing the way cancer is treated and diagnosed by allowing for early detection and effective targeted therapeutics at the earliest stages of the disease [79]. Multifunctional nanoparticles incorporate the imaging probe with the therapeutic agent and targeting ligand. These applications can recognize cancer cells, deliver drugs to the targeted tissue, aid in reporting the outcome of therapy, and provide real-time assessment of therapeutic efficacy

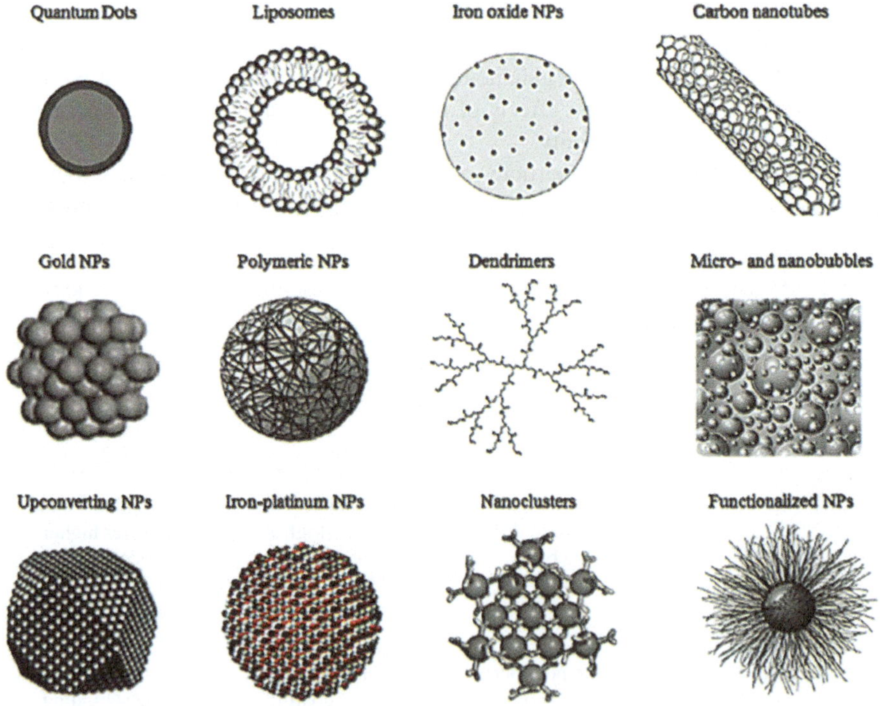


**Fig. 28** Common nanoparticle platforms for medical imaging [84]

[79]. Another major application is the ability to monitor intracellular changes to help prevent precancerous cells from becoming cancerous. Multifunctional nanoparticles can also be used to monitor the effectiveness of the drug in use by detecting changes in the tumour over time and can help to limit the dosage of drugs administered to the patient, as one drug will have many functions [79]. Figure 29 shows a multifunctional nanoparticle that has the capability of targeting, treating, and imaging cancerous tumours.

Tumour-targeted imaging devices are currently in clinical testing and have many promising applications [86]. There are also some issues that must be addressed before nanoparticles for cancer imaging can be used in clinical applications. These issues include [86]:

1. Identifying targeting molecules that are highly expressed in tumour cells, but can also be found in low levels in normal cells
2. Producing stable and high-affinity targeting molecules, which can be produced in large quantities
3. Developing imaging probes that are able to give off strong signals to improve cancer diagnosis, but has low toxicity to normal healthy tissues/cells
4. Increasing the retention time of imaging probes that allow for longer periods of blood circulation.

Overall nanotechnology for cancer imaging requires further research and study before being used in clinical applications, but has many promising benefits. As imaging is used to detect cancerous cells and track the progress of cancer treatments, it is essential that the methods in place are optimized for the highest level of

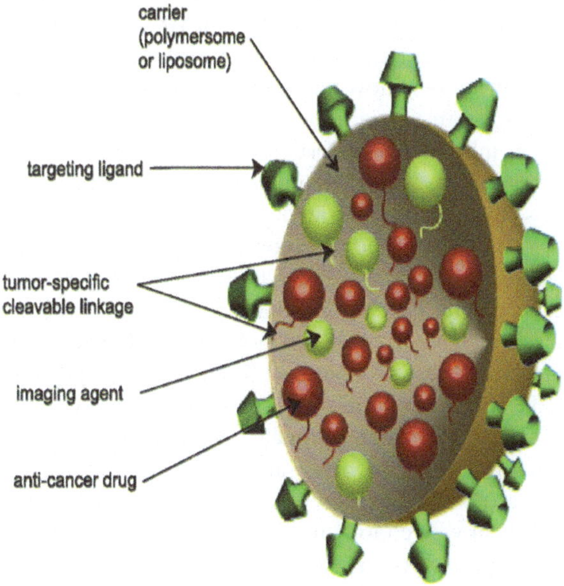


**Fig. 29** Nanoparticle with both diagnostic and therapeutic functions [79]

success. Nanoparticles provide a new and innovative way of tagging cancer cells to lead to earlier detection and better diagnosis and treatment.

## 5 Nanoparticle Cytotoxicity

In cancer imaging, therapy, or treatment within biological systems, the unanticipated behaviour of nanoparticles is a major concern requiring extensive research. As matter enters the nanoscale, the physical properties differ from those expected of that material on larger scales. In cancer applications, the goal is to develop nanoparticles that are toxic to cancer cells but do not damage healthy cells.

One major risk is caused by the exceptionally high surface area to volume ratio of nanoparticles. A one-gram sample of nanoparticles with diameters of 30 nm may have a combined surface area exceeding 200 $m^2$. Therefore, any risk is further compounded by the vast biological surface in contact with these particles. Another concern is intrinsic risks associated with the use of nanoparticles constructed with biologically toxic materials. Quantum dots are often built with cadmium, which may lead to the release of toxic-free cadmium ions if the polymeric shell deteriorates. Some materials which have traditionally been accepted as inert and/or biocompatible are toxic when utilized on the nanoscale. Nanoparticles of silver as well as those of industrial smoke have been found to be highly toxic when on the nanoscale compared to the microscale. There have been limited studies on possible reactions between artificially introduced nanoparticles and biological nanoparticles such as ribosomes. This becomes a concern when nanoparticles penetrate the cell [22]. Concerns surrounding size are a major risk to evaluate when using nanotechnology in cancer applications. For example, particles smaller than 100 nm can permeate into intestinal cells, while particles larger than 500 nm are unable to enter the circulatory system [2].

New studies show the size of inhalable nanoparticles, such as those used for lung cancer treatment, is important in relation to the blood–brain barrier [2]. Figure 30 shows inhalable nanoparticles passing through the blood–brain barrier.

As pictured in the figure above, undesired nanoparticles under a certain size can pass through this barrier leading to the possibility of brain toxicity and interaction with nerves and receptors in the nasal cavity. In cancer therapy, nanostructures designed to cross the blood–brain barrier may be toxic to cerebral endothelial cells. Experiments have found that cationic nanostructures when administered in high densities have toxic interactions with endothelial cells, while anionic and neutral nanostructures are nontoxic. It has been found that Au-NPs, even at the scale of 2 nm, are nontoxic in rats in part due to their anionic nature [2].

One example of inadequate biocompatibility limiting the use of potential nanoparticles in cancer revolves around the use of nanodiamond structures. The effect of these nanodiamonds was studied on glioblastoma cells (U-87 MG) and human umbilical vein endothelial cells (HUVEC). When administered, the results obtained from a transmission electron microscope showed that the HUVEC cells died at high doses (100 μg/milliliter). Comparatively, the U-87 MG cells were unaffected by the

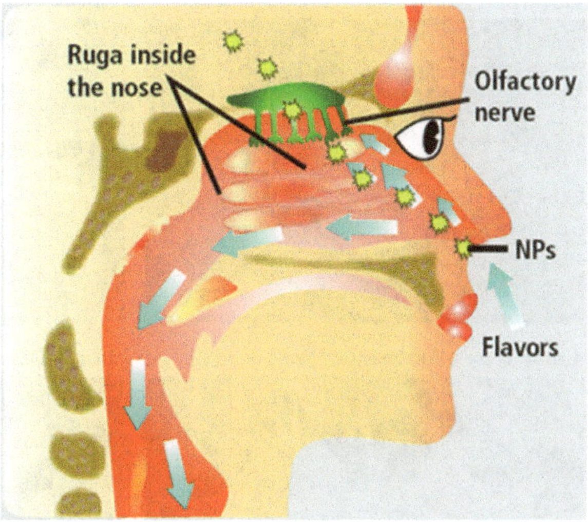


**Fig. 30** Passage of nanoparticles from the nose to the cerebral system via the cribriform plate ([2], p. 1120)

nanodiamonds. The increase in U-87 MG cell viability has been attributed to lower nanodiamond adsorption [63].

Reports have been made that there is insufficient clear evidence that particle hazards in biological systems substantially increase once they reach sizes under 100 nm. Thus, researchers are suggesting that rather than taking 100 nm as a hazard threshold, conventional information about hazards, information about particle toxicity from the macroscale should be applied on the nanoscale [24]. Figure 31 shows an example comparing the hazards of nanoscale particles to macroscale materials.

The above figure shows an example where particle toxicity on rodent macrophages is more dependent on shape than on size. In a 2009 experiment by Hamilton et al., $TiO_2$ cytotoxicity was evaluated in the form of 60–200 nm spherical particles, 0.8–4 μm fibres, and 15–30 μm fibres. It was found that both the spherical and shorter fibres were completely encapsulated by the cell through phagocytosis (seen in the upper right image). The longer length fibres were found to be toxic by distorting the cell (seen in the lower right image). The conclusion reached was that in this case 15–30 μm fibres were extremely toxic while the shorter particles and spherical nanoparticles of the same material were nontoxic [24]. The studies raise the important fact that tiered hazard classification must be implemented to establish a basis for risks associated with nanoparticle applications in biological systems.

In 2008, the FDA and EMEA have reported no significant observed health risks associated with the use of nanoparticles as therapeutic agents. Further development into targeting agents and particle functionalization will help ensure that nanoparticles take the most direct route to target as possible, and reduce unnecessary biological contact. Further development is being done to reach the ideal situation where hollow nanoparticles will not release any of the payloads until inside the target cell, then release the drug at a controlled rate. Tiered testing methods are being implemented

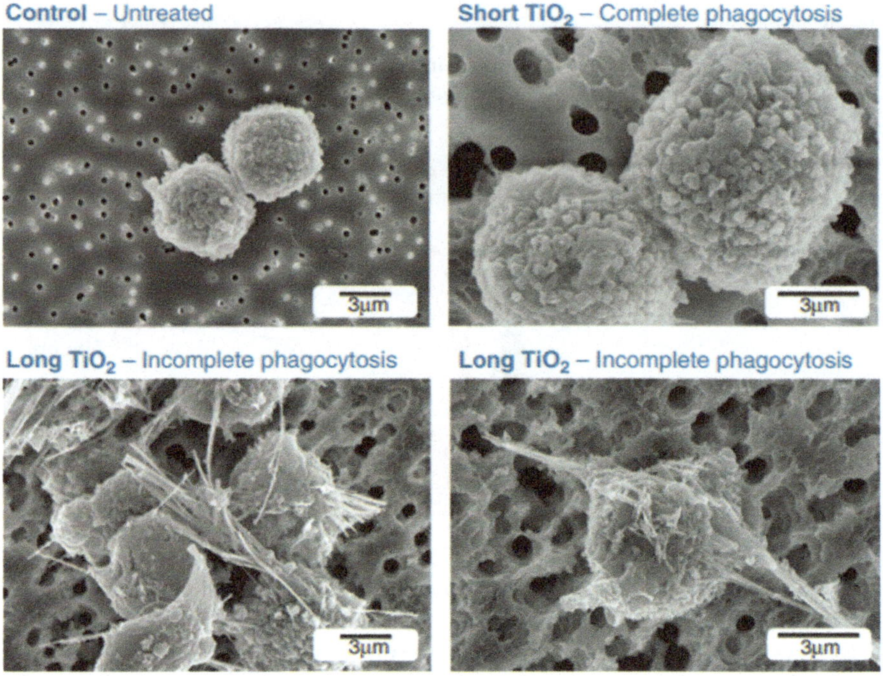


**Fig. 31** Effect of particle shape/length on macrophage uptake ([24], pp 728)

to further understand the impact of different nanoparticle characteristics—such as size, composition, and porosity—on biological systems [22].

## 6 Conclusion

The benefits of nanotechnology in cancer research are numerous. The small size of the nanoparticles makes it possible for them to interact with the cell organelles of the malignant cells, which is rather impossible in any other treatment mode. The active and passive drug delivery and controlled drug release makes it possible for the nanoparticles to target the malignant cells at the exclusion of healthy cells. Besides simply treating cancer, the usefulness of magnetic nanoparticles in cancer imaging, and thermal ablation makes NPs ideal for multimodal applications.

The use of nanoparticles in cancer is not completely free of side effects. The greatest risk in using NPs for cancer treatment is its carcinogenicity and toxicity, while some clinical trials have associated NPs with asthma, bronchitis, Alzheimer's, Parkinson's disease, and blood clots (Gmeiner and Ghosh 2014). However, these effects are insignificant compared to traditional chemotherapy and can be easily overcome and controlled. Another major obstacle in applying NPs in cancer treatment is the cost associated with it. However, with further research and development in technology, this can be overcome.

The latest research in the application of nanotechnology in cancer research suggests the application of NPs in cancer preventative measures. This can be accomplished by synthesizing nanobodies with genetic material that can be introduced into vulnerable cells shielding them against mutations caused by UV radiation (Mansoori et al. 2007). Moreover, the application of NPs in blocking the nutrient channels for malignant cells is also being studied. Further development in areas of nanoarrays, nanosensors, liposomes, monoclonal antibodies, and improved nanoparticles (dendrimers, diamondoids, gold-based nanoparticles, magnetic nanoparticles, and quantum dots) can make significant progress in the field of nanoparticles application in cancer research.

## References

1. Admad M et al (2012) Nanometric gold in cancer nanotechnology: current status and future prospect. J Pharm Pharmacol: 634–651
2. Ai J, Biazar E, Jafarpor M, Montazeri M, Majdi A, Aminifard S, Mandana Z, Akbari H, Rad H (2011) Nanotoxicology and nanoparticle safety in biomedical designs. Int J Nanomed 6:1117–1127
3. Alexis F, Pridgen E, Molnar L, Farokhzad O (2008) Factors affecting the clearance and biodistribution of polymeric nanoparticles. Mol Pharm 5(4). http://pubs.acs.org.ezproxy.lib.ryerson.ca/doi/abs/https://doi.org/10.1021/mp800051m
4. Arvizo R, Bhattacharya R, Mukherjee P (2010) Expert Opin Drug Deliv 7:753
5. Ayub A, Nazir S, Hussain T, Rashid U, MacRobert A (2014) Nanomaterials in combating cancer: therapeutic applications and developments. Nanomed Nanotechnol Biol Med 10:19–34
6. Barenholz Y (2012) Doxil-the first FDA approved nano-drug: lessons learned. US National Library of Medicine National Institute of Health. http://www.ncbi.nlm.nih.gov/pubmed/22484195
7. Bertrand N, We J, Xu X, Kamaly N, Farokhzad O (2014) Cancer nanotechnology: the impact of passive and active targeting in the era of modern cancer biology. Adv Drug Deliv Rev 66:2–25
8. Bhati W, Vishwa A (2013) Nanotechnology method comparison for early detection of Cancer. I.J. Intell Syst Appl 03:58–65. https://doi.org/10.5815/ijisa.2013.03.06
9. Bhati W, Vishwa A (2013) Nanotechnology method comparison for early detection of cancer. Int J Intell Syst Appl 5(3):58–65. Web
10. Blanco E, Shen H, Ferrari M (2015) Principles of nanoparticle design for overcoming biological barriers to drug delivery. Nat Biol 33(9). http://www.nature.com/nbt/journal/v33/n9/full/nbt.3330.html
11. Brannon-Peppas L, Blanchette JO (2004) Nanoparticle and targeted systems for cancer therapy. Adv Drug Deliv Rev 56:1649–1659
12. Brown DM, Wilson MR, MacNee W, Stone V, Donaldson K (2001) Size-dependent proinflammatory effects of ultrafine polystyrene particles: a role for surface area and oxidative stress in the enhanced activity of ultrafines. Toxicol Appl Pharmacol 175:191–199
13. Burns A, Ow H, Wiesner U (2006) Fluorescent core-shell silica nanoparticles: towards "Lab on a Particle", architectures for nanobiotechnology. Chem Soc Rev 35(11):1028–1042. Web
14. Cagdas M, Sezer A, Bucak S (2014) Liposomes as potential drug carrier systems for drug delivery. Intech. http://www.intechopen.com/books/application-of-nanotechnology-in-drug-delivery/liposomes-as-potential-drug-carrier-systems-for-drug-delivery#article-front

15. Chaudhari K, Ukawala M, Manjappa A, Kumar A, Mundada P, Mishra A, Mathur R, Monkkonen J, Murthy R (2012) Opsonization, biodistribution, cellular uptake and apoptosis study of PEGylated PBCA nanoparticle as potential drug delivery carrier. Pharm Res 29:53–68. http://link.springer.com.ezproxy.lib.ryerson.ca/article/10.1007%2Fs11095-011-0510-x
16. Chen J et al (2005) Gold nanocages: bioconjugation and their potential use as optical imaging contrast agents. Nano Lett 5:473–477
17. Chen Y, Liu L (2012) Modern methods for delivery of drugs across the blood-brain barrier. Adv Drug Deliv Rev 64(7):640–665
18. Clancy L, Goodman P, Sinclair H, Dockery DW (2002) Effect of air pollution control on death rates in Dublin, Ireland: an intervention study. Lancet 360:1210–1214
19. Clementi C, Miller K, Mero A, Satchi-Fainaro R, Pasut G (2011) Dendritic poly(ethylene glycol) bearing paclitaxel and alendronate for targeting bone neoplasms. Mol Pharm 8:1063–1072
20. Cuenca AG, Jiang H, Hochwald SN et al (2006) Emerging implications of nanotechnology on cancer diagnostics and therapeutics. Cancer 107:459–466
21. Daniel M, Astruc D (2004) Gold nanoparticles: Assembly, supramolecular chemistry, quantum-size-related properties, and applications toward biology, catalysis, and nanotechnology. Chem Rev 104(1):293–346. https://doi.org/10.1021/cr030698+
22. Debbage P, Jaschke W (2008) Molecular imagining with nanoparticles: giant roles for dwarf actors. Histochem Cell Biol 130:845–875
23. Demers LM, et al (2002) Direct patterning of modified oligonucleotides on metals and insulators by dip-pen nanolithography, 296, 1836–1838
24. Donaldson K, Poland CA (2013) Nanotoxicity: challenging the myth of nano-specific toxicity. Curr Opin Biotechnol 24(4):724–734. https://doi.org/10.1016/j.copbio.2013.05.003
25. Donaldson K, Stone V, Tran CL, Kreyling W, Borm PJA (2004) Nanotoxicology. Occup Environ Med 61:727–728
26. Hanahan D, Robert A (2011) Weinberg, hallmarks of cancer: the next generation. Cell 144(5):646–674. ISSN 0092-8674
27. Dow Chemical Company (2014) Chemical companies; dow electronic materials to build first, large-scale quantum dot manufacturing operation in Korea. Nanotechnol Wkly 142.
28. Durr S, Janko C, Lyer S, Tripal P, Schwarz M, Zaloga R, Alexiou C (2013) Magnetic nanoparticles for cancer therapy. Nanotechnol Rev 2(4):395–409
29. El-Sayed IH (2010) Nanotechnology in head and neck cancer: the race is on. Curr Oncol Rep 12(2):121–128. https://doi.org/10.1007/s11912-010-0087-2
30. Gentile F, Ferrari M, Decuzzi P (2008) The transport of nanoparticles in blood vessels: the effect of vessel permeability and blood rheology. Ann Biomed Eng 36:254–261
31. Fabbro C, Ali-Boucetta H, Da Ros T, Kostarelos K, Bianco Al, Prato M (2012) Targeting carbon nanotubes against cancer. R Soc Chem 48(33):3911–3926
32. Fakhoury M, Takechi R, Al-Salami H (2014) Drug permeation across the blood-brain barrier: applications of nanotechnology. Br J Med Med Res: 547–556
33. Ferrari M (2005) Cancer nanotechnology: opportunities and challenges. Nat Rev Cancer 5:161–171
34. Ferrari M (2010) Frontiers in cancer nanomedicine: directing mass transport through biological barriers. Trends Biotechnol 28(4):181–188
35. Fritz J, et al (2000) Translating biomeolecular recognition into nanomechanics, 288, 316–318
36. Gao X et al (2004) In vivo cancer targeting and imaging with semiconductor quantum dots. Nature Biotechnol 22(8):969–976. Web
37. Ghasemi Y, Peymani P, Afifi S (2009) Quantum dot: magic nanoparticle for imaging, detection and targeting. Acta Biomed 80:156–165
38. Goodall S, Jones M, Mahler S (2014) Monoclonal antibody-targeted polymeric nanoparticles for cancer therapy—future prospects. J Chem Technol Biotechnol 90:1169–1176. http://journals1.scholarsportal.info.ezproxy.lib.ryerson.ca/details/02682575/v90i0007/1169_mapnfctp.xml

39. Grossman JH, McNeil SE (2012) Nanotechnology in cancer medicine. (Cover story). Phys Today 65(8):38–42
40. Gunasekera UA, Pankhurst QA, Douek M (2009) Imaging applications of nanotechnology in cancer. Target Oncol 4(3):169–181. https://doi.org/10.1007/s11523-009-0118-9
41. Jawaid AM, Chattopadhyay S, Wink DJ, Page LE, Snee PT (2013) "A". ACS Nano 7(4):3190–3197. https://doi.org/10.1021/nn305697q
42. Jiang W, Kim BY, Rutka JT, Chan WC (2007) Advances and challenges of nanotechnology-based drug delivery systems. Expert Opin Drug Deliv 4(6):621–633. https://doi.org/10.1517/17425247.4.6.621
43. Johnston HJ, Hutchison G, Christensen FM, Peters S, Hankin S, Stone V (2010) (2010), A review of the in vivo and in vitro toxicity of silver and gold particulates: particle attributes and biological mechanisms responsible for the observed toxicity. Crit Rev Toxicol 40:328
44. Jokerst JV, Lobovkina T, Zare RN, Gambhir SS (2011) Nanoparticle PEGylation for imaging and therapy. Nanomedicine (lond) 6(4):715–728
45. Hamilton RF, Wu N, Porter D, Buford M, Wolfarth M, Holian A (2009) Particle length-dependent titanium dioxide nanomaterials toxicity and bioactivity. Part Fibre Toxicol 6:35
46. Hayashi K, Nakamura M, Miki H, Ozaki S, Abe M, Matsumoto T, Sakamoto W, Yogo T, Ishimura K (2014) Magnetically responsive smart nanoparticles for cancer treatment with a combination of magnetic hyperthermia and remote-control drug release. Theranostics 4(8)
47. Kang B, Mackey MA, El-Sayed MA (2010) Nuclear targeting of gold nanoparticles in cancer cells induces DNA damage, causing cytokinesis arrest and apoptosis. J Am Chem Soc 132(5):1517
48. Karathanasis E, Suryanarayanan S, Balusu S, McNeeley K, Sechopoulos I, Karellas A, Annapragada A, Bellamkonda R (2009) Imaging nanoprobe for prediction of outcome of nanoparticle chemotherapy by using mammography. Radiology 250:398–406
49. Kreyling WG, Semmler-Behnke M, Seitz J, Scymczak W, Wenk A, Mayer P, Takenaka S, Oberdorster G (2009) Size dependence of the translocation of inhaled iridium and carbon nanoparticle aggregates from the lung of rats to the blood and secondary target organs. Inhal Toxicol 21(Suppl 1):55–60
50. Kroll A, Pillukat MH, Hahn D, Schnekenburger J (2009) Current in vitro methods in nanoparticle risk assessment: limitations and challenges. Eur J Pharm Biopharm 72:370–377
51. Lamprecht A, Yamamoto H, Takeuchi H, Kawashima Y (2005) Nanoparticles enhance therapeutic efficiency by selectively increased local drug dose in experimental colitis in rats. J Pharmacol Exp Ther 315:196–202
52. Leroux J-C, Allemann E, De Jaeghere F, Duelker E, Gurny R (1996) Biodegradable nanoparticles—from sustained release formulation to improved site specific drug delivery. J Control Release 30:339–350
53. Loo C, Lowery A, Halas N, West J, Drezek R (2005) Immunotargeted nanoshells for integrated cancer imaging and therapy. Nano Lett 5(4):709–711
54. Yezhelyev MV, Al-Hajj A, Morris C, et al (2007) In situ molecular profiling of breast cancer biomarkers with multicolor quantum dots. Adv Mater 19(20):3146–3151. Web
55. Malam Y, Loizidou M, Seifalian A (2009) Liposomes and Nanoparticles: nanosized vehicles for drug delivery in cancer. Trends Pharmacol Sci 30(11). http://www.sciencedirect.com.ezproxy.lib.ryerson.ca/science/article/pii/S0165614709001370
56. Marega R, Karmani L, Flamant L, Nageswaran PG, Valembois V, Masereel B, Bonifazi D (2012) Antibody-functionalized polymer-coated gold nanoparticles targeting cancer cells: an in vitro and in vivo study. J Mater Chem 22(39):2135–21312. https://doi.org/10.1039/c2jm33482h
57. Matsumura Y, Maeda H (1986) A new concept for macromolecular therapeutics in cancer chemotherapy: mechanism of tumoritropic accumulation of proteins and the antitumor agent smancs. Can Res 46:6387–6392
58. Mühlfeld C, Gehr P, Rothen-Rutishauser B (2008) Translocation and cellular entering mechanisms of nanoparticles in the respiratory tract. Swiss Med Wkly 138:387–391

59. Nie S, Xing Y, Kim GJ, Simons JW (2007) Nanotechnology applications in cancer. Annu Rev Biomed Eng 9:257–288
60. Nima Z, Mahmood M, Karmakar A, Mustafa T, Bourdo S, Xu Y, Biris A (2013) Single-walled carbon nanotubes as specific targeting and Raman spectroscopic agents for detection and discrimination of single human breast cancer cells. J Biomed Opt. https://doi.org/10.1117/1.JBO.18.5.055003
61. Oerlemans C, Bult W, Bos M, Storm G, Nijsen J, Hennick W (2010) Polymeric micelles in anticancer therapy: targeting, imaging and triggered release. Pharm Res 27(12):2569–2589. http://www.ncbi.nlm.nih.gov/pmc/articles/PMC2982955/
62. Osaka T, Nakanishi T, Shanmugan S, Takahama S, Zhang H (2009) Effect of surface charge of magnetite nanoparticles on their internalization into breast cancer and umbilical vein endothelial cells. Elsevier Coll Surf B 71(2):325–330
63. Pagano A, Honore S, Esteve M, Braguer D (2012) Nanodrug potential in cancer therapy: efficacy/toxicity studies in cancer cells. Int J Nanotechnol 9(3/4/5/6/7):502
64. Pan Y, Neuss S, Leifert A, Fischler M, Wen F, Simon U, Schmid G, Brandau W, Jahnen-Dechent W (2007) Size-dependent cytotoxicity of gold nanoparticles. Small 3(11):1941–1949
65. Peng, C-W, Li Y (2010) Application of quantum dots-based biotechnology in cancer diagnosis: current status and future perspectives. J Nanomater 2010(2010):1–11. Web
66. Poland CA, Byrne F, Cho WS, Prina-Mello A, Murphy FA, Davies GL, Coey JM, Gounko Y, Duffin R, Volkov Y et al (2012) Length dependent pathogenic effects of nickel nanowires in the lungs and the peritoneal cavity. Nanotoxicology 6:899–911
67. Prajapati PM, Shah Y, Sen DJ (2010) Gold nanoparticles: a new approach for cancer detection. J Chem Pharm Res 2(1):30–37
68. Jain RK (1999) Transport of molecules, particles, and cells in solid tumors. Annu Rev Biomed Eng 1:241–263
69. Reynolds AR, Moein Moghimi S, Hodivala-Dilke K (2003) Nanoparticle-mediated gene delivery to tumour neovasculature. Trends Mol Med 9:2–4
70. Rhyner MN, Smith AM, Gao X, Mao H, Yang L, Nie S (2006) Quantum dots and multifunctional nanoparticles: new contrast agents for tumor imaging. Nanomed 1(2):209–217
71. Roberts MJ, Bentley MD, Harris JM (2012) Chemistry for peptide and protein PEGylation. Adv Drug Deliv Rev 64:116–127
72. Roux E, Francis MF, Winnik FM, Leroux JC (2004) Stimuli responsive liposome-polymer complexes: towards the design of intelligent drug carriers. In: Svenson S (ed) Carrier based drug delivery systems. CS Symposium Series 879, American Chemical Society, Washington, DC, pp 26–39
73. Rutishauser R, Mühlfeld C, Blank F, Musso C, Gehr P (2007) Translocation of particles and inflammatory responses after exposure to ine particles and nanoparticles in an epithelial air-way model. Part Fibre Toxicol 4:9–15
74. Salmaso S, Caliceti P (2013) Stealth properties to improve therapeutic efficacy of drug nanocarriers. J Drug Deliv 2013
75. Sanvicens N, Marco MP (2008) Multifunctional nanoparticles—properties and prospects for their use in human medicine. Trends Biotechnol 26(8):425–433
76. Sawant R, Torchilin V (2010) Liposomes as 'smart' pharmaceutical nanocarriers. Soft Matter 6:4026–4044
77. Sengupta S, Sasisekharan R (2007) Exploiting nanotechnology to target cancer. Br J Cancer 96:1315–1319
78. Shi C, et al (2008, 2009) Quantum dots-based multiplexed immunohistochemistry of protein expression in human prostate cancer cells. Eur J Histochem EJH 52.2:127–34. Web
79. Sinha R, Kim GJ, Nie S, Shin DM (2006) Nanotechnology in cancer therapeutics: bioconjugated nanoparticles for drug delivery. Mol Cancer Ther 5(8):1909–1917. https://doi.org/10.1158/1535-7163.MCT-06-0141
80. Slingerland M, Guchelaar H-J, Gelderblom H (2012) Liposomal drug formulations in cancer therapy: 15 years along the road. Drug Disc Today 17:160–166

81. Strebhardt K, Ullrich A (2008) Paul Ehrlich's magical bullet concept: 100 years of progress. Nat Rev Cancer 8:473–480
82. Sun Y et al (2002) Template-engaged replacement reaction: a one-step approach to the large-scale synthesis of metal nanostructures with hollow interiors. Nano Lett 2:481–485
83. Sutradhar K, Amin M (2013) Nanotechnology in cancer drug delivery and selective targeting. ISRN Nanotechnol 2014. http://search.proquest.com.ezproxy.lib.ryerson.ca/docview/1503653656/fulltextPDF?accountid=13631
84. Toy R, Bauer L, Hoimes C, Ghaghada KB, Karathanasis E (2014) Targeted nanotechnology for cancer imaging. Adv Drug Deliv Rev 76:79–97. https://doi.org/10.1016/j.addr.2014.08.002
85. Toy R, Hayden E, Shoup C, Baskaran H, Karathanasis E (2011) The effects of particle size, density and shape on margination of nanoparticles in microcirculation. Nanotechnology 22(2011):115101. http://iopscience.iop.org.ezproxy.lib.ryerson.ca/article/10.1088/0957448 4/22/11/115101/meta;jsessionid=6918F0A405162504A3701210EA0BB28D.c3.iopscience.cld.iop.org
86. Wang X, Yang L, Chen Z, Shin DM (2008) Application of nanotechnology in cancer therapy and imaging. CA A Cancer J Clinic 58:97–110
87. Watson P, Jones AT, Stephens DJ (2005) Intracellular trafficking pathways and drug delivery: fluorescence imaging of living and fixed cells. Adv Drug Deliv Rev 57:43–61
88. Xia Y et al (2011) Gold nanocages: from synthesis to theranostic applications. Acc Chem Res 44:914–924
89. Yacobi NR, Phuleria HC, Demaio L, Liang CH, Peng CA, Sioutas C et al (2007) Nanoparticle effects on rat alveolar epithelial cell monolayer barrier properties. Toxicol in Vitro 21:1373–1381
90. Zhong Y, Meng F, Deng C, Zhong Z (2014) Ligand-directed active tumor-targeting polymeric nanoparticles for cancer chemotherapy. Biomacromolecules 15:1955–1969. http://pubs.acs.org.ezproxy.lib.ryerson.ca/doi/pdf/https://doi.org/10.1021/bm5003009

# Chapter 12
# Nanotechnology for Food and Packing Application

**Pranta Barua, Adnan Hossain Khan, and Nazia Hossain**

## 1 Introduction

Nanotechnology is a developing innovation that allows us to produce materials at the atomic level into functional structures with nano-meter measurements. This application will make food items less expensive, and creation is more effective and more reasonable by utilizing less water and synthetic substances. Novel agrarian and food security frameworks, malady treatment conveyance strategies, instruments for subatomic and cell science, sensors for microbe identification, and natural insurance are instances of the significant applications that nano-innovation may have in the science and designing of farming and food frameworks. It can reform the worldwide food framework. To avoid biohazards and other metal, glass, toxic plastic and other components, physical screening test or statistics-based screening can be performed by using X-ray. Moreover, different biosensors, insert of green nanocomposite for packaging, automation are being used rapidly. With following technological applications, contamination can be detected fast, which is user-friendly and robust. However, this fast step will help manufacturers take precautions immediately and stop total food processing from endangering living beings and from the total financial loss of the food industries.

Nanomaterials are continuously being used in the food business because of the extent of front line functionalities to the packaging materials. Nanotechnology-enabled food packaging can generally be isolated into three elementary classes [58]. They are improved, dynamic, keen/smart packaging. It is basic to assess conceivable development of packaging constituents into food and review their viable risk for a comprehensive threat assessment regarding buyer security. Nevertheless, relatively

P. Barua (✉) · A. H. Khan
Department of Electrical and Electronic Engineering, Chittagong University of Engineering and Technology, Kaptai Highway, Chattogram 4349, Bangladesh

N. Hossain
School of Engineering, RMIT University, Melbourne VIC 3001, Australia



M. Mujawar et al. (eds.), *Nanotechnology for Electronic Applications*, Materials Horizons: From Nature to Nanomaterials, https://doi.org/10.1007/978-981-16-6022-1_12

few reviews have been distributed in regard to the impact of nanomaterials dependent on body intake or the conceivable coordinated effort of nanomaterial-based food contact materials (FCMs) with the food section [58]. Nanotechnology, for the most part, contains manufacture, portrayal, and control of nano-extend (<100 nm) atoms. The use of nanotechnology in polymers includes the project, assembling, handling, and use of polymer materials loaded up with nano-particles and the plan, assembling, handling, and use of polymer materials loaded up with nano-particles and gadgets of nano range (Paul et al. 2008).

Nanotechnology derived food packaging materials are the largest category of current nanotechnology applications for the food sector [25]. The rising consumer worries about food quality and medical advantages are actuating the scientists to discover the way that can upgrade food quality while upsetting least the dietary benefit of the item. The interest of nanoparticle-based materials has been extended in the food business a similar number of them contain principal segments, and besides found to be non-toxic [50].

They are ended up being sound and hygienic at high temperatures and weights [52]. Nanoparticles have more than average characteristics for exemplification and conveyance efficiency than customary epitome structures. Nano-exemplifications cover fragrances or tastes, control interchanges of dynamic trimmings with the food organization, control the appearance of the diligent administrators, ensure availability at a target time and unequivocal rate, and shield them from moistness, heat, manufactured, or common corruption during getting ready, storing, and utilization, and besides show closeness with various blends in the framework. Besides, these movement systems have the ability to invade significantly into tissues because of their smaller gauge and accordingly license capable transport of dynamic blends to target objections in the body [34, 64, 69].

In food microbiology, nano-sensors or nano-biosensors are utilized to discover microorganisms in preparing plants or in food material, size of available food constituents, cautioning purchasers, and wholesalers on the well-being status of food. The nano-sensor functions as a marker that reacts to changes in ecological conditions, for example, stickiness or temperature away rooms, microbial defilement, or item debasement. Different nanostructures like slight movies, nanorods, nanoparticles, and nanofibers have been inspected for their potential applications in biosensors [10, 36, 30, 32]. Other than a ton of favorable nanotechnology circumstances to the food business, security issues related to nanotechnology cases to the food business, and security issues related to the nanomaterial can not be disregarded. Numerous scientists examined security concerns to nanomaterial giving accentuation on the chance of nanoparticles moving from the bundling material into the food and their effect on customer's wellbeing. Albeit a material is being considered GRAS (for the most part regarded as sheltered) substance, additional studies must be brought to examine the danger of its nano-partners because the physicochemical properties in nano-states are not the same as that are in macro-state. Additionally, these nanomaterials' nominal size may expand the danger for bioaccumulation inside body organs and tissues [11, 31, 51]. Nanotechnology in food packaging systems is categorized into some sectors. They are outlined in Fig. 1.

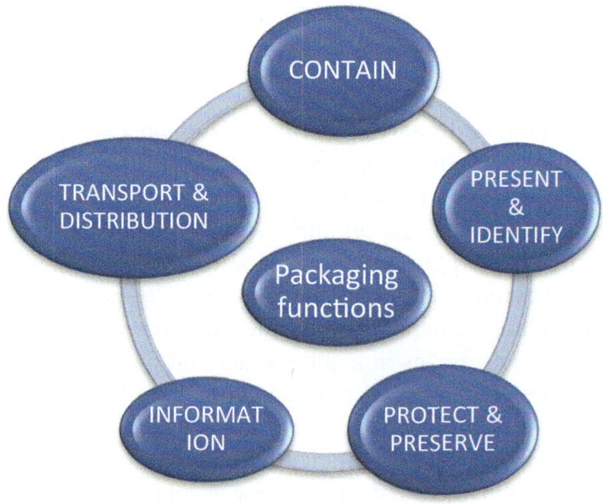


**Fig. 1** Fundamental processing steps of packaging

It can similarly be expected to share disinfectant, cell fortifications, blends, tastes, and healthy food to extend the period of the ease of use in nano-wrapping. The innovative nanotechnology things for food packaging were in the pipeline. Some foe of microbial movies to improve the period of reasonable convenience of food and dairy things have recently been entered the market. Novel food packaging advancement is by far the most reassuring benefit of nano-development in the food business soon, as portrayed starting late by Dasgupta et al. as to nanotechnology in the agricultural and food area one of the fastest creating fields in nano-research [17]. This chapter aims to highlight and review recent updates, current technologies, and future perspectives of nanotechnology in food packaging application.

## 2 Required Technological Advances in the Food Industries

Food industry orbs any company engaged in producing, processing, manufacturing, selling, and serving food, beverage, and dietary supplements [43]. The food industry has recently undergone a revolutionary change due to Industrial whirl, Industry 4.0. 'Industry 4.0' has become a potential factor in directing the innovative production processes that are partially or entirely automated by technical supports used to run the creative production processes [23]. Technology plays a vital role in that case. So, the demand for technological advancement in the food sector is also increasing. Align with that, manufacturing models are altering through robotics, machine learning, and artificial intelligence. The reasons behind technological demand in the food industry can be given as follows:

- The expectation of healthy products
- A pleasing appearance of the products

- Efficient production of the products
- Less touch of hands
- Low production cost.

A notable digital transformation is happening in the food industry. The meaning of digitalization does not confine to transfer from analog to digital; rather, it presents a communication between the prepared interfaces, the data interchanges, the business procedures, and the administrations [8]. Smart technologies, e.g., the internet of things (IoT) platforms, advanced robotics systems, significant data methods, and cloud computing play a pivotal role in the food industry. The smart and automated production systems can provide real-time data, which helps the workers to make accurate decisions in a concise amount of time. The industries rely more on team-oriented works because of the combination of smart and artificial intelligent tools. Teamwork will become more critical between the workers and subsystems (Fast Company 2019). An efficient workforce is required in this aspect. Due to increase automation, the force will become accountable for more complex loads. The demand for social skills will also rise due to the welcome of leading technologies [14], Han et al. [29]. Food-tech in the industries mainly focuses on high quality at low cost (ING Econonics 2019).

For short terms:

Robotics—Mainly focus on the Processing of food and packaging
Data Technology—Big data analysis, Artificial Intelligence, RFID chips
Novel Processing Techniques—High-pressure Processing, Steam blanching process.

For long terms:

Nanotechnology—nano-sized food ingredients
3D printing—personalizing in production
Cell technology—use of stem cells to produce meat and fish.

To automate, robots are advisable in the food industry. Figure 2 showcased the International Federation of Robotics (IFR), and there is an increment of robot sales in the food industry with time. This robot power helps to explain this in case of heavy workload. As the food manufacturers focus more on efficiency based on per unit cost; relying on the automated system increases, relying on the computerized system increases.

## 3 Application of Nanotechnology in Food Packaging

Analysts have begun to investigate the potential outcomes of "smart" packaging. Instead of merely shielding food from going off for as far as might be feasible, bundling with implanted splendid materials and versatile nano-gadgets could adequately control the atmosphere inside the bundling and prepared customers the

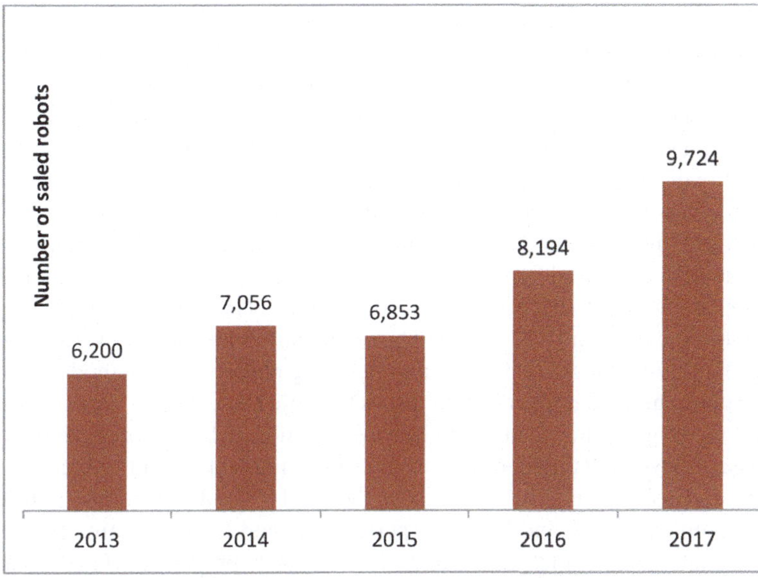


**Fig. 2** Worldwide robot selling depiction (IFR) (ING Economics 2019)

food has begun to rot. Auto-cooling packaging is proposed, which would use a compound or actual cycle, for instance, disappearing of gas for maintaining the internal temperature down inside the packaging. Controlled systems could similarly use a pitiful film solar cell to control cooling using a thermoelectric material. This would reduce the prerequisite for tremendous extension refrigeration along the nimble chain, notwithstanding how vague it is to be cash sparing a bit of leeway for this circumstance (Lawley 2008). Self-repairing polymers have been under investigation by researchers for a couple of years now, and a couple of models have appeared accessible. Usage of these materials as an outer layer in food bundling could allow little infiltrates and tears in the wrapping, decreasing wastage due to hurt bundling. One intriguing headway is using nano-particles to conceal changing plastic packaging, which showed when food is going off. There are two standard instruments to achieve this—substance and physical. The compound part uses an engineered marker that changes tone. RFID (Radio Frequency Identification) marks have been used for specific years now. They work a great deal of like a normalized tag, particularly perceiving many things. Yet, they do far off of up to a few meters inside seeing gases emanated when food oxidizes [39].

A few associations are creating Smart Packaging frameworks. For instance, alongside specialists at Rutgers University in the US, Kraft nourishments are building up an "electronic tongue" for incorporation in bundling. This comprises a variety of nanosensors that are amazingly delicate to gases provided by food as it ruins, causing the sensor strip to change shading, giving an away from sign of if the food is new. Bayer polymers has built up the Durethan KU2-2601 packaging film, which has lightweight, improved strength, and more warmth safe than those right now available. The primary role of food wrapping films is to keep the substance from drying out and shield them from dampness and oxygen. The new film is known as a "mixture framework" that is advanced with a more considerable amount of silicate nano-ingredients. These significantly decrease the passageway of oxygen and different gaseous compounds, and the leftover of dampness, keeping food from ruining.

Agro-Micron has built up the Nano-Bioluminescence Detection Spray, which contains a glowing protein designed to connect to the outside of organisms, such as Salmonella and Escherichia coli. When bound, it radiates a clear sparkle, hence permitting simple discovery of defiled food or drinks. The more extreme the sparkle is, the higher the bacterial defilement. The organization means to advertise the item under the name Bio-Mark and is presently planning new shower methods to apply in sea cargo containerized transportation just as to battle bioterrorism. In a comparative methodology to guarantee sanitation, European Union (EU) scientists in the 'Good Food Project' have built up a convenient nanosensor to identify synthetic compounds, microbes, and poisons in food. This evades the need to send tests to research centers (which is both expensive and lengthy), permitting food to be investigated for well-being and quality at the ranch, abattoir, during transport, preparing, or at the bundling plant. The undertaking is likewise building up a gadget utilizing DNA biochips to distinguish microorganisms, a strategy that could again be applied to decide the presence of various types of hurtful microscopic organisms in meat or fish or growths influencing natural product. Likewise, the scheme has plans to develop microarray sensors that can be used to determine pesticides on foods grown from the ground only as those that will screen natural conditions at the ranch. These have been authored "Acceptable Food sensors" [9]. The principle nanomaterials utilized in purchaser items are silver nanoparticles (AgNPs), titanium dioxide (TiO2), and silicon dioxide (SiO2) NPs. SiO2, NPs were used to build up a light, stable, thermally safe plastic (Durethan) with gas obstruction obligations to forestall food decay and drag out period of usability [3, 19].

Biosensors have been accounted for that can recognize microorganisms (for example, E. coli, Salmonella), contaminants (for example, hefty metals, lingering veterinary anti-infection agents, leftover pesticides, illicit added substances, mycotoxins, bisphenol A, and others), allergens, and supplements (for example, cell reinforcements) with potential to conquer the constraints of expensive and arduous research center strategies [2, 6, 20, 41, 44, 67, 71, 72]. Nanomaterials that are regularly utilized in these stages include (i) gold nano particles (AuNPs), silver nano particles (AgNPs) that have surface stage reverberation (SPR) properties and high conductivity empowering them to be used in ocular and phytochemical sensors, (ii) attractive NPs, for example, iron nano-particles ($Fe_2O_3$NPs) that give attractive capacities to the proficient detachment and improvement of the focused on analytes, and (iii) carbon nanotubes and graphene-based nanomaterials that enhance electrochemical signals because of their high electrical conductivity. The accompanying segment examines the usage of different nanomaterials in phytochemical and ocular(bio)sensors for food investigation and gives instances of utilization [35, 57, 68].

# 4 Latest Advancement of Nanotechnology in the Global Food Industries

As the number of mouths to be fed is increasing rapidly, there is a way to meet up the food demand with rapid foois a way to meet up the food demand with quick food production alignment with proper nutrition values. Nanotechnology can play an exigent induction in this situation. So nanotechnological demand in the food industry is flourishing. Several financial experts predicted this field to be boomed by 2025 [28]. The pasteurization process, which was invented in the eighteenth century, can be considered as the first footstep in the food nanotechnology field. The applications of nanoscale materials in sensing and detection purposes are practical. Besides, nanoparticles possess a large surface area, which results in leading chemical and biological activity than large particles of the same chemical composition [27, 42].

## *4.1 Food Safety*

Food safety is a concerning issue among people all over the world. Healthy food tends to a healthy life. Food microbial diseases cause 20 million infected people annually. To preserve the nutrients in food for a long time, there is no option to ensure proper food management. Foods are required to protect from physical, chemical, and biological contamination by an act of processing, handling, packaging, etc. [70]. In food industries, nanotechnology can be worked for finding out bacteria, producing strong flavors in food, proper color, and ensuring safety to improve barrier properties [56].

## *4.2 Food Processing*

The process that converts raw ingredients into food or related types intend to extend shelf life is called food processing. It includes the removal of toxins, preclusion from pathogens, and better consistency of foods. The main benefit of processed food is that it is less prehensile to quick spoilage than raw foods and best suited for distant deliveries. Nowadays, the effectiveness of food processing depends more on the incorporation of nanotechnology. Nano capsule delivery systems play a crucial role in food processing and handling properties, which are facilitated by emulsions, biopolymer-based materials, and others into foods [1]. Nano-drop is a kind of nano-sized structural lipid that works as a transporter of nutritional elements and inhibits cholesterol movement from the digestive system into the bloodstream [24].

### 4.3 Food Packaging, Bio-degradability, and Nutritional Supplements

The packaging is a momentous stage in the food industry. To keep the food quality intact, the packaging is a convenient option to opt for. Besides, nutrition information and marketing of the product also rely on the packaging it. Different nanomaterials like silver nanoparticle, titanium nitride nanoparticle, nano-titanium dioxide, nano zinc oxide, and nano-clay are commenced as working additives to food packaging. Nanotechnology-related food packaging can be categorized as [26, 70].

Improved packaging—Nanoparticles and polymer chains are mixed, intending to develop the gas barrier features and also temperature, humidity resistance of packaging. United States Food and Drug Administration supports the usage of nanocomposite in the food industry.

Active packaging—Nanomaterials are conducive to interlude directly with food to allow better preservation of the product. Nanomaterials like nano-silver, nano-titanium dioxide, nano magnesium oxide, nano copper oxide, and carbon constructed nanotubes can provide antimicrobial characteristics. At present, silver nanoparticles are being used more rapidly as antibacterial agents in food packaging.

Smart packaging—It senses any kind of biochemical or microbial alteration in food. This kind of packaging is also used to detect specific gases that can damage food and tracing food safety. Companies like Nestle and British Airways use chemical sensors to detect color change [33].

Different types of nano-sensors are used in food packaging like array biosensors, nano-cantilevers, nano-test strips, and others Nano-sensor. Nano-sensors can detect the rotten foods can detect the rotten foods, and packaging can change the color of itself to alert the consumers. Nanotechnology provides efficient packaging and storage of fresh foods for a long time. It can detect toxins, pesticides, and spoilage [70].

Environmental pollution is a concerning issue globally. Packaging system, specifically, which is non-degradable, plays a harmful role in the ecological system. But this problem is come over by nanotechnology-based packaging materials composed of synthetic nanoparticles that are biodegradable. These nanoparticles are attained from animals and plants and metal oxide, and carbon nanoparticles are used [38].

The market value of nano-food is thriving. Nano-ceuticals and Nutrition-be-nanotech are commercial denominations of such types of supplements. Nano-cochleates are used for nutrient delivery to cells without altering the taste and color of food. Supplementary prospect entangles encapsulation techniques where probiotics and other products are aimed for the living system with the aid of iron and zinc nanostructured capsules [4, 56].

## 5 Future Perspective

The possible use of nanotechnology in food packaging can be predicted to be vast and wide. Although a lot of analysis has shown benefits of using nanoparticles for developing food checking, quality and freshness, studies of their harmfulness, future, and debasement happen in their earliest stages. No later than their maximum capacity may be acknowledged, more examinations should experiment and a careful year-round investigation of nanoparticles in food and the climate. Comprehension of their dangers needs improvements in a few zones. To start with, techniques should be created to portray and quantify nanomaterials' essential features and their biodegradation in food. Traditional approaches have restricted application due to their significant expense, failure to gauge a massive amount of tests, and food frameworks' fineness. Further examinations are then expected to assess the harmfulness of nanoparticles, either mixed as fixings to food or in contact with food (for example, nanomaterial in bundling). After that, price, biologically acceptable, and scale-up should be additionally tended to for fruitful usage. Future undertakings are relied upon to ensure long stretch quality and limit recognizing parts in field conditions, similarly as an endorsement of the advancement. Architects of food nanotechnology should like manner center around industry and customer affirmation and capacity to pay. The development of nanomaterials from packaging into food and the atmosphere is a further concern that requires a predominant understanding. The issues referenced above should be surveyed as a significant part of the improvement cycle. Four, the usage of nanotechnologies in food misses the mark on a regulatory framework and makes the client reluctant to use nano-engaged food. Fifth, standards and a predominant course of action and depiction of nano-things similarly as naming are required to address client demands and authoritative issues and ultimately animate business gathering. These topics are the subjects of colossal degree gatherings and discussion conversations. The Institute of Food Technologists (IFT) has standard issues and is encouraging influential communities discussing the use and likely utilization of nanotechnologies in food. Such activities are depended upon to be extended later on [15, 40].

Nanotechnology in the food business is depended upon to altogether increment in an impetus to $20.4 billion by 2020, with a couple of creating domains for headway, reports. Future applications could consolidate immobilizing impetuses to improve their efficiency and reuse and using nanoscale structures to make new food textures. In the latest couple of years, the upsides of applying nanotechnology have been driving the headway of contemporary and world-class materials for the food region. Thus, the applications at laboratory yet also on a cutting edge scale have dramatically extended. Over the latest twenty years, the number of conveyances and licenses developed 40 and 90%, exclusively. These numbers show the potential and energy of pros and associations of nanotechnology in the food territory and the increased ability to use nanotechnology-based things in a couple of food cycles and applications. Besides, the number of associations focusing their inventive creative work on nanotechnology-based things is more than 1,000—a number that should fill in the next years [15]. Another uncommon open door for nanotechnology is in smart packaging. Quick food packaging can screen and offer a hint of the packaged food's image and likewise guarantee their security, not merely during amassing and transportation for industry and retailers yet notwithstanding the purchasers. A bit of the model the sensors-engaged RFID (radiofrequency unmistakable detailed verification) marks and pointers using nanotechnology to teach about the quality or novelty of the squeezed food things.

An example in the food business is the probability to set example in the food business's probability of setting up modified sustenance plans with on-demand prosperity necessities and license buyers to pick food things subject to their potential advantages safely. This is possible with the post and improvement of food things where nanotechnology can have critical work. The usage of nano-encapsulation can be used to make sure about and pass on down to earth blends, improving quality, bioavailability, and known structures. It is possible to nano-encapsulation a couple of blends, for instance, supplements, minerals, malignant growth avoidance specialists, and polyunsaturated fats, and achieve better security (decrease mix and effect of environmental conditions, for instance, light and temperature), improve organoleptic properties (low influence of flavor and concealing) and augmentation bioavailability (high digestion in the human gut with controlled conveyance) when differentiated and the current systems [65].

In outline, nanotechnology offers innovative responses to improve quality, produce, secure, and separate food things. In any case, stresses over toxicity have raised different issues for prosperity possibilities and regular impact. This study gave a survey of late advances and potential applications, including risks and preferences of these developments. Attempts should accumulate later on to ensure nanotechnology's secured service in the food territory and its possible repercussions.

## 6 Environmental and Economic Prospect

Bio-based packaging is biodegradable packaging films applied to food things to control sogginess move or possibly gas exchange to improve prosperity and protect the energizing and material quality [16]. These packaging materials are believed to be more characteristic friendly than the other standard packaging films. The unequivocal brand name that perceives biodegradable packaging films from other packaging courses of action is suitable for decaying through the movement of living creatures [61]. This packaging type is generally observed to be more biological neighborly as the breakdown things are to a great extent typical, i.e., carbon dioxide, biomass, and water. Bio-based packaging does not (or less) use oil subordinates to convey the materials, yet uses inexhaustible sources, upon expulsion energy can be recovered by incineration. Under appropriate conditions (suddenness, temperature, and oxygen openness), this plastic prompts brokenness or disintegration of the plastics with no harmful or damaging earth development [22].

Novel food packaging advancement is by far the most promising favorable position of nanotechnology in the food business shortly, as depicted starting late by [66] concerning its application in the agro-food zone one of the speediest creating fields in nano-research. Food bundling and noticing are critical points of convergence of food industry-related nanotechnology creative work [21]. The primary headway in food packaging is dynamic and smart packaging that pledges to improve disinfection and quality and upgrades thing period of sensible ease of use [12]. On the other hand, in bundling adventures, the most significant bit of materials used is non-degradable oil-based plastic polymer materials, which in this way, address a troublesome issue on the overall biological [33]. In such a manner, by strategies for the correct assurance of materials and packaging headways, it is possible to keep the thing quality and originality during the time required for its commercialization and use [45, 63]. By and by, the most read bio-nano composites fitting for packaging applications are starch and derivates, i.e., polylactic destructive, poly(butylene succinate), polyhydroxybutyrate, and aliphatic polyester. Like this, bio-based packaging can similarly be used simultaneously for improved and dynamic packaging.

Polylactic corrosive is an eco-obliging bioplastic with excellent biocompatibility, vulnerable durability, slow degradation, hydrophobicity, and nonappearance of responsive side-chain packs limit its application [58]. Like this, the fitting of its properties to show up at end-customers' demands is required. Various has been represented the course of action of polylactic destructive/earth nanocomposite materials [18, 46, 59]. Polyhydroxybutyrate has been the subject of comprehensive examinations as a usually welcoming polymeric material, which is the most acclaimed polyhydroxyalkanoate used in food packaging. Polyhydroxybutyrate is a polymer with a place with the polyesters class of premium as bio-decided and biodegradable plastics [60].

The bundling industry contributes by and large to the world economy; right around 55–65% of $130 billion was spent on food and savor packaging in the USA [13]. Nanotechnology remains an astounding portrayal of rising advances in this century [5]. A couple of cash related pros foreseen that this advancement would wind up being impacted by 2025 [28]. In 2008, nanotechnology referenced over $15 billion when all is said in done creative work cash (open and private) and utilized more than 400,000 specialists in general [47, 49]. Nanotechnologies are anticipated to be worth $3 trillion over the general economy by 2020, and nanotechnology business endeavors all in all may require no under six million specialists to help them before the decade is over [37]. It is challenging to consider the veritable cash related capacity of nanotechnology, until this point, it has been utilized in different fields, for example, PC gadgets, correspondence, energy plan, courses of action, and the food industry, likely obliging novel administrative strategies [7, 62]. Cash related and social affirmations and chances of nanotechnologies are fundamental for people, considering the effects they would have on everybody, going like the individual fulfillment and creation of abundance. Foodborne disorders absolute $152 billion in prosperity related costs each year in the US, particularly for bacterial defilements achieved by Salmonella, Escherichia coli, Listeria monocytogenes, and Campylobacter spp [48, 53, 54, 55].

## 7 Conclusions

Nanotechnology is spreading its potentiality in all applications, including our daily usages. The food packaging system is now more organized and perfect than ever before because of the nanotechnology application. The article showcased some recent and latest outcomes of nanotechnology implementation in the food industry. Considering the facts and future perspective, it is almost sure that nanotechnology will triumph in the food packaging industry as within a short period it has achieved more urgency than other technologies because of the quality maintenance of food in every corner of all processes. Moreover, robot technology is now being addressed, which is a blessing in the food industry that can be used to sustainable food production efficiently and securely. The manuscript highlighted some technical advancements also, which reflects the prosperous future of the food industry.

## References

1. Abbas KA, Saleh AM, Mohamed A, MohdAzhan N (2009) The recent advances in the nanotechnology and its applications in food processing: a review. J Food Agric Environ
2. Anirudhan TS, Athira VS, Chithra Sekhar V (2018) Electrochemical sensing and nano molar level detection of Bisphenol-A with molecularly imprinted polymer tailored on multiwalled carbon nanotubes. Polymer (guildf). https://doi.org/10.1016/j.polymer.2018.05.052
3. Ashraf SA, Siddiqui AJ, Elkhalifa AEO et al (2021) Innovations in nanoscience for the sustainable development of food and agriculture with implications on health and environment. Sci Total Environ
4. Avella M, De Vlieger JJ, Errico ME et al (2005) Biodegradable starch/clay nanocomposite films for food packaging applications. Food Chem. https://doi.org/10.1016/j.foodchem.2004.10.024
5. Bachmann G, Holtmannspötter D, Korte S, Zweck A (2001) Angels on a pinhead: new research networks for nanotechnology. Foresight
6. Ben Messaoud N, Ghica ME, Dridi C et al (2017) Electrochemical sensor based on multiwalled carbon nanotube and gold nanoparticle modified electrode for the sensitive detection of bisphenol A. Sens Actuators B Chem. https://doi.org/10.1016/j.snb.2017.06.160
7. Blind K (2008) Regulatory foresight: methodologies and selected applications. Technol Forecast Soc Change. https://doi.org/10.1016/j.techfore.2008.02.004
8. Bogner E, Voelklein T, Schroedel O, Franke J (2016) Study based analysis on the current digitalization degree in the manufacturing industry in Germany. In: Procedia CIRP
9. Bounie D, Arcot J, Cole M et al (2020) The role of food science and technology in humanitarian response. Trends Food Sci Technol
10. Bouwmeester H, Dekkers S, Noordam MY et al (2009) Review of health safety aspects of nanotechnologies in food production. Regul Toxicol Pharmacol. https://doi.org/10.1016/j.yrtph.2008.10.008
11. Bradley EL, Castle L, Chaudhry Q (2011) Applications of nanomaterials in food packaging with a consideration of opportunities for developing countries. Trends Food Sci Technol. https://doi.org/10.1016/j.tifs.2011.01.002
12. Brody AL (2003) "Nano, Nano" food packaging technology. Food Technol
13. Brody AL, Bugusu B, Han JH et al (2008) Innovative food packaging solutions. J Food Sci
14. Bughin J, Hazan E, Lund S et al (2018) Automation and the workforce of the future | McKinsey. McKinsey Global Institute

15. Bujang A, Rahman FA, Omar SRS (2020) Nanotechnology in the food processing and packaging: an overview of its halalan tayyiban aspect. Malaysian J Consum Fam Econ
16. Cerqueira MA, Pinheiro AC, Silva HD et al (2014) Design of bio-nanosystems for oral delivery of functional compounds. Food Eng Rev
17. Cha DS, Chinnan MS (2004) Biopolymer-based antimicrobial packaging: a review. Crit Rev Food Sci Nutr. https://doi.org/10.1080/10408690490464276
18. Chang JH, An YU, Sur GS (2002) Poly(lactic acid) nanocomposites with various organoclays. I. Thermomechanical properties, morphology, and gas permeability. J Polym Sci Part B Polym Phys. https://doi.org/10.1002/polb.10349
19. Chaudhry Q, Scotter M, Blackburn J et al (2008) Applications and implications of nanotechnologies for the food sector. Food Addit Contam Part A Chem Anal Control Expo Risk Assess
20. Cheung W, Shadi IT, Xu Y, Goodacre R (2010) Quantitative analysis of the banned food dye sudan-1 using surface enhanced raman scattering with multivariate chemometrics. J Phys Chem C. https://doi.org/10.1021/jp908892n
21. Dasgupta N, Ranjan S, Mundekkad D et al (2015) Nanotechnology in agro-food: from field to plate. Food Res Int
22. Del Nobile MA, Conte A, Buonocore GG et al (2009) Active packaging by extrusion processing of recyclable and biodegradable polymers. J Food Eng. https://doi.org/10.1016/j.jfoodeng.2008.12.022
23. Didenko NI (2020) Concept and mechanism for managing the digital transformation of Arctic target subspaces and spheres of vital activity. In: IOP conference series: materials science and engineering
24. Dingman J (2008) Nanotechnology: its impact on food safety. J Environ Health
25. Duncan TV (2011) Applications of nanotechnology in food packaging and food safety: Barrier materials, antimicrobials and sensors. J Colloid Interface Sci. https://doi.org/10.1016/j.jcis.2011.07.017
26. Emamhadi MA, Sarafraz M, Akbari M et al (2020) Nanomaterials for food packaging applications: a systematic review. Food Chem Toxicol. https://doi.org/10.1016/j.fct.2020.111825
27. Garnett MC, Kallinteri P (2006) Nanomedicines and nanotoxicology: some physiological principles. Occup Med (Chic. Ill)
28. Glenn JC, Gordon TJ (1997) The Millennium Project: 1997 state of the future—implications for actions today. Technol Forecast Soc Change. https://doi.org/10.1016/s0040-1625(97)00114-5
29. Han ES, Goleman D, Boyatzis R, Mckee A (2019) Blueprint for sectoral cooperation on skills: towards an eu strategy addressing the skills needs of the steel sector
30. Helmke BP, Minerick AR (2006) Designing a nano-interface in a microfluidic chip to probe living cells: challenges and perspectives. Proc Natl Acad Sci U S A
31. Jain A, Ranjan S, Dasgupta N, Ramalingam C (2018) Nanomaterials in food and agriculture: an overview on their safety concerns and regulatory issues. Crit Rev Food Sci Nutr
32. Jianrong C, Yuqing M, Nongyue H et al (2004) Nanotechnology and biosensors. Biotechnol Adv. https://doi.org/10.1016/j.biotechadv.2004.03.004
33. Kuswandi B, Wicaksono Y, Jayus et al (2011) Smart packaging: sensors for monitoring of food quality and safety. Sens Instrum Food Qual Saf. https://doi.org/10.1007/s11694-011-9120-x
34. Lamprecht A, Saumet JL, Roux J, Benoit JP (2004) Lipid nanocarriers as drug delivery system for ibuprofen in pain treatment. Int J Pharm. https://doi.org/10.1016/j.ijpharm.2004.03.018
35. Lin X, Ni Y, Kokot S (2013) Glassy carbon electrodes modified with gold nanoparticles for the simultaneous determination of three food antioxidants. Anal Chim Acta. https://doi.org/10.1016/j.aca.2012.12.036
36. Luo S, Zhang P, Gao D (2020) Preparation and properties of antimicrobial poly(butylene adipate-co-terephthalate)/TiO2 nanocomposites films. J Macromol Sci Part B Phys. https://doi.org/10.1080/00222348.2020.1712045

37. Malanowski N, Zweck A (2007) Bridging the gap between foresight and market research: integrating methods to assess the economic potential of nanotechnology. Technol Forecast Soc Change. https://doi.org/10.1016/j.techfore.2007.05.010
38. McClements DJ (2019) Future foods: How modern science is transforming the way we eat
39. Mihindukulasuriya SDF, Lim LT (2014) Nanotechnology development in food packaging: a review. Trends Food Sci Technol. https://doi.org/10.1016/j.tifs.2014.09.009
40. Mustafa F, Andreescu S (2020) Nanotechnology-based approaches for food sensing and packaging applications. RSC Adv
41. Neethirajan S, Weng X, Tah A et al (2018) Nano-biosensor platforms for detecting food allergens—new trends. Sens Bio-Sens Res
42. Nel A, Xia T, Mädler L, Li N (2006) Toxic potential of materials at the nanolevel. Science (80–)
43. Nestle M (2013) Food politics: how the food industry influences nutrition and health
44. Oh SY, Lee MJ, Heo NS et al (2019) Cuvette-type lspr sensor for highly sensitive detection of melamine in infant formulas. Sensors (Switzerland). https://doi.org/10.3390/s19183839
45. Pang MM, Choo HL, Buys YF (2020) Plastics in food packaging. In: Reference module in materials science and materials engineering
46. Pluta M, Galeski A, Alexandre M et al (2002) Polylactide/montmorillonite nanocomposites and microcomposites prepared by melt blending: structure and some physical properties. J Appl Polym Sci. https://doi.org/10.1002/app.11309
47. Roco MC (2011) The long view of nanotechnology development: the national nanotechnology initiative at 10 years. J Nanoparticle Res
48. Roco MC, Harthorn B, Guston D, Shapira P (2011) Innovative and responsible governance of nanotechnology for societal development. J Nanoparticle Res
49. Roco MC, Mirkin CA, Hersam MC (2011) Nanotechnology research directions for societal needs in 2020: summary of international study. J Nanoparticle Res
50. Roselli M, Finamore A, Garaguso I et al (2003) Zinc oxide protects cultured enterocytes from the damage induced by Escherichia coli. J Nutr. https://doi.org/10.1093/jn/133.12.4077
51. Savolainen K, Pylkkänen L, Norppa H et al (2010) Nanotechnologies, engineered nanomaterials and occupational health and safety—a review. Saf Sci. https://doi.org/10.1016/j.ssci.2010.03.006
52. Sawai J (2003) Quantitative evaluation of antibacterial activities of metallic oxide powders (ZnO, MgO and CaO) by conductimetric assay. J Microbiol Methods. https://doi.org/10.1016/S0167-7012(03)00037-X
53. Scallan E, Hoekstra RM, Mahon BE et al (2015) An assessment of the human health impact of seven leading foodborne pathogens in the United States using disability adjusted life years. Epidemiol Infect. https://doi.org/10.1017/S0950268814003185
54. Scharff RL (2012) Economic burden from health losses due to foodborne illness in the United States. J Food Prot. https://doi.org/10.4315/0362-028X.JFP-11-058
55. Scharff RL (2010) Health-related costs from foodborne illness in the United States. Sci. York
56. Sekhon BS (2010) Food nanotechnology—an overview. Nanotechnol Sci Appl
57. Sharpe E, Frasco T, Andreescu D, Andreescu S (2013) Portable ceria nanoparticle-based assay for rapid detection of food antioxidants (NanoCerac). Analyst. https://doi.org/10.1039/c2an36205h
58. Silvestre C, Duraccio D, Cimmino S (2011) Food packaging based on polymer nanomaterials. Prog Polym Sci
59. Sinha Ray S, Maiti P, Okamoto M et al (2002) New polylactide/layered silicate nanocomposites. 1. Preparation, characterization, and properties. Macromolecules. https://doi.org/10.1021/ma011613e
60. Sinha Ray S, Yamada K, Okamoto M et al (2003) New polylactide/layered silicate nanocomposites. 3. High-performance biodegradable materials. Chem Mater. https://doi.org/10.1021/cm020953r
61. Siracusa V, Rocculi P, Romani S, Rosa MD (2008) Biodegradable polymers for food packaging: a review. Trends Food Sci Technol

62. Sozer N, Kokini JL (2009) Nanotechnology and its applications in the food sector. Trends Biotechnol
63. Stewart CM, Tompkin RB, Cole MB (2002) Food safety: new concepts for the new millennium. Innov Food Sci Emerg Technol. https://doi.org/10.1016/S1466-8564(02)00010-3
64. Ubbink J, Krüger J (2006) Physical approaches for the delivery of active ingredients in foods. Trends Food Sci Technol
65. Vanderroost M, Ragaert P, Devlieghere F, De Meulenaer B (2014) Intelligent food packaging: the next generation. Trends Food Sci Technol
66. Vroman I, Tighzert L (2009) Biodegradable polymers. Materials (Basel)
67. Wang C, Hu L, Zhao K et al (2018) Multiple signal amplification electrochemiluminescent immunoassay for Sudan I using gold nanorods functionalized graphene oxide and palladium/aurum core-shell nanocrystallines as labels. Electrochim Acta. https://doi.org/10.1016/j.electacta.2018.05.061
68. Wang J, Wu X, Wang C et al (2016) Facile synthesis of Au-coated magnetic nanoparticles and their application in bacteria detection via a SERS method. ACS Appl Mater Interfaces. https://doi.org/10.1021/acsami.6b07528
69. Weiss J, Takhistov P, McClements DJ (2006) Functional materials in food nanotechnology. J Food Sci
70. Wesley SJ, Raja P, Raj AA, Tiroutchelvamae D (2014) Review on—nanotechnology applications in food packaging and safety. Int J Eng Res. https://doi.org/10.17950/ijer/v3s11/1105
71. Yan X, Li H, Su X (2018) Review of optical sensors for pesticides. TrAC—Trends Anal Chem
72. Zhu Y, Cai Y, Xu L et al (2015) Building an aptamer/graphene oxide FRET biosensor for one-step detection of bisphenol A. ACS Appl Mater Interfaces. https://doi.org/10.1021/acsami.5b00199
73. How AI is changing the way companies are organized. https://www.fastcompany.com/3068492/how-ai-is-changing-the-way-companies-are-organized. Accessed 18 July 2019
74. Food tech: technology in the food industry. ING Economics Department April 2019
75. Nanotechnology in packaging: a revolution in waiting. (2008). Food Eng Ingred 33(3):6

The manufacturer's authorised representative in the EU is Springer Nature Customer Service Centre GmbH, Europaplatz 3, 69115 Heidelberg, Germany. If you have any concerns regarding our products, please contact ProductSafety@springernature.com

Printed and bound by CPI Group (UK) Ltd, Croydon, CR0 4YY
30/09/2026
02225295-0005